This volume is affectionately dedicated to those inspired men and women who have devoted themselves to the protection of the lives and property of their friends and neighbors since the time of Amenhotep — the locksmiths of the world. There are few higher purposes, few causes more worthy, and none more deserving of recognition. So, my fellow locksmiths, this is for you, individually and collectively.

Complete Course in

Professional Locksmithing

Robert L. Robinson

Professional / Technical Series

Nelson-Hall nh Chicago

ISBN 0-911012-15-X

Library of Congress Catalog Card No. 73-174584

10 9 8 7 6 5 4 3 2

Contents

Contents continued

Preface

Locks protect the lives or property of almost every person in the world today, whether directly by locking his home against intruders, or indirectly by locking switches which could launch the nuclear missiles and destroy us all. Most of these billions of people own locks for the personal protection of their own individual persons and property. Even those who live in such abject poverty that they do not have a lock of their own, and may never have one, have a vital interest in the use of locks by industry and the government of the country in which they live.

The billions of locks in the possession of individual persons are of the greatest interest to locksmiths because they are not only more numerous, but also are more evenly distributed throughout the human race. As the rising tide of lawlessness seemingly engulfs whole nations, with crimes against property and humanity on the increase, the need for citizens to protect themselves is an acute and pressing matter. Most people look to locks for protection.

This is where the locksmith comes in. He can help people buy the right kind and type of lock for their individual protection requirement, service the locks they now have, arrange for the replacement of existing locks with some of a more adequate type, and make any keys needed to implement new or existing locks. Meeting these ever-increasing demands for lock services, and thereby helping to protect the countless millions of people who depend upon their locks for protection, is the special province of the locksmith. The knowledge that a locksmith can accomplish this vital function is vastly satisfying, and so is the knowledge that he is an independent businessman — a solid and respected member of the community .

Increasing demands upon locksmiths make the locksmithing business a lucrative one. A profession which provides satisfaction with a sense of accomplishment and reasonable remuneration is quite extraordinary and worthy of being anyone's choice as a career. Within reasonable limits, the hours are what you make them, and the work is not physically demanding in comparison to most other occupational fields. In fact, many kinds of physical disability are of little or no handicap in locksmithing.

It is the purpose of this book to provide needed information for the person who is handy with tools and who wishes to become a proficient and competent locksmith. The author hopes that it will likewise help many of those who are already locksmiths to become better ones. Hopefully, this will result in new standards of excellence in a field where adequate information is all too often lacking.

To achieve these objectives, the author has resorted to empirical logic based on a scientific method of analysis. We shall go into why and how locks behave. On a part-by-part basis, we shall build the operation of simple parts into operating principles, which will enable us to understand the many types and manufactures of locking devices which have existed in the past or which may exist in the future. For this reason, the author asks that the book be read in the order in which the various subject matter is discussed. To do otherwise would be to lose the empirical logic on which a particular subject matter is based. In any kind of analysis, no steps are skipped, and this book is no exception; it is important to begin right at the start and progress through it step by step if the greatest possible benefit is to be gained. An index is included for reference purposes, but it should not be used in the initial study of this material.

Acknowledgments

The author wishes to express his grateful acknowledgement to a number of firms for their generous contributions to the material appearing in this text. The following have made known their devotion to the progress of the locksmithing profession and the lock-making industry in a very tangible and instructive manner:

American Lock Co.
Curtis Industries, Inc.
Eaton Yale & Towne, Inc.
Falcon Lock Co.
Glen's Key Lock & Safe Co., Salt Lake City, Utah
Intermountain Lock & Supply Co.
Red's Gateway Key & Cycle, Pocatello, Idaho
Sargent & Co.
Sargent & Greenleaf, Inc.
Schlage Lock Co.

Introduction

ALTHOUGH the first use of door locks is lost in the mists of time, various means of locking doors are known to have been used by the Assyrians, Egyptians, and Greeks. Although the Egyptians are credited with the invention of a pin tumbler lock and key, it was not until the Roman period that locks evolved to something more nearly resembling the modern form. The Romans, using a notched boltwork copied from the Greeks, put the boltwork in an iron case, added wards, and used a true key.

Other Roman innovations included the use of the spring-loaded bolt, spring detents, the addition of spring operated pin tumblers to the Egyptian design, and the first padlocks. These early locks were mostly of the warded type and were produced in a wide range of designs and varied widely in craftsmanship. The broadly based Roman commerce of about the time of Julius Caesar led to a substantial demand for locks and keys by the prosperous merchants, government, and officials of Rome. There is every indication that locksmithing was a flourishing trade. The mechanical knowledge of these early lockmakers was such that many of the Roman innovations such as spring-loaded boltwork, spring detents, warded and paracentric keyways are very much in evidence to this day.

After the fall of Rome in the fifth century A.D., before the onslaught of people whom the Romans had contemptuously referred to as "barbarians," commerce and trade vanished in the crashing ruins of civilization. The warring, squabbling factions of what had been the proud Roman Empire had few valuables to protect and locksmithing declined to the point where it was a most uncommon trade. Locks made during this period were usually ordered by the nobility and by those few merchants who had achieved the considerable prosperity required to warrant so expensive a purchase.

Eventually, as the scattered fragments of the Roman Empire began to regroup into more or less stable nations, commerce began to revive in a limited way and more people came to buy locks. Even though nations achieved a degree of stability in those times, the hazards of individual living were considerable. These very hazards also contributed to the demand for locks, and between these two factors, enough locks were made to keep lockmaking from becoming a lost art.

The Crusades brought about a strong and persistent demand for locks to protect valuables while the owner was away. The locksmith, in much the same circumstances as in the Roman period, began to prosper once more. Locks of the time of the Crusades were of the same warded type which had been used by the Romans. Although there was a tendency in this period for locks to become occasionally massive and very strongly sprung, it was primarily a revival period for an almost forgotten craftsmanship.

During the Crusades and in the following centuries, an exquisite artistry and craftsmanship developed which was nowhere more evident than in the making of locks. Keys for the better locks of this period were delicately formed and had the artistic merit of a fine piece of Cellini* gold work. One can readily see that the best locks of this time usually found their way into the hands of the reigning monarch and the extremely wealthy.

This high degree of craftsmanship was attained as a result of the development of an apprenticeship system which finally became the Guild System. The Guild required a locksmith to serve a specified period as an apprentice; at the end of that period he had to pass a rigid examination. After passing his examination, he was known as a *journeyman locksmith*. In order to become a *master locksmith*, a journeyman was then required to design and produce a lock (known as a *masterpiece*) which would meet the strictest judgement of the master locksmith for whom he worked. Such a masterpiece was judged on operating smoothness, functional design, the skill with which it was produced, and its artistic merit.

With the advent of indentured servitude and the gradual corruption of the Guild System, lock craftsmanship began to decline and, although some excellent locks were still produced, the general tendency was for them to become more crudely made.

This decline in quality continued until the invention in 1784 by an Englishman, Josep Bramah, of the lock which bore his name. The Bramah lock utilized a bolt with multiple notches into which were fitted other notched bolts. These multiple bolts, when their notches were aligned by the key, would permit the lock to rotate. This was the first lock mechanism of any importance to make use of a rotating element in the lock itself. It was considered unpickable for many years until it was finally picked by an American, Hobbs, in 1851.

About the time of the invention of the Bramah lock in England, events were taking place in America which would have profound significance for lockmaking. The young republic was just getting established and there were growing fears that England, suspecting that the republic was about to succeed, would make a determined effort to regain her former colonies while she could. Anticipating an invasion, the young government called on the redoubtable Eli Whitney, giving him a contract in 1798 for 10,000 muskets to be delivered to the fledgling army by 1800.

Later, Whitney was called to Washington to ex-

plain why, with his contract time so nearly expired, he had delivered only 500 muskets. Whitney took 10 unassembled muskets with him when he went to answer the committee's summons. He spread the thoroughly mixed parts of all 10 before the surprised committee and asked one of the committeemen to choose enough parts from the lot to assemble a musket. Whitney then assembled the parts and fired the weapon for the committee. This was an astonishing feat in those days, for it was the inception of the interchangeable parts concept.

Eli Whitney had been very busy during the period when he had produced so few muskets — busy establishing sizes, tolerances, shapes, and standards of many kinds, as well as building the *machine tools* required to produce a complex device such as a musket to his standards. This was the beginning of mass production and was to have a far-reaching effect on lockmaking; its effect would be manifested in the design and quality of locks, as well as in bringing the cost of good locks within the purchasing range of most people. This reduction in the price of a good lock is indeed a remarkable development when one considers that, prior to that time, a reliable lock often commanded a price of $500. Such a sum would be about 10. to 20 times the cost of comparable protection today, but even this is not a fair comparison unless one considers also the lessening purchasing power of the dollar. If one contemplates that the three pounds of pork chops which today cost us $2.25 or more could be had for 25¢ in 1933, it would seem that $500 would "buy a lot of beans" in Revolutionary Boston.

The gradual development of the machine tool industry and the laborious evolution of a system of standards and tolerance contributed in large measure to the development, by Linus Yale Jr., in the middle of the nineteenth century, of the detachable pin tumbler cylindered lock. Mr. Yale's lock was very different from previous locks in that the change in keys from one lock to another did not depend upon a change in position or shape of some part of the lock, but upon a simple, precisely measured difference in the cuts made on the same shape and size of key material or *key blank*. The only change made in order to accomplish differing of keys in the cylinder mechanism itself was in a single dimension of the tumblers, which — except for length — were exactly alike from one lock to the next.

Since Mr. Yale's lock cylinder was detachable

*Benvenuto Cellini, 1500 – 1571; generally considered to be one of the foremost goldsmiths of all time. Although much of his work was gold based, he worked with equal facility in silver and bronze.

without the need to disassemble the rest of the lock to get at it, and was not an integral part of the rest of the lock (or locking mechanism), the same cylinder or similar cylinders could be used in locks having widely different modes of operation (or function). This versatility, together with offering good protection, has led to the pin tumbler mechanism becoming one of the most widely copied, adopted, and adapted mechanisms since the invention of the wheel.

About a century before the Yale lock was developed in America, the lever lock evolved in England. Like the pin tumbler lock, the lever lock keys differed from one another by precise measurement of the cuts. The dimensioning and usually, the configuration of the tumblers varied according to the depth of cut of the key. The tumblers were placed inside of and as an integral part of the locking mechanism, thus lacking the versatility of the pin tumbler lock.

As a result of this lack of versatility and the difficulty of changing the combination in event of a lost key, the lever lock did not retain the tremendous popularity enjoyed by some other types, although it has a substantial and enviable place in the family of locks. A lever lock in inexpensive form offers good protection for many things such as desks, cabinets, and lockers. More expensive precision-made lever locks fill the need for maximum security lock uses in banks, prisons, and other special use requirements.

The disc tumbler lock is a product of the twentieth century technique of die-casting, which made its intricate shape economically practical to produce. It makes use of tumblers which are usually stamped from sheet metal, spring driven and aligned fully within the plug by the key. The disc tumbler lock fills the protective gap between the pin tumbler lock and warded locks. It is economy priced and, in the usual form, offers moderate protection, although some disc tumbler variations offer good protection. Combinations may usually be changed with a reasonable degree of facility, and this lock's versatility is, in some respects, even greater than that of the pin tumbler mechanism. These factors have led to a widespread acceptance of the disc tumbler lock in homes, automobiles, industry, and business.

Mortise Lock and Panic Exit Device Construction

I

The Mortise Passage Latch

THE mortise lock unit is heavier — not only in weight, but also in the size and strength of the operating parts — than most other types of locks. Its larger size permits design variations which can contribute to its durability and general reliability. Since it is set into (or mortised into) the door, it acquires a measure of extra strength from that of the door itself. Being mortised into the door, it is also somewhat less vulnerable to burglarious attack than are surface mounted locks.

For these reasons and others which are inherent in specific designs and design variations, the mortise lock types find their greatest use where the heaviest type of service is encountered or where the highest degree of reliability is required. There are problems and deficiencies associated with their use, of course, but such problems are usually less frequent and less severe than with most other types of locks.

The passage latch is one of the most basic of the mortise functions. It provides the basis for most of the other mortise functions. Despite the superficial simplicity of the passage latch design, it is composed of a rather surprising number of pieces and parts,* all of which contribute in an important degree to the proper functioning of the unit. In order to place this interdependence of the various parts in proper perspective, it is necessary to consider the mode of operation of the unit as a whole, together with its physical layout and the objectives which it is intended to accomplish.

_____
*The passage latch may consist of as high as 40 or more individual pieces, including screws.

Referring to Figure 1, we find a latch which is retracted into the case by turning the knob on either side of the door. When the knob is released, it returns to its original position and the latch is again extended. If the door is pushed closed without turning the knob, the strike plate pushes the latch into the case and holds it there until it comes even with the strike plate pocket. The latch then springs out to enter the pocket, thus latching the door.

Each of the parts contributing to this function will be discussed separately in order to establish normal construction practices and the requirements

Figure 1. A mortise passage latch. The cap has been removed to expose the working parts.

Figure 2. Another version of the mortise passage latch.

of proper functioning. In order to perform any needed repairs or required service properly, it is necessary to understand the construction of the parts thoroughly, and to know what the contributory factors may be.

The Latch Bolt Assembly

Figures 1 and 2 show the latch with the cap removed to expose the works in the case. Note the *latch bolt assembly;* this will be the first part we will discuss and it is shown separately in Figure 3. It consists of the *latch bolt,* a pin used to attach it to the *shaft,* a *tail piece,* a compression type helical *coil spring,* and a *thrust washer.*

The Latch Bolt

The latch bolt may be cast, forged, or cut off from

Figure 3. Latch bolt assemblies.

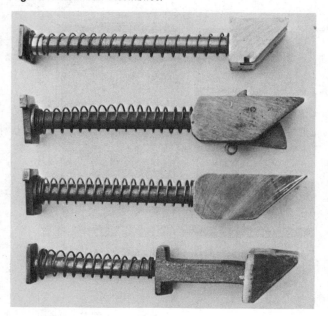

a preshaped extruded bar and may be made of cast iron, cast steel, mild steel, brass, or bronze. Those which are cast or forged are occasionally made with the shaft integral in one piece with the bolt. It is considerably more rare to find the bolt, shaft, and tail all formed in one piece as this precludes the use of a thrust washer. The bolt itself comes in a variety of shapes and sizes, the older designs usually being considerably smaller and generally made of steel or iron, or occasionally brass, and frequently constructed integrally with the latch shaft.

The tapered or "meeting" surface of the old style latch bolt, as seen in Figure 4, has a sharper taper than that of the larger latch bolts in order that it may protrude far enough from the door to enter the strike pocket properly. As a result, it is somewhat more difficult for the edge of the strike plate to push the latch bolt back into the lock case than is a larger

Figure 4. Latch bolt with convex curved meeting edge. Note that a curved meeting edge such as this one means that the strike plate will contact the meeting edge at a different angle at every position on the meeting edge. Contrast this with the meeting edge of more conventional latch bolts as seen in Figures 5 and 6.

latch bolt with a taper which need not be so sharp in order to extend an equal or greater distance.

The configuration of the tapered surface is also of interest; the meeting edge of the old style latch bolt may be curved, and frequently is; but it is rarely domed. It meets the strike plate edge squarely across its full width except for the bevel usually found at the corners where the meeting edge adjoins the top and bottom surfaces of the latch bolt.

This bevel is provided in case the hammering action of the strike plate edge on the meeting edge of the latch bolt causes the meeting edge to expand, raising a "blister" at that point on the top or bottom surface of the latch bolt, thereby preventing

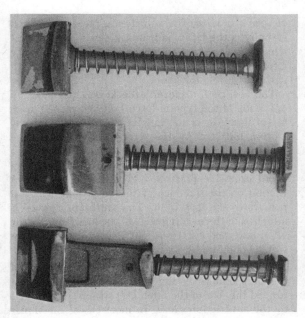

Figure 5. Latch bolts with domed surfaces on the meeting edge.

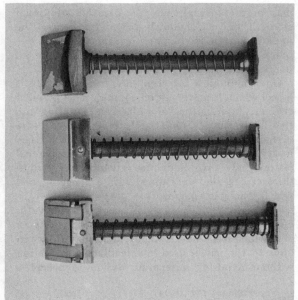

Figure 6. Various types of meeting surfaces used in latch bolt construction. Top to bottom: domed, flat, and plastic insert.

it from moving freely through the hole in the front of the case in which it operates.

The newer, large latch bolt frequently uses a domed surface on the meeting edge, reducing the area of friction with the strike plate while eliminating the need for the beveled edge.

An increasing number of manufacturers are using a latch bolt with one or two strips of plastic set lengthwise into the meeting edge. The inherent lubricity of these plastic strips and their low coefficient of friction results in a more smoothly operating and more durable latch unit. A few of the latch bolts with the plastic strip inserts retain the domed configuration, but the majority are essentially flat, with the plastic strips being used as a replacement for the domed structure in keeping the face of the meeting edge away from the strike plate edge. Various types of meeting surfaces may be seen in Figure 6.

The latch bolt toe, or small tip end of the latch bolt, as seen in Figure 7, is cut off square with the flat latching surface and consists of a small flat area some 3/32 in. to 1/8 in. wide, the full height of the latch bolt. All edges and corners joining this surface are left square.

The toe of a latch bolt is flatted for several reasons: It is safer if someone should happen to bump against it. It is less subject to distortion and

battering than a sharp edge would be, and does not so quickly damage the strike plate surface over which it travels. It provides a surface where a curved or beveled meeting edge may join and blend in with the latching edge without producing a toe which would not otherwise be straight, but would reflect a cross section of the meeting edge where the latching edge intersected it.

As may be seen by referring to Figure 3, the *heel*, or shaft end, of the latch bolt may be cut off square with merely a bevel at the points where the meeting edge and latching edge adjoin it in order to keep these square corners from snagging on the case. It

Figure 7. Latch bolt toe. Note that this portion of the latch bolt is perfectly flat. See text.

may protrude deeply into the case for additional strength, or very shallowly in order to reduce the mass which the latch spring must move to provide quick latching instead of allowing the door to bounce back when it is slammed shut.

One compromise between these two conditions is a heel which is partly cut off square with a pyramid shape rising from the flat surface and pointing into the latch shaft. Some latch bolts which were designed so that they could be used with bit key locks* or locks having rather unusual mechanical features have a pyramidal rise extending deeply into the lock case and are capped with a flange against which other parts may work.

The feature which most latch bolt heels have in common is that somewhere on the heel there is an area which has a flat or slightly undercut surface in order that it may be used in a deadlatch application without requiring a separate design to accept a

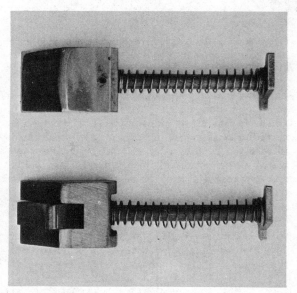

Figure 8. An antifriction latch bolt (bottom) shown in comparison with its equivalent in a plain latch bolt (top).

variation in function. A few deadlatch functions work off the latch tail; consequently, those designs do not require this flat surface on the heel.

Other latch bolts are of the *antifriction* type and are composed of several pieces. A typical antifriction latch bolt consists of the *antifriction piece;* a *pivot pin* by which the antifriction piece is attached to the lock front in such a way that it is free to swing in a restricted arc; the *latch bolt yoke,* a form

*A lever tumblered lock, with the levers housed in the lock case, or a warded lock.

of yoke straddling the antifriction piece and which actually is the latch bolt; and a *thrust pin,* against which the antifriction piece presses to push the latch bolt back into the case when the door is closed without turning the knob.

The antifriction piece is the key to the mode of operation of the type of latch bolt shown in Figure 8. Its function is to transfer the direction of the motion initiated by the strike plate edge on the meeting edge of the latch bolt to a direction at a right angle to the motion of the door, in order to avoid excessive pressure and the resultant higher friction on the flat face of the latch bolt where it penetrates the lock front. The antifriction piece is attached to the meeting edge of the lock front by the pivot pin, and rotates about it in an arc segment which is limited by the travel of the thrust pin.

The thrust pin is fixed firmly between the tangs of the latch bolt yoke, and directly absorbs the thrust toward the rear of the latch bolt which is generated by the antifriction piece. In most cases, the thrust pin engages and travels in a slot at the back, or heel, of the antifriction piece. In a variation of this design, the thrust pin operates in a hole, rather than in a slot, in the antifriction piece. This serves substantially the same purpose, but is intended to avoid damage to the small protruding tang which the slotted method leaves.

In either design, it is noteworthy that the configuration of the latch bolt yoke at the bottom of the slot and the shape of the heel of the antifriction piece must be compatible, or the heel of the antifriction piece will engage that surface and produce a powerful levering action against the thrust pin. The meeting edge of the latch bolt yoke is customarily flat, with a small bevel at those edges where it adjoins horizontal surfaces, as this surface customarily makes little if any direct contact with the strike plate edge.

The *roll-back latch bolt* is pivot pinned to the meeting edge of the lock front in much the same fashion as the antifriction piece, except that in this case, the entire latch bolt travels in a segmented arc.

The meeting edge of this type of latch bolt customarily has a concave curve and the opposite, or latching, edge has a convex curve, which engages a correspondingly curved pocket in the strike plate. The edges of both of these surfaces are beveled slightly to avoid the latch bolt sticking in the lock front in the event of any distortion of the edges. Latch bolts of this type are ordinarily used only in metal doors with metal frames, as shrinkage or swelling of door or frame is likely to cause the strike to be either excessively loose or so tight as to fail to latch altogether.

The roll-back latch bolt, because of its rotary motion, requires that the latch bolt shaft be free to move with respect to the latch bolt, or its motion, too, would be rotary; therefore, the heel of the latch bolt is hollowed out and connection to the shaft is made with a pivot pin. This pivot pin is customarily set through an intermediate piece (for strength) which, in turn, is firmly attached to the shaft.

The Latch Tail

With the exception of the roll-back type of latch bolt, the reversing characteristics (establishing the "hand" of the latch) of the entire latch bolt assembly are often established in the *latch tail*. In reversible latches, the latch tail is always symmetrical in order that it may always present the same configuration, top and bottom, regardless of the direction the latching surface of the latch bolt faces. The latch tail comes in a wide variety of configurations, some of which are shown in Figure 9, and which are largely dependent upon the type of mechanism with which it must function.

The latch tail is designed (as are the other parts of the latch bolt assembly,) in such a way that it may fulfill a number of functions which may or may not all be used in any one assembly or application. This is simply good design practice, and affects economies in production and avoids the need for storage of various types of latch tails embodying minor variations from one usage to another.

Except for those which are cast or forged integrally with the latch bolt shaft, most latch tails are stamped from sheet metal. The metal is ordinarily from 1/16 in. to 1/8 in. thick, depending on the strains involved in the uses to which it may be subjected. It should be wide enough to fill the space between the case and cap sufficiently to avoid slap in the latch bolt assembly and to avoid jamming of any of the pieces which may operate with or from the latch tail, and yet be free enough to move very easily along the area it traverses between the case and cap.

Its vertical height should be sufficient to provide a good bearing surface for all of the parts which function against the latch tail in retracting the latch bolt, even though some of those other parts may become somewhat worn. The tail may be attached to the shaft by riveting or by welding. If the attachment is made by welding, the latch tail may be positioned along the shaft by a simple shoulder

Figure 9. Latch tail configurations. Note the undercut edges on the one at the left. These latch tails are all symmetrical, hence reversible.

turned on the end of the shaft, and its angular position held during welding by a welding jig (a specialized clamp to hold it in place).

In those instances where the tail piece is riveted to the shaft, as is usually done, the hole in the tail piece is customarily of an eccentric shape, with the end of the shaft shaped to match. This keeps the tail piece from rotating on the shaft, even though the riveting may loosen slightly in use.

A simple means of accomplishing this is by cutting one or more flats on the latch shaft end where the tail is installed. This may be done on one, two, or four sides of the shaft, with the hole in the tail piece being punched to conform to this configuration. The tail piece is then installed and the protruding end of the latch shaft is pressure riveted over the tail piece, resulting in a firm, wobble-free fit.

At other times, the tail is not fixed in such a way as to prevent its tendency to rotate, being more or less of a free floating arrangement, kept in alignment by the parallel surfaces of case and cap. Such an arrangement is not in widespread use due to its susceptibility to wear. This wear, when it becomes sufficiently pronounced, will permit the tail piece to tip excessively and eventually results in the failure of the internal mechanism to fully withdraw the latch bolt into the case and clear the strike plate pocket.

The Latch Bolt Shaft

The latch bolt shaft, unless it is cast or forged integrally with the latch bolt, is commonly made of mild steel, and in most modern construction is 1/4 in. in diameter. It is long enough to enclose, between the heel of the latch bolt and the latch tail, all of the parts essential to the operation of the latch in both extended and retracted positions, yet short enough to permit the latch bolt to retract fully into the case with the toe even with the front.

The shaft is attached to the latch bolt by a variety of methods, most of them involving a certain amount of preshaping of the front end of the shaft, as may be seen in Figure 10. In one example of this, the latch bolt is drilled to the shaft diameter; the shaft is straight knurled into a spline, pressed into the latch bolt to the proper depth, and pinned into position with a straight pin.

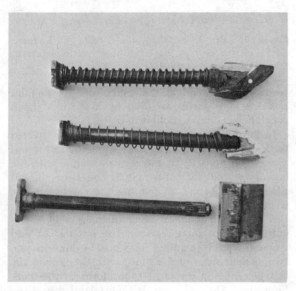

Figure 10. Methods of attaching the latch bolt to the latch shaft. Top: Latch bolt is forged or cast over a knob on the end of the shaft. Center: Latch bolt is forged or cast over notched end of the latch shaft. Bottom: Latch shaft is splined, forced into a hole in the latch bolt and pinned in place.

Another example has the front end of the latch shaft with a groove turned in (lathe cut) and the latch bolt either cast or forged around it. Others drill and tap the latch bolt, thread the front end of the shaft (keeping the threaded portion slightly shorter than the depth of the tapped hole), and screw them very tightly together. Still other methods may include forming a bolt head or other

eccentric knob on the front of the shaft, making it larger in size than the body of the shaft, and casting or forging the latch bolt over the front of the shaft.

The Latch Bolt Spring

The *latch bolt spring*, which actuates the latch return function, is a compression type helical coil spring with the latch bolt shaft thrusting back and forth through the hollow center of the coil. The effective inside diameter of this coil is normally anywhere from 1/32 in. to 1/16 in. greater than the diameter of the latch shaft itself. This allows free motion of the latch shaft within the spring, and yet it fits closely enough to allow the latch shaft to serve as an alignment aid in keeping the coil from kinking as it is compressed.

The forward end of the spring thrusts directly against the latch bolt, while the back end of the spring thrusts against the alignment yoke which is part of the lock case. This thrust may be imposed directly upon the alignment yoke, but usually it is done indirectly with a thrust washer interpositioned between the alignment yoke and the spring itself, as may be seen in Figure 3. Instances where the thrust washer is not used sometimes require the back end of the latch spring to be flared, involving a compound spiral in the spring; that is, the spring increases in diameter as it increases in length.

The spring may be made of spring steel, stainless steel, or of bronze. The general trend in modern manufacture is to make it from stainless steel, because of the superior corrosion resistance of that material.

Since there is no appreciable change in the leverage involved, the compression spring is very uniform and smooth of action through the length measured by the amount of straight line thrusting force against the latch toe which is required to fully retract the latch into the case when working against the spring.

In locks of the European pattern (and some of the older American-made locks), where the spring operates not only the latch bolt, but also serves to return the knobs to their normal or "at rest" position, the strength of this spring and the tension it generates must be considerably greater than in the type usually found in the United States, which is referred to by British lockmakers as the "easy action pattern." The strength of these latch springs is set by the British Standards Institution* at 2-1/2 to 3

---

*British Standards Institution, 2 Park Street, London, W.1., in their standard, B.S. 2088, "Performance Tests for Locks."

pounds for mortise and rim locks, plain action. For mortise and rim locks, easy action (as in most American manufacture), this is set at 1-1/2 to 2 pounds.

Easy action locks of American manufacture are usually somewhat under these figures, or tend toward the low edge; the result of this light tension is a very easy closing of the door without turning the knob. If, however, the latch bolt assembly is too heavy in point of physical mass or weight to be readily moved by the spring tension, the latch may not extend quickly enough to catch the latch pocket in the strike plate, resulting in the door bouncing open again before the latch bolt can engage the strike.

The Thrust Washer

The *thrust washer* provides a flat, smooth surface for the spring to push against, and prevents the spring from pushing against the split yoke, which is the latch shaft guide. This avoids entangling the spring between the yoke and the latch shaft, which would cause them to jam. In addition to this, the thrust washer provides a flat, square-set surface for the spring thrusting against it, thereby reducing the kinking tendency of the latch spring.

The thrust washer is commonly made in two basic patterns, both of which are shown in Figure 3. One of these is merely an ordinary flat washer, intended merely to provide a flat surface for the spring end to thrust against. The hole in this washer must fit the shaft rather closely in order to avoid the spring end becoming entangled between the washer and shaft; its outside diameter must be such that it will fit freely between the case and cap of the latch without tipping or otherwise causing the shaft to bind.

The second type of thrust washer, shown in its operating position in Figure 14, may best be described as a short segment of tubing which has been flanged at the end in contact with the shaft guide yoke. This type of thrust washer is intended to

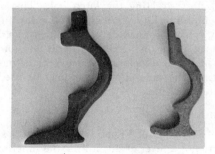

Figure 11. Rocking latch levers.

provide more positive means of ensuring that the end of the spring does not entangle either the washer or the guide yoke and wedge the shaft. In addition, the short end of tubing tends to hold the flange more nearly at right angles with the shaft, providing more uniform compression of the coil spring. The objective of this is to reduce the kinking of the latch spring to a minimum, thus keeping it from rubbing the latch shaft, wearing the spring thin, and reducing its tension or causing it to break.

The Latch Lever

The function of the *latch lever* is fulfilled by one of three types of devices in the various manufactures of locks; the *rocking latch lever*, the *compound latch lever*, or the *shoe*, which is not a lever at all, but fills the function usually ascribed to one.

The Rocking Latch Lever

The fulcrum of the *rocking latch lever* of Figure 11 is frequently a protrusion from the lower wall of the case and is undercut in either a square or rounded manner so as to engage the heel of the latch lever in such a fashion that the lever is free to rock through its full limit of travel. The heel is protruded far enough beyond the side surface opposite the toe to engage the fulcrum undercut of the case to full depth in any position within the lever's limit of travel. A variant of this arrangement uses a pivot post, or stump (a British term), as a fulcrum, with a hole drilled through the heel of the latch lever to fit loosely over the pivot post.

The toe of the rocking latch lever is opposite the heel and serves as a bearing surface on which the latch lever return spring may function to return the latch lever to the "at rest" position, allowing the latch bolt to extend without having to drag a resisting latch lever along with it, thereby wasting the power of the latch bolt spring. This is the basis of the "easy action" latch operation which is so prevalent in American production. The latch lever toe is customarily rounded in such a way that the direction of thrust of the spring is as nearly as possible directly toward the "at rest" position of the latch lever, regardless of the position of the latch lever at any given moment. This not only tends to return the latch lever to the "at rest" position smoothly, but also tends to make uniform the strength of the return thrust by largely avoiding the creation of a "moment of forces" situation.

The *cam bearing surfaces* of the rocking type of latch lever are located directly above and below the center line of the cam; however, the lower bearing surface (nearer the heel or fulcrum) ordinarily has its center appreciably nearer the cam center line than does the upper. This helps make the turning of the knob require more nearly the same turning force and angle of rotation in either direction from the "at rest" position.

Balancing of the mechanical advantages of knob and lever spring by this method requires the use of an asymmetric cam. The cam bearing surfaces are usually flat, but are occasionally found in a slightly concave form; in this case, the corresponding surfaces of the cam assume a convex shape which tends to maintain uniform knob return force as the meeting surfaces of cam and latch lever become worn. The curved portion of the latch lever between the cam bearing surfaces serves merely to provide clearance for the body of the cam while connecting the bearing surfaces.

The uppermost part of the latch lever is the *latch bolt retracting tang,* a nominally rectangular protrusion extending between the latch shaft guide yoke, which is attached to, or part of, the case, and the latch tail. In length, this portion usually extends to just about flush with the top of the latch tail when in the "at rest" position. The corner which engages the latch tail in the retracted position is frequently rounded to provide smoother operation; however, this rounding is usually to a very small radius, amounting to no more than removing the sharp corner.

Latch levers which are intended solely for passage latch use ordinarily conform to this description; however, latch levers which are for multiple use (in other functions such as the spring latch or deadlatch) may have protrusions of various shapes to enable them to work with the additional parts required for other functions, and may occasionally extend above the latch tail. These variations in the retracting tang will be discussed in some detail.

The Compound Latch Lever

The *compound latch lever* is made up of two levers, the *prime lever* and the *floating lever.* The floating lever is positioned under the prime lever and has a hole at the top which fits over a pivot pin. The pin is attached to, or is part of, the case in a position well above both the center line and the upper arm of the cam. This position is customarily the same distance from the center of the cam as is the pivot pin for the prime lever. From this position, the floating lever extends downward beyond the center line of the cam, then turns sharply inward toward the cam's lower arm. A bearing surface for

the cam is formed at this point. At a point usually near the center line of the cam, a pin is attached to the floating lever, protruding upward through a slot in the prime lever. It is this pin which transfers the power of the spring or cam from one lever to the other and causes them to act in unison as the cam is rotated in one direction or the other.

The prime lever is pivoted on a pin near the bottom of the case. The pivot pin frequently does double duty as a post for the latch lever spring. The prime lever extends upward from this point, past the lower cam arm to the coupling slot, which is oblong, roughly parallel to the length of the lever, and just long enough to accommodate the travel of the coupling pin. The slot is centered directly

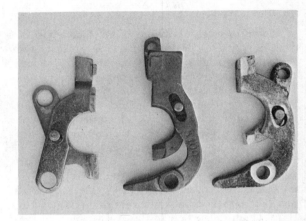

Figure 12. Compound latch levers. Note that both the prime lever and the floating lever are drilled for pivot pins.

opposite the coupling pin, usually near the center line of the cam. Above this point, the prime lever is extended toward the upper cam arm and a cam bearing surface is provided.

The cam bearing surface, on both the prime and floating levers, is usually flat. Upward, beyond the upper cam bearing surface, the prime lever assumes a shape for retracting the latch bolt. This shape, although ordinarily *effectively* rectangular, in cross section, is likely to include features such as a detent lug for an unlocking lever or a pin or protrusion from which a deadlocking lever may be actuated in order that the same latch levers may be used in a deadlatch function.

The Latch Shoe

The *latch shoe* of Figure 13, although a sliding part and not a lever at all, fulfills the function of a latch lever. It is a roughly U-shaped part, usually of cast or forged brass, and operates between the latch shaft guide yoke and the bottom of the case. Cam bearing surfaces are flat and located inside the U at the beginning of the curve of the loop. Further along the curve of the lower side of the shoe and on the outside is another flat surface against which the *shoe spring*, a compression-type helical coil spring, operates.

Figure 13. Latch shoe. Left: top view. Right: underside. Note the flat surface at the bottom of the underside view; this is the thrusting surface for the shoe spring.

Opposite the surface provided for the shoe spring, in a position corresponding to the tail of the small letter u, a tang is extended to engage the latch bolt and retract it. This tang does not engage either the full width or the center of the latch tail as does a more conventional latch lever, but engages only a corner of the latch tail as shown in Figure 14. The underside of the shoe has a groove in the middle of the curve which rides a raised rail cast into the lock case in order to keep the shoe from tipping out of position and failing to completely draw the latch bolt. If this groove becomes too wide or the straight sides of the shoe get too far out of parallel, the latch bolt will not be completely drawn into the case.

Figure 14. The latch shoe, showing its engagement with the latch tail. Note the flanged tube type of thrust washer, shown in the operating position.

Figure 15. Action of the shoe spring and its positioning. Refer also to the underside view in Figure 13.

The Shoe Spring

The *shoe spring* is a compression-type helical coil spring, and operates between the flat surface provided for it on the shoe and a corresponding surface on the rear of the lock case, in the position shown in Figure 15.

Latch Lever Springs

The *latch lever spring* is commonly one of the types seen in Figure 16. It may consist of simply a flat spring, either perfectly straight or with one, two, or three bends or curves in it. These bends or curves control the positioning of the spring and its direction of thrust, and they adjust the amount of force which is applied by a spring of given stiffness and dimension. This flat type of latch lever spring is used with latch levers having a toe against which the spring can operate. The toe end of the spring may have a slight curve at the tip, bringing the end upward and away from the toe of the latch lever. This design prevents a sharp edge or corner of the spring from gouging into the latch lever toe, which would cause the latch lever to remain in the retracted position.

The opposite, or fixed, end of the latch lever spring is engaged in protrusions of the case and is frequently bent or curved in such a way as to keep it properly in place. Depending on the location of those protrusions and their nature, and the amount of tension derived from the type of material used in the spring and its cross-sectional dimensioning, another bend or curve may occur at or near the middle of the spring in order to change the direction

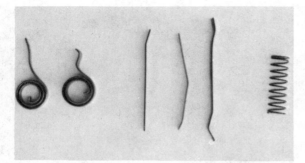

Figure 16. Latch lever springs. Left to right: two spiral springs, three flat steel springs, a shoe spring.

of the spring thrust or to increase the amount of tension applied by the spring to the latch lever, or both.

The other type of latch lever spring in common use is also made of flat spring stock, but is wound into a spiral with a small tang bent to protrude into the hollow center of the spiral. This type fits over a post which is part of the case and which has been notched to receive the protruding tang of the spring, thus fixing it in a position to generate tension as the spiral is more tightly wound. Tension is exerted on the latch lever by the terminal end of the flat spring stock from which the spiral is wound. The terminal end is bent so as to protrude straight out from the mounting post on which the spring is mounted, rather than tangentially from it.

The thrust exerted by this type of spring is controlled by the type and dimensioning of the material from which it is made, and by the angle of departure of the terminal end relative to the positioning tang, as well as by the number of coils in the spiral. It is adjustable by bending the terminal end to suit the thrust requirements engendered by the type and dimensioning of the latch lever and other components.

The spiral type of latch lever spring may be used in conjunction with latch levers having a toe in much the same manner as a flat spring is used, or with pivoted types of single- or double-acting latch levers, and using a peg, pin, or protrusion from the main body of the latch lever to thrust against. If the lever is pivoted, the spring is usually mounted on an extension of the same post which is the pivot post, in an arrangement such as shown in Figure 1.

Cams (or Hubs)

The *cam*, working against the latch lever, begins the transition from the rotary motion of the knob to the straight line draw of the latch bolt. The

camming action is usually provided by two extending arms projecting from the body of the cam; although in rather isolated instances, only one arm is provided. Those with two arms are *double acting cams*, as they may turn in either direction to operate; those with only one projecting arm are *single acting cams*, as they may be turned in only one direction to operate the latch bolt. Most passage latches use double acting cams; however, some (mostly those fitted with lever type handles) may use the single acting cam.

The arms of the cam vary in length, shape, and positioning on the body of the cam according to the requirements of the latch lever with which they must operate. If the cam is used with a rocking type of latch lever, the upper arm of the cam, being further from the pivot, must be longer than the lower one in order to maintain more uniform knob turning pressure. It must also keep the angular

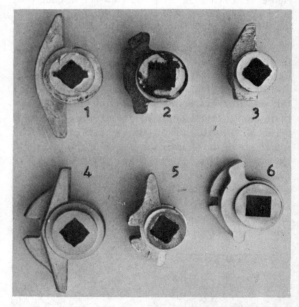

Figure 17. Cams 1, 2, 3, and 5 are one-piece cams. Number 3 is single-acting. Numbers 4 and 6 are two-piece cams. Two-piece cams are shown disassembled in Figure 18.

rotation of the knob constant in both directions despite the difference in distance from the latch lever's pivot point (fulcrum). A slightly rounded corner at the contact surface of the cam arm tip usually results in smoother action between the cam and the latch lever.

A few cam arms are undercut next to the body of the cam so that only the rounded tip of the cam arm contacts the lever, with the objective of ensuring that the required turning force in actuating the latch lever is constant in any position of cam travel. Cams with arms of equal length are usually intended for use with either compound latch levers or a shoe, as these types are designed to require both equal thrust and angular rotation regardless of the direction of rotation of the cam.

The cam body is normally round with smaller diameter shoulders turned concentric with the body in order to provide bearing surfaces which enter holes provided in the case for that purpose. A hole is provided longitudinally through the center of the cam body to accommodate the spindle, or shaft, by which turning force is exerted on the cam. This hole, if the spindle is solid, is ordinarily square. However, some spindles are made in longitudinal sections, and holes for this type of spindle are rectangular, permitting the expanding action inherent in securing knobs to such spindles.

Cams for use with a passage latch are usually of one-piece construction, but it is by no means infrequent that two-piece cams such as those seen in Figure 18 are alternatively used. Although the two-piece cams are intended for use in locking units wherein one knob locks and the other is always free, it is sometimes more economical to produce the extra part for a two-piece cam than to manufacture and stock two types of cams for use with the same latch lever, what with a two-piece cam being able to fill both purposes adequately. Essentially, the two-piece cam is much like a one-piece cam which has been cut in the center, perpendicular to its rotational axis, leaving a half-thickness pair of cam arms on each of the resulting pieces. Indeed, if it were not for the thickness of the saw cut, a two-piece cam could be made in this way.

Since the primary reason for the existence of two-piece cams is to allow one knob to lock, a locking arrangement is provided in the form of a notched hump on one or both halves of the cam, centered on the cam body's axis, in order that a *cam dog* or *stop* may enter one-half of the two-piece cam, immobilizing it while leaving the other half free to turn.

In many cases, the two halves are not coupled together, alignment between the two being maintained by the bearing surfaces of the cam and the spindle. In other instances, the halves are coupled by each half being provided with a cam bearing surface on each side of the half cam, with a short length of tubing between the shoulders of the adjoining bearing surfaces. Others are coupled by the use on one cam half of a bearing surface somewhat smaller in diameter than that which

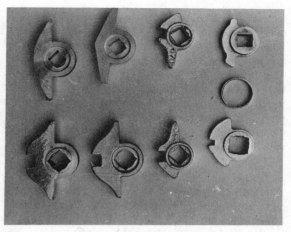

Figure 18. Two-piece cams. Note coupling sleeve which fits between the two cam halves at the right-hand side of the illustration. Pairing of the cam halves is vertical throughout.

bears on the case; this bearing surface, in turn, enters into a shallow matching recess on the mating surface of the other cam half. With this type of coupling, the cam half carrying the bearing surface has the spindle hole drilled out from the mating side to a depth somewhat greater than the depth of the shoulder of the mating bearing, in order that the swivel point of the knob spindle may be centered in the case without turning or attempting to turn the opposite half of the cam.

THE FRONT

The *front*, or *forend* (British term), is that part which is exposed at the edge of the door through which only the latch bolt operates (in the passage latch function). It nevertheless serves several purposes: It supports the latch bolt through the entire distance of its travel. It holds the latch case in position in the door. It provides a convenient means of attaching the latch firmly to the door. It provides the bevel which permits the latch case to be parallel to the inner and outer surfaces of the door, even though the door edge is beveled, to permit opening and closing without binding.

The front is usually one of two types, *plain* or *armored*. The *plain front* is of one-piece construction, and has ears attached to it for mounting the front to the latch case, as well as holes for screwing the front to the edge of the door. Depending on the way in which it is attached to the lock case, a plain front may be *handed, reversible,* or *adjustable*. An adjustable type front is attached to the top and bottom edges of the lock case, usually by three screws. Two of these screws are close to the base of the tang, which is drilled and tapped to receive them. The holes in the case through which these two screws pass fit the screws rather closely. One tang has provision for another screw further away from the base of the tang; the corresponding hole in the edge of the case is enlarged to allow the screw to move about a bit before it is tightened, permitting the front to swivel through a small arc, the center of which is the two inner screws.

A *handed front* is not attached to the case ends, but rather to the body of the case, which has special shoulders in the two open corners, through which the attaching screws (or rivets) pass to the front's attaching tangs. With this method of attachment, the bevel of the front is established by a taper on either the case shoulder or the tang of the front, or both; thus, the latch is not reversible.

A *reversible front* may be achieved in two ways, one of which is much like the handed front. It requires that the latch bolt come from the exact center of the front; the tangs are tapered in opposite directions at the contact surface with the case shoulder (which is flat). To reverse the front, it is necessary to turn it end for end with respect to the case, attaching it with opposite sides of the front tangs in contact with the case shoulders. This system will work only with the passage latch function, and then only with one having the latch bolt in the exact center of the front.

The second means by which a reversible front may be achieved is similar to that for the adjustable front. One tang is pivoted on a single screw, while the other is pivoted on one screw and provision is made in both the case and that tang for two additional screws. The holes for the two additional screws do not both come into alignment at the same time; thus, bevel of the front is dependent upon which of the two pairs of holes is used for the bevel-fixing screw. As an alternative, two holes may be used in each tang. If the second hole is used in the top tang, the bevel is set in one direction; if, however, it is the bottom tang, which is used for the bevel-fixing screw, the bevel of the front is established in the opposite direction.

An *armored front* serves the same purpose as the plain front, but is made in two pieces, an *inside front* and an *outside front* (or *scalp*). The armored front is nearly always of the adjustable type, and is more versatile than the plain front, as the inside front may be used for more than one function without having to be modified in any way. As may be seen in Figure 19, it is precut to accept addition-

al parts whether or not those parts are used. If the parts are not used, then the holes through which they would operate are blanked off by the outside front. This results in production economies, as the

Figure 19. An inside front cut for multiple function use. Note the unique pivot pin near the bottom tang. It is used to attach the tang to the lock case. The flange of this pivot pin is positioned between the case and the tang.

outside front (various functions of which are shown in Figure 20) is merely stamped from sheet metal, and for this reason is much more easily produced in various types for the different functions than is either the inside front or the plain front.

The screws which mount the latch unit to the door are set in countersunk holes in the ends of the inside front and are covered when the outside front is installed; the cylinder set screw which is present in other than passage latch functions is also covered by the outside front, reducing its vulnerability. The outside front is attached to the inside front by two screws (usually #8-32) located just centerward from the mounting screws or just outward from them along the centerline of the front.

The Case

The *case* is one of the most important yet most neglected portions of any latch or lock unit. Although it is true that it fills no active function, it fills a number of passive functions such as positioning, aligning, limiting travel of various parts, and providing bearing surfaces for active parts to operate on. These passive functions are not only numerous, but exacting, as may be seen by an examination of Figure 2.

As in the past, mortise cases continue to be made from cast iron; however, there is a small but growing tendency to stamp them from sheet metal, punching them to receive the various protruding posts, etc., necessary to hold and align the working parts, which are then riveted to the case. This method provides greater accuracy and smoother bearing surfaces for the working parts to move on. With the advent of automatic machinery, this type of production is becoming more and more competitive with the cast iron cases which can be molded in one operation into rather complex shapes. The two major disadvantages of cast iron cases as compared to sheet metal cases are the brittleness of cast iron, and the rough, comparatively abrasive surface on which the working parts must move.

The case is much more than a mere box in which is stuffed some assorted parts. The portion of the case immediately behind the latch bolt forms a bearing surface and guide for the latch bolt as it is drawn in, and keeps it from slapping from side to side. Further back in the case is a yoke, either attached to, or part of, the case. The yoke forms a guide for the shaft of the latch bolt assembly. This same part also serves as a thrust member against which the latch bolt spring may push to extend the latch bolt automatically. Immediately behind this yoke, the case, or rails raised from the case, form a bearing surface for the latch tail.

Another part of the case forms a working surface for the latch lever. At times this surface is merely a flat part of the case; other applications have a raised ridge, curved to correspond to the radius described by the latch lever at that point, thus reducing the area of friction between the case and latch lever. Those cases in which a shoe is used instead of a latch lever have a raised guide rail to keep the shoe properly aligned with the latch bolt assembly. Holes drilled in the case and cap serve both as bearing surfaces for the cam and a means of keeping it in alignment.

Since a latch case is seldom produced solely for use in the passage latch function, it is likely to contain various posts, stops, and holes to accept other parts, making possible the use of the same basic case for a number of functions when supplied with compatible parts. Posts are supplied, drilled, and tapped to accept screws, usually two, for attaching the cap.

The Cap

The *cap* of a mortise unit is similar in construction to the case, lacking, of course, the three turned-up edges (or sides) of the case. It is also similar in function, with two important differences: no working parts are attached to or mounted on the cap; also, the cap is provided with raised portions, posts, etc., as shown in Figure 1, which have the purpose of keeping thin parts properly in place and aligned within the case. These posts and raised portions serve at the same time as bearing surfaces for those parts, so both their length and smoothness is important.

As in the instance of the case, a cap is seldom produced for use exclusively in the passage latch function, and is, as in Figures 1 and 2, likely to contain features of construction which are superfluous to the passage latch function. Even beyond this, in some designs of case and cap for multiple purpose use, eccentricities of design may require the inclusion in a passage latch of parts which serve only a passive purpose. However, these parts are vital to another function and appear in the passage latch merely because their omission would cause a change in tolerances, limits of travel, alignment, or mode of operation to such an extent that it would operate imperfectly, if at all.

Figure 20. Outside fronts for various functions. Left to right: deadbolt; combined deadbolt and springlatch; passage latch; deadlatch. These outside fronts do not mate with the inside front of Figure 19, but do mate with other multiple purpose inside fronts.

The Mortise Springlatch

THE mortise springlatch shown in Fig. 21 varies from the corresponding passage latch in that one knob may be locked while the other is always free, and a lock cylinder has been added to provide for unlocking the unit from the *locked side*. This function requires the use of the two-piece cam, rather than the solid cam.

THE STOPWORKS

The locking in position of one side of the cam is accomplished by the addition of a stopworks mecha-

nism to the passage latch function. Customarily this stopworks consists of four pieces: a *dogging stop;* a *floating stop;* a *stop lever;* and a *stop spring.* These four active parts, in conjunction with passive functions supplied by various guides, supports, and posts (which are part of, or attached to, the case, cap, and front) customarily comprise the stopworks. Unlocking the mortise springlatch is accomplished by the addition of a *mortise lock cylinder,* a *cylinder set screw* or *cylinder clamp* (two different devices), and a *latch unlocking lever,* or by direct action of the *cylinder cam* upon an extension of the stop lever.

Figure 21. A mortise springlatch. Note the cylinder set screw at the right of the cylinder hole.

Figure 22. The four operating parts of a stopworks mechanism. Top to bottom: stop spring, dogging stop, stop lever, floating stop.

16

The Dogging Stop

The *dogging stop* has four essential features: a round push button which protrudes through the front, providing a means for setting the stop into the locked position; a provision for mounting or receiving the stop spring, thus holding the stops in the

Figure 23. The stopworks of Figure 22 in operating position, locking the outside cam.

position in which they are set; a post, pin, or notch to receive the stop lever; and a stop bar to engage the locking half of the cam.

The Floating Stop

Essential features of the *floating stop* include a round push button, which is similar to that of the dogging stop and protrudes through the front in a like manner, and serves the purpose of moving the stopworks into the unlocked position; a pin or notch to receive the stop lever; and it may, as shown in Figure 23, include a flat surface serving as a positioning and alignment guide.

The Stop Lever

The *stop lever* is pivoted, near its center, on a pin or post which is attached to, or part of, the case. Since the dogging stop is supported on both ends, it is essential that the stop lever *not* be firmly affixed to the dogging stop, but rather, be able to adjust to the varying distance between the stop lever pivot pin and the point of contact of the stop lever with the dogging stop. In the design illustrated by Figure 24, this is accomplished by running the end of the stop lever into or through a notch in the dogging stop, allowing the stop lever to slip back and forth through the notch as the movement of the dogging stop between the locked and unlocked positions changes the distance between the notch and the

Figure 24. Another version of the stopworks mechanism. Note the extension of the stop lever through a notch in the dogging stop. In this design, the stop spring is mounted on the dogging stop and moves along with it.

stop lever pin. In the type illustrated in Figure 23, the stop lever has a notched end which fits over a pin on the dogging stop, with the pin sliding back and forth in the notch as the distance requirement changes.

The Stop Spring

The stop spring, in the design shown in Figure 24, is mounted in two notches in the dogging stop, and slides back and forth with the dogging stop as it is moved between the locked and unlocked positions, causing a hump near the tip of the stop spring to move back and forth across a corresponding hump on the side of the stop yoke. This maintains the position of the stopworks in either the locked or unlocked position.

In the types of Figure 23, the stop spring is fitted over a post which is part of, or attached to, the case. This stop spring has two arms projecting from it, one arm being straight and the other having a hump near the tip of the arm. The straight arm thrusts against an extrusion from the case, while the humped arm thrusts against a corresponding hump

Figure 25. This stopworks mechanism utilizes a two-armed stop spring, of which both arms perform a retaining function. Note that both ends of the stop lever are in a floating coupling with the floating stop and the dogging stop. Illustration courtesy of Eaton Yale & Towne, Inc.

which is part of the dogging stop; or in alternate designs, the humped arm may engage one of two notches (also part of the dogging stop), depending on the position of the dogging stop. Figure 25 illustrates a similar principle in which both arms of the spring are functional in positioning the stopworks, one arm operating with the floating stop, and the other functioning with the dogging stop.

The Stop Yoke and Other Case Requirements

The stop yoke, which is part of the case, is set quite close to the locking side of the cam in order to gain a mechanical advantage. The purpose of this is to keep the dogging stop from becoming bent or the stop yoke from being broken in the event excessive turning pressure is applied to the locked cam, as in an attempt to force the lock.

The cap and/or case may also incorporate such extrusions as are necessary to keep the floating stop, stop lever, and stop spring securely in place, and yet allow them to function freely and smoothly. Excessive length of any of these extrusions would cause the stopworks to bind or the spring to malfunction, and result in the stopworks slipping out of the position in which it had been placed. If these extrusions are too short, the stop spring may become dislodged and fall out of position; or the stop lever may become disengaged at one or more points.

The front contains two holes through which the push buttons are activated. These holes are at the bottom of recesses much like drilled holes which have been countersunk. This construction is neces-

sary in order that the push buttons will not project past the face of the front and bind on the strike plate when the door is closed.

A variation in the stopworks design has the dogging stop made in two pieces, with the stop bar detachable and reversible; in this variation, the stop bar covers only one-half of the depth of the lock case at the point of engagement with the cam, and both cams are notched for the stop bar to enter. This arrangement is used primarily in locks having a single acting latch lever since the cam arms in many of these locks are of unequal length, hence they are not reversible.

Operating Elements of the Cylinder

Although the pin tumbler mortise lock cylinder is one of the most important constituents of modern mortise locking devices, it is also one of the most versatile, and therefore most complex components. To discuss it in depth at this point would only detract from the subject matter at hand. For this reason, the pin tumbler cylinder, except for its relationship with other parts, will be relegated to a section of its own. It is sufficient, at this point, to indicate that the mortise cylinder has a threaded body, or *cylinder hull*, containing a *plug* which is rotatable by use of the key. To the rear of this plug is attached a *cylinder cam* which varies in shape and size to be compatible with the design of the mechanism in the lock case in which it is intended to operate.

The mortise cylinder is affixed to the lock case by screwing it into a threaded hole which is provided in the case and/or cap for that purpose. It is locked firmly in place by the use of a *cylinder set screw* or *cylinder clamp*, either of which protrudes through the edge of the threaded hole in the lock case and into a notch in the sides of the cylinder. This prevents removal of the cylinder from the lock until the cylinder set screw or cylinder clamp has been released. Ordinarily, this may only be accomplished through the lock front, or, in the case of an armored front, after first removing the outside front.

The Cylinder Set Screw

The cylinder set screw is usually a long machine screw similar to that partially visible in Figure 21, and has a head which is of the fillister, or modified fillister, type. The head is flush, or nearly flush, in the upper portion of the front. It usually travels along a groove indented into the lock case or cap nearly to the portion which has been threaded to receive the cylinder, and where the case usually thickens to provide extra strength for the cylinder mount. It is into this portion of the case or cap that the cylinder set screw is threaded — right on through the cylinder mounting threads to give access to the cylinder hull's longitudinal groove.

The Cylinder Clamp

The cylinder clamp, as may be seen in Figure 26, is essentially a two-piece device consisting of a roughly Y-shaped yoke, with the sides of the Y paralleled and free to slide back and forth in notches in the case and cap, extending from the cylinder hole toward the front of the lock. The base of the Y is drilled and tapped (from the end), customarily with a *left hand thread*, to receive the threaded end of a rod with a groove. This groove fits into a yoke which is part of the lock case, preventing the rod from moving back and forth while being free to rotate.

With the rod unable to move back and forth, the yoke is drawn out of the cylinder groove as the threaded end of the rod is screwed deeper into the yoke; conversely, as the rod is unscrewed out of the yoke, the yoke is forced into tighter contact with the cylinder. This makes the reason for the left-hand thread apparent since the yoke tightens as the rod is unscrewed. While a left-hand thread unscrews by turning it to the right, long habit causes people to think of everything turning right to tighten. It is manifestly confusing to have anything but a left-hand thread in this application. A variation of this arrangement has the arms of the Y joined together, but with a pivot pin attaching them to the base of the Y.

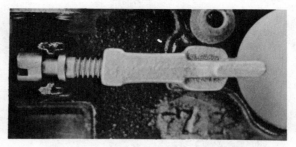

Figure 26. Cylinder clamp is shown in position in the lock case. Note also the attachment of the top tang of the inside front to the lock case.

THE LATCH UNLOCKING LEVER

The *latch unlocking lever* is a relatively simple device, as it consists merely of a simple lever having a pivot hole (or fulcrum) near the cylinder hole in the lock case. The pivot hole is fitted over a pin which is attached to, or is part of, the lock case. The pin may be a long one, nearly the same as the depth of the lock case, in which case (if the latch unlocking lever is, as is usual, made of sheet metal) a sleeve to take up the unused portion of the pin is either attached to the latch unlocking lever or is slipped onto the pin over the latch unlocking lever, as in Figure 27. This sleeve helps to hold the lever in place and in position to contact those parts, and only those parts with which it is supposed to function.

Figure 27. Latch unlocking lever with attached spacer tube. Note that the shape of the lever requires the lock cylinder cam to sweep from the rear toward the front of the lock in order to retract the latch bolt. The round pin just above the latch bolt allows the lever to be used in a deadlatch function if desired.

Figure 28. Latch unlocking lever on a short pin. A pin attached to the cap which contacts the lever just below the pivot pin is used to keep the lever in position. Note that the shape of the lever requires that the cylinder cam sweep from the front toward the rear of the lock in order to retract the latch bolt.

If the latch unlocking lever pin is short, as in Figure 28, it is necessary for an extrusion or applied pin to protrude from the cap into close proximity with the latch unlocking lever in order to keep it in position. The lower tang of the latch unlocking lever may be a simple rectangular shape, one corner of which thrusts directly against the latch tail in much the same manner as does the latch lever; or it may be notched to rest against a stop provided on the latch lever to avoid excessive free motion of the latch unlocking lever between the latch shaft yoke and the latch tail. This leaves the latch unlocking lever free to be operated independently by the key, yet causes it to be caught up by and moved along with the latch lever as the knob is turned.

The pin which is often affixed to the latch unlocking lever in a position usually just above the latch tail serves no purpose in the springlatch function, but enables the same latch unlocking lever to be used in a deadlatch function as well as in the springlatch function. Frequently, the latch unlocking lever has the corner of the tang which contacts the latch tail in the process of drawing the latch bolt slightly rounded in order to smoothen the interaction of these parts.

The upper end of the latch unlocking lever (nearest the lock cylinder) customarily has a right angle bend which extends nearly the full inside depth of the lock case. It is against this bent tang on the latch unlocking lever that the cylinder cam thrusts to rotate the latch unlocking lever through a segment of an arc, thereby drawing the latch bolt.

The latch unlocking lever may be shaped to draw the latch bolt from one direction of rotation of the cylinder cam by using a straight lever, as in Figure 27; or it may have the upper arm of the lever offset to one side, as in Figure 28, to permit drawing the latch bolt by rotating the cylinder cam in the opposite direction. In the latter event, both the cylinder arm and the latch drawing arm of the lever are effectively on the same side of the pivot pin, which is offset to one side of the cylinder.

LESSON 3

The Mortise Deadlatch

T HE *mortise deadlatch* of Fig. 29 is very similar
in its mode of operation to the mortise springlatch;
the basic difference between the two being that the
mortise deadlatch has a latch bolt which is protected against end pressure on the latch bolt when
the door is closed and locked. Thus, when a mortise
deadlatch is in good operating condition, a knife
blade, screwdriver, strip of plastic, or similar object
inserted between the lock front and strike plate will
not succeed in forcing back the latch bolt and
unlocking the door. This feature is made possible by
the addition of an *auxiliary latch bolt* which does
not enter a pocket in the strike plate, but merely
rides the surface of the strike plate, usually in a
position above the latch bolt.

Figure 29. Mortise deadlatch utilizing the addition of parts
to create the deadlatching function from the basis of the
springlatch parts.

The depression of the auxiliary latch bolt by the
strike plate causes certain internal parts of the
mortise deadlatch to move into position behind the
latch bolt, securing it against end pressure after it
has fully extended into its pocket in the strike plate.
At the same time, the movement of internal parts is
usually arranged in such a manner that the push
buttons are also secured against end pressure,
keeping them safe from being tampered with by a
bent wire or similar tool inserted between the lock
front and strike plate.

THE AUXILIARY LATCH BOLT ASSEMBLY

The auxiliary latch mechanism of Figure 30
consists of an *auxiliary latch bolt assembly* resembling a scaled-down version of the main latch bolt
assembly, complete with a guide yoke on the inside
of the case for the *auxiliary latch shaft,* and against
which the auxiliary latch spring thrusts, exactly as
is the case with the main latch bolt assembly. The
auxiliary latch bolt occupies much less space along
the length of the front than its other proportions
would indicate in comparison with the main latch
bolt assembly. It generally has a frontal height of
about one-third that of the main latch bolt.

Ordinarily, the auxiliary latch bolt does not
extend quite as far through the lock front as does the
main latch bolt, since the auxiliary latch bolt rides
the surface of the strike plate rather than penetrating it. The auxiliary latch mechanism is activated by the operation of the strike plate on the
auxiliary latch bolt as the door is closed, and has
the purpose of causing another part to function, as

21

Figure 30. Another version of the mortise deadlatch. Note the resemblance of the auxiliary latch bolt assembly to the latch bolt assembly.

Figure 31. Use of the auxiliary latch tail (top left) to control the functioning of the deadlocking lever, shown in the locked position, as it secures the latch tail rather than the latch bolt.

Figure 32. Auxiliary latch bolt module (at right). This illustration shows the deadlocking mechanism in the disengaged position, although the stopworks is set in the locked position.

shown in Figure 31, rather than being operated by another internal part.

THE AUXILIARY LATCH BOLT MODULE

The variation of the auxiliary latch mechanism seen in Figure 32 utilizes a three-piece module rather than an integrated assembly. It consists of the *auxiliary latch bolt*, the *auxiliary latch lever*, and the *auxiliary latch lever spring*.

The Auxiliary Latch Bolt

The auxiliary latch bolt in this type of mechanism may resemble an isosceles or equilateral triangle, the sides of which are all in the form of a convex curve, with the curve of the inner side of the triangle being of somewhat shorter radius than those of the two sides which protrude through the front. That side of the two which protrudes through the front and which meets the strike plate may have a concave, rather than a convex curve, as shown in Figure 33, to permit positive retraction of the main latch bolt by the strike plate before functioning of the deadlocking mechanism causes the main latch bolt to be incapable of being retracted by the strike plate.

Along the perimeter of the inner curve of this auxiliary latch bolt are six protrusions resembling three pins which have been driven completely through the inner curve, perpendicular to the curvature and approximately evenly spaced along it. The two outer pins of the group fit into notches in the lock front which have been provided to receive them. In the case of a solid, or one-piece front, these are merely indentations in the back side of the lock front. If an armored front is used, these notches may go entirely through the inside front, allowing the pins to come to rest against the inner surface of the outside front (which serves as a stop for them). As the door is closed and the meeting edge of the auxiliary latch bolt comes into contact with the strike plate, the auxiliary latch bolt is rotated about, or pivoted on that pin or pair of protrusions which is nearest the meeting edge of the auxiliary latch bolt. The outer pin, as a result, is pushed back out of its recess and serves no purpose at that time.

After the door is closed, and as it is being opened, the auxiliary latch bolt may simply maintain the position established as the door was closed and slide along the surface of the strike plate in this position; or, if there is sufficient friction between the strike plate and the auxiliary latch both, the auxiliary latch bolt may flip over to the opposite position with the opposite pin becoming the pivot pin for the opening of the door. In any case, when the door is opened and the auxiliary latch bolt clears the edge of the strike plate and springs back out into the "at rest" position, that pin, or pair of protrusions, opposite that which has been serving as a pivot pin, then becomes a second stop, preventing the auxiliary latch bolt from traveling too far through the lock front.

The center pin, or pair of protrusions, when used with a unit having an armored front, serves to keep the auxiliary latch bolt in position during the assembly of the lock, and helps keep it from being lost completely out of the lock in the event the outside front becomes loose or detached from the inside front.

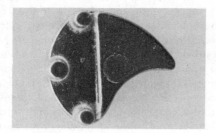

Figure 33. Modular type latch bolt. Note the concave curved meeting edge, and the protrusions along the curve of the edge, which remains inside the lock case.

The Auxiliary Latch Lever

The second part of this three-piece mechanism is the *auxiliary latch lever*. At times this part is made of cast brass or bronze, but it is frequently formed or stamped from ferrous sheet metal, taking the form shown in Figure 34. In either mode of construction,

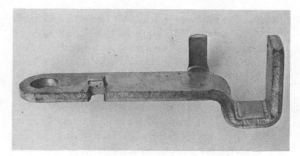

Figure 34. An auxiliary latch lever.

there is a bar which extends virtually the full thickness of the depth of the lock case and is positioned directly behind the innermost curvature of the auxiliary latch bolt, maintaining a tension against it at all times in order to restore the auxiliary latch bolt to its "at rest" position as the door is opened. The body of the auxiliary latch lever, or that portion which is between its two ends, occupies much less of the depth of the case. It uses almost, but not quite, half of the inside depth of the case.

A short distance from the thrust bar, along the body of the lever, is a protrusion, either cast in, or in the form of a bent-out tang, extending most of the way from the latch lever body to the cap. It is this protrusion which controls the movement of the *deadlocking lever* in accordance with the position of the auxiliary latch bolt as the door is opened and closed. The auxiliary latch lever is pivoted at the end opposite the one which meets the auxiliary latch bolt. A hole is provided at that point, enabling it to work freely on a pin or post which is provided on the lock case. This portion of the auxiliary latch lever, if it is of sheet metal, is merely the thickness of the sheet metal; if, however, it is cast, there may be an integral "tube" of the parent metal which may be nearly equal in length to the remaining depth of the lock case.

Somewhere along the body of the auxiliary latch lever, usually between the pivot hole and the protrusion which operates the deadlocking lever, will be another protrusion or groove in which the *auxiliary latch lever spring* may be engaged, except in those cases where the spring is bent with a U shape at the tip to hook directly over the body of the auxiliary latch lever.

The Auxiliary Latch Lever Spring

The auxiliary latch lever spring is customarily of the constrictive type of torsional coil spring; that is, as the spring is flexed away from its "at rest" position, the diameter of the coil tends to constrict, or become smaller. The coil of the spring may be mounted on the same pivot post and directly on top of the auxiliary latch lever, provided that the lever is sufficiently thin to accept the length of the coil on the same pivot pin.

If the auxiliary latch lever is of the type which has a "tube" at that point to provide a broad bearing surface for the pivot, the coil of the spring may be sufficiently large in inside diameter to fit over this "tube." Failing either of these two conditions, the spring may be mounted on a separate post adjoining the auxiliary latch lever. Since this is a torsional type of coil spring, it requires two arms: one for positioning it and keeping the whole coil from rotating on the post, and one for exerting tension on the moving part (in this case, the auxiliary latch lever).

The stationary arm of the spring may be in a location that permits it to thrust against a wall of the lock case, or against a pin or protrusion which is attached to or is part of the lock case. This completes the requirements of the spring for exerting and maintaining tension against the auxiliary latch lever, while allowing it to move away from the lock front with the movement of the auxiliary latch bolt, yet producing sufficient thrust to return both parts to their "at rest" position.

The Deadlocking Lever

The *deadlocking lever* is the device which implements the commands given by the auxiliary latch bolt; that is, when the auxiliary latch bolt is depressed by the strike plate, the deadlocking lever drops into a position in the line of travel of the main latch bolt, securing it against end pressure when it is fully extended into the strike pocket. In addition to this, an arrangement is usually provided for deadlocking the push buttons at the same time, if they are in the locked position. If the three-piece auxiliary latch mechanism is used, the deadlocking lever receives its commands from the auxiliary latch lever; if a unitized auxiliary latch bolt assembly is used, those commands are received from the auxiliary latch tail, as is shown in Figures 31 and 35.

The deadlocking lever is usually stamped from sheet metal, although large numbers of them are cast. The deadlocking lever has a number of features which are essential to its operation: a pivot pin or pivot hole which maintains the deadlocking lever in position in the lock case, yet allows it sufficient movement to perform its function; a shape which coordinates the deadlocking lever's movements with those of the auxiliary latch bolt mechanism; a tang or hook to engage the latch bolt assembly; a protrusion or notch which, in coordination with a complementary pin or protrusion on the latch unlocking lever, operates to disengage the deadlocking lever from the main latch bolt assembly as the lock is unlocked by the key; and usually a shoulder or notch to provide a means of locking the stopworks when the auxiliary latch bolt is depressed, either directly by the deadlocking lever, as in Figure 31, or indirectly by the addition of a *stopworks locking lever,* as in Figure 32. (The deadlocking mechanism of Figure 35 uses another means of locking the stopworks. See also Figure 39.)

Although not always vital, it is common practice for the deadlocking lever to be provided with a means by which it may be spring operated or spring supplemented in a gravity-drop type of operation. Those systems which do not use a spring supplement in a gravity-drop type of operation frequently have a deadlocking lever whose balance is heavily upset in the direction favoring a positive drop.

The deadlocking lever may be said to have one of three modes of operation: *gravity-drop,* with only the force of gravity to move it into position; *spring supplemented gravity-drop;* or *spring operated.* The pivot point (or fulcrum) of the deadlocking lever is usually sufficiently remote from the center of gravity and in a direction from the center of gravity to make a totally spring operated deadlocking lever unnecessary. This avoids sole dependence on a spring which could weaken or break in adverse conditions.

Many types of deadlocking levers which would function efficiently in a pure gravity-drop condition do have supplemental springing due to the fact that conditions of operation are not always optimum. The presence of foreign matter or congealed oil or grease inside the lock case may interfere with the normal operation of a gravity-drop. Supplemental springing, light enough to avoid interfering with the operation of the auxiliary latch bolt, frequently helps to overcome these and other conditions such as a rough interior in the lock case, or a lock which has been improperly installed.

A common method of achieving the deadlocking of the latch bolt is by dropping a part of the end of the deadlocking lever close behind and in the path of travel of the latch bolt. It may be accomplished by providing the deadlocking lever with an end which is roughly in the shape of an inverted J, with the curve of the J riding on top of a protrusion on the auxiliary latch bolt lever as in Figure 32; or, it may ride on the upper surface of the auxiliary latch

Figure 35. Control of the deadlocking lever by the auxiliary latch tail. Notice that the upward extension of the latch lever serves a dual purpose as a latch unlocking lever. Illustration courtesy of Eaton Yale & Towne.

tail. In either case, the deadlocking lever contacts this control point on the auxiliary latch mechanism at a point which is very near the tip of the curve of the J when the auxiliary latch bolt is extended, coming nearer the center of the curve (allowing the deadlocking lever to drop) as the auxiliary latch bolt is further depressed into the lock front.

The sequencing of this series of operations is very important. As the door is pushed closed, the first thing which should occur is the depression of the latch bolt into the lock by the meeting edge of the strike plate. After this depression of the latch bolt has begun, and before it reaches the point where it would be intercepted by the deadlocking lever, the depression of the auxiliary latch bolt into the lock is started by the strike plate, and the deadlocking lever begins to descend toward the engagement position. The deadlocking lever does not reach the engagement position, however, until the latch bolt has already passed the intercept position; therefore, the deadlocking lever merely rides on the top of the latch bolt until the latch bolt springs out into the strike plate pocket. Then, and only then, does it drop behind the latch bolt and assume its function of preventing end pressure on the latch bolt from allowing the door to be opened.

The fact of the obstruction of the path of travel of the main latch bolt by the deadlocking lever creates a problem in the unlocking of the lock with the key in that the deadlocking lever must be lifted from the path of travel of the latch bolt before the latch unlocking lever can draw the latch bolt. This is accomplished by means of a pin or protrusion on the latch unlocking lever (usually just above the latch tail) which engages a tapered surface on the edge of the deadlocking lever to push it out of the path of travel of the latch bolt as the latch unlocking lever rotates through its segment of arc to draw the latch bolt. The tapered surface may be in the form of a hump on the edge of the deadlocking lever, as in Figure 32, or it may be in the form of a notch moving onto the pin as the deadlocking lever is lowered into the engagement position, as in Figures 31 and 35.

Here again, sequencing of these operations is critical. The lifting of the deadlocking lever must begin with the first movement of the latch unlocking lever and must have lifted the deadlocking lever enough to clear the path of travel of the latch bolt *before* it reaches the intercept point.

It is well to recall at this point that only one knob is locked and that this locking is accomplished by the use of a two-piece cam. Since the cam on the unlocked side of the lock is free to turn at all times, it may be used at any time to operate the latch lever. Because the latch lever is incapable of drawing the latch bolt with the deadlocking lever in the engagement position, a means must be provided for the latch lever to move the deadlocking lever, just as does the latch unlocking lever. This is usually done through a coupling of the operation of the latch lever in such a way that the latch unlocking lever moves along with the latch lever to raise the deadlocking lever.

Figure 36. Showing how the latch lever actuates the latch unlocking lever.

This requires that if the latch unlocking lever tang occupies a position on top of, or under, the latch lever tang, a protrusion must be provided on that edge of the latch lever tang which is nearest the latch shaft yoke on the case. A notch is sometimes provided in the latch unlocking lever to receive the protrusion, as in Figure 36; the latch unlocking lever may be thinned in compensation for the room taken by the protrusion.

Another method of accomplishing this requires the interpositioning of the latch unlocking lever between the latch lever and the latch tail. Such an arrangement requires that the latch unlocking lever be used, even in the passage latch function, in order to take up the slack between latch lever and latch tail which its omission would generate, or that another acceptable means of taking up that slack be used.

Another arrangement extends the tang of the latch unlocking lever on downward past the latch tail sufficiently to be picked up by the latch shoe just as it begins to move to draw the latch bolt. It is well to point out that the latch unlocking lever may be operated without its having to operate the latch lever as well, since turning the key would be excessively hard if the latch lever spring were also to be overcome. This does not apply to latch unlocking

levers which are combined with the latch lever as in Figure 37, since the mechanical advantage of the cylinder cam at that distance from the fulcrum is considerable.

Although almost any type of springing may be used to supplement the gravity operation of deadlocking levers, the one most frequently used is of the torsional constricting coil type. This may be mounted on a tube which is part of the latch unlocking lever, with one arm bearing on the deadlocking lever and the other arm bearing on the lock cylinder end of the latch unlocking lever, thus providing an independent return function for both of these parts. Alternatively, the *deadlocking lever spring* may be mounted on a separate post, or pin, attached to or part of the lock case; or it may be mounted on the deadlocking lever itself, with one end engaging the deadlocking lever and the other arm thrusting against some extruded part of the case, as in Figure 35.

The lock case and cap requirements, other than those previously mentioned, for the operation of the deadlocking lever are comparatively simple. A pivot pin, or a tube (to accept a pivot pin on the deadlocking lever), is raised from the inside surface of the case. A pin or protrusion on the cap may be required at or near this point to keep the dead-

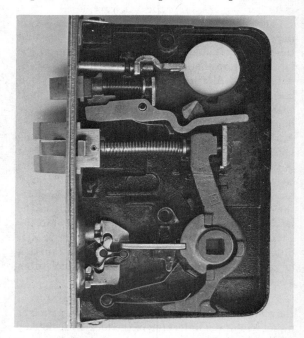

Figure 37. Illustrating a combined latch lever and latch unlocking lever in the deadlatch function. Illustration courtesy of Eaton Yale & Towne, Inc.

locking lever engaged with its pivot. Various other pins or protrusions are added to help maintain the alignment and restrict the travel of the deadlocking lever.

Further help in maintaining alignment of the deadlocking lever may be provided by such parts as the auxiliary latch lever, or the latch unlocking lever, or both. Optimum positioning of the deadlocking lever is usually in the exact center of depth of the lock case to avoid any side strain which might cause it to bend under the pressure of an attempt to force the latch bolt. These positioning aids and factors must accomplish their purpose and yet allow the deadlocking lever to function freely. They must be neither too loose nor too tight against the lever and must present a smooth surface at all points of contact with the deadlocking lever.

Guarding the Push Buttons

One of the means by which the push buttons are guarded against end pressure (when the door is locked and closed) is an arrangement whereby a stair-stepped type of notch is provided on the bottom edge of the deadlocking lever to engage a *stopworks locking lever*. Figure 38 (top) shows how this type of stopworks locking lever is engaged in the deepest portion of the notch as the deadlocking lever descends onto it when the door is closed (if the push buttons have been set previously in the locked position). This locks the push buttons in position for as long as the deadlocking lever remains in the intercept position. Since opening the door by use of key or knob raises the deadlocking lever, and the auxiliary latch bolt mechanism keeps it raised after the door is open, the push buttons are free to operate any time that the door is open. Operation of the push buttons (moving them into the unlocked position) moves the stopworks locking lever off to one side where it cannot be engaged by the notch in the deadlocking lever, and into an adjoining position on the deadlocking lever. The deadlocking lever is held in position and kept from dropping into the intercept position as the door is closed.

The stopworks locking lever is commonly stamped from sheet metal. By means of a pivot hole near the center, it mounts on a shouldered pin or post which is part of, or attached to, the lock case. One end of the lever is provided with a slot or open notch to receive a pin which is part of the dogging stop in ordinary practice. The other end is bent in a right angle extending nearly the full depth of the lock case in order to provide positive engagement with the deadlocking lever when it drops to the intercept

Figures 38a and 38b. A stopworks locking lever. This lever is shown with the push buttons in the locked position and the deadlocking lever (top) ready to descend and immobilize the stopworks locking lever. The illustration below shows the push buttons in the unlocked position and the stopworks locking lever used to keep the deadlocking lever from engaging the latchbolt.

position. This arm of the stopworks locking lever must be long enough to engage the notch in the deadlocking lever; and the notch must be deep enough to allow the deadlocking lever to move freely and far enough to ensure positive interception of the latch bolt.

Other forms of the deadlocking lever include one which deadlocks the latch tail rather than the latch bolt. The arrangement shown in Figure 31 uses a hook which is part of the deadlocking lever and drops down over the top edge of the latch tail. The latch tail is usually grooved on the back side to keep the hook from sliding off the latch tail when end pressure is applied to the latch bolt. In this example, operation of the deadlocking lever is controlled by an arm of the lever which extends upward to a point just ahead of the lower end of the auxiliary latch tail. Locking of the push buttons is achieved by a third arm of the lever, projecting toward the lock front, which moves to intercept the travel of the floating stop end of the stop lever.

Another deadlocking lever includes the use of an expanding coil spring in a spring-supplemented operation. One end of the coil is engaged in a hook which is part of the deadlocking lever, and the other end is slipped over a post or pin which is part of the lock case. Other variations include a completely spring-operated deadlocking lever, which is moved against the force of gravity by a flat spring inserted in a slot in the deadlocking lever, with the free end of the spring resting on a protrusion of the inside of the lock front, or on a case protrusion. This type of deadlocking lever rises from below the latch bolt assembly to the intercept position, in which case the auxiliary latch bolt mechanism may be found to be below the main latch bolt.

One of the more common variations of the mortise deadlatch, affecting its mode of operation, is one in which no floating stop or stop lever is used. Only the dogging stop is used, and its push-button end is shortened so that it does not protrude through the lock front. If an armored front is used, the inside front has holes for push buttons, but the outside front does not. If a solid or one-piece front is used, the hole for the dogging stop is drilled part way through the front from the inside.

In either of these situations, no indication of the existence of a push button appears on the outside surface of the lock front. This frequently requires that the push button on the end of the dogging stop be shortened somewhat. The dogging stop for this mode of operation is usually notched to receive a *dogging stop operating lever,* which is operated by

the lock cylinder. Often, the dogging stop operating lever merely replaces the latch unlocking lever and locks or unlocks the locking knob cam according to the direction in which the key is turned; in this instance, it is necessary that the pin which moves the deadlocking lever out of the intercept position for the unlocked cam be on the latch lever tang.

At other times, both the dogging stop operating lever and the latch unlocking lever are used. This often necessitates the use of a lock cylinder on the inside of the door to operate the dogging stop operating lever, and another lock cylinder on the outside of the door to operate the latch unlocking lever. Since such a lock is not reversible without changing these parts, the lock is handed unless a *coordinating lever* is added.

The coordinating lever may be considered as a lever with only one arm which is wide enough to rest on top of both the latch unlocking lever and the dogging stop operating lever, with the result that both are moved at the same time, resulting in one cylinder operation. For this mode of operation, the dogging stop operating lever is equipped with another lug (or tang) so that rotation of the cylinder in the opposite direction changes the position of the dogging stop. An alternative to the use of the coordinating lever would be the positioning of one tang of the dogging stop operating lever behind the tang of the latch unlocking lever. Either of these methods would allow single cylinder operation and maintain reversability.

The method of protecting the push buttons illustrated in Figure 39 utilizes a spring operated deadlocking lever to obstruct the return of the floating stop to the unlocked position. It is moved out of the intercept position by the toe of the latch lever when it is actuated by either the free knob or the key.

Yet other variations of the mortise deadlatch (and certain other mortise functions) provide for operation of the latch lever by *latch lever lift blocks* which, in turn, are operated by *thumb levers* pivoted on the lock trim. The latch lever lift blocks replace the cams, but thrust against the latch lever in a slightly different way since their motion is either a straight line motion, as in Figure 40, or (in a few cases) very nearly so. The lock case is notched so that the lower edges of the latch lever lift blocks are exposed for a distance slightly greater than that which they are required to move in fully retracting the latch lever, and, in the "at rest" position, come even — or nearly even — with the bottom end of the lock case. A channel (or two parallel channels) is

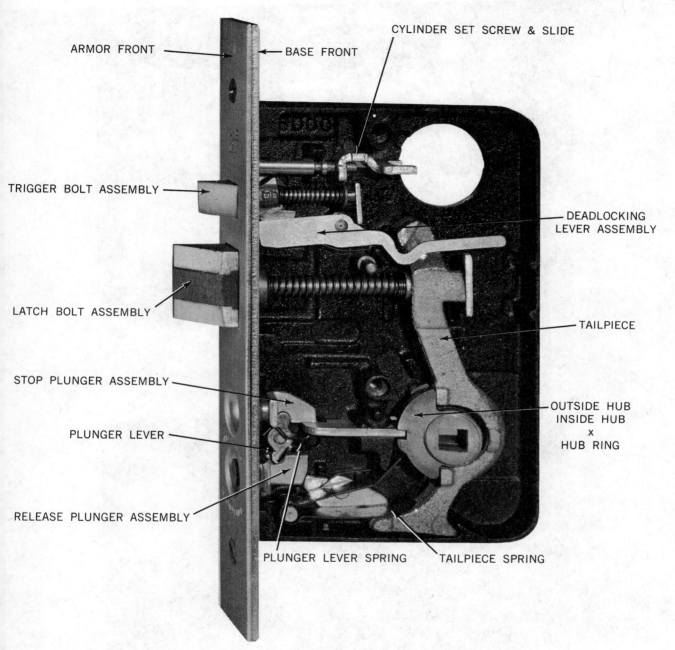

ARMOR FRONT — BASE FRONT

CYLINDER SET SCREW & SLIDE

TRIGGER BOLT ASSEMBLY

DEADLOCKING LEVER ASSEMBLY

LATCH BOLT ASSEMBLY

TAILPIECE

STOP PLUNGER ASSEMBLY

PLUNGER LEVER

OUTSIDE HUB
INSIDE HUB
x
HUB RING

RELEASE PLUNGER ASSEMBLY

PLUNGER LEVER SPRING — TAILPIECE SPRING

Figure 39. Push buttons guarded by an independent lever (at the bottom of the lock case, just left of center). Note the variance in terminology of a number of these parts from that used by the author and certain other manufacturers. Terminological differences for corresponding parts has long been one of the greatest problems a locksmith has in ordering replacement parts, unless he has a parts catalog for the particular lock. Illustration courtesy of Eaton Yale & Towne, Inc.

Figure 40. Latch lever lift blocks. This illustration shows the use of latch lever lift blocks in a mortise deadlatch. Note the use of a coil spring on the dead-locking lever and the operation of the dogging stop operating lever.

provided in the lock case to function as a guide for these blocks as they are operated.

In a conventional arrangement, the latch lever lift block which is farthest from the lock front thrusts directly against a rocking type of latch lever and is always free to operate it. The *free latch lever lift block* operates independently of the *locking latch lever lift block* and may be moved into the "retract" position without disturbing the locking latch lever lift block. The locking latch lever lift block in this conventional arrangement thrusts against a projection of the free one, causing both to operate together. A notch or slot is provided in the locking latch lever lift block into which a dogging stop may enter, locking that block into position. Springing of the latch lever lift blocks is normally accomplished by bringing the arm of a spring (flat or torsional constricting coil) to bear on the free latch lever lift block. Alternatively, the latch lever spring is provided with sufficient thrust against the latch lever for it to thrust, in turn, against the free latch lever lift block with sufficient force to return both latch lever lift blocks to the "at rest" position.

Occasionally, the latch lever lift blocks are attached to a pair of levers which are pivoted inside the case near the lock front, eliminating the need for the guide channels in the case, and causing the latch lever lift blocks to move in a path slightly deviant from the straight line of the channeled blocks.

One of the levers overlaps the other with an arm extending from its side; this is the *free lift lever*. The *locking lift lever* is locked in place by the dogging stop in much the same manner as is the locking latch lever lift block. In all other respects, the operation of the lift levers is similar to that of the latch lever lift blocks; that is, the free lever operates the latch lever directly and is never dogged. The free lift lever is the one which is sprung and is operated by the locking lift lever.

Another variation of the mortise deadlatch is one in which both a cam and a lift lever (or latch lever lift block) are used. Except for the fact that dogging may occur on either the cam or the lift lever, the principle is the same as has been discussed under those parts where they occur separately.

These, then, are a few of the variations which one may expect to find in the mortise deadlatch function. Since the mortise deadlatch is capable of more variations than other types of mortise function, other variations not discussed here do occur and will be encountered from time to time; however, the principles which have been discussed apply to them also and a little thought should resolve them.

The Mortise Deadbolt

THE mortise deadbolt differs from the other mortise functions in that the operation of its locking function is completely independent of the operation of the latch bolt and its associated mechanism. Locking of the mortise deadbolt is done by a bolt, ordinarily rectangular in cross section, which comes just flush with the outside of the lock front in the unlocked position. In the locked position, the bolt extends on through the lock front into a rectangular hole in the strike plate which has been especially prepared to receive it. This bar, or bolt, is the *deadbolt* from which the name of the lock function is derived.

THE DEADBOLT

Essential features of the deadbolt include the deadbolt itself, an arrangement whereby the travel of the deadbolt in and out of the lock case is limited in the extent of its travel; a *detent*, or other means of holding the deadbolt in the position in which it is placed (either locked or unlocked); a spring to operate this detent; and, in those cases where the deadbolt is not operated directly by the cylinder cam, a *deadbolt cam* is provided in an intermediary function between the cylinder cam and the deadbolt.

The Bolt

The bolt itself — that is, the portion which protrudes through the lock front — is usually made of brass or bronze, either bar stock or cast. If it is cast, as in Figure 42, the material is uniform throughout the entire deadbolt. If, however, it is made of bar stock, the deadbolt consists of two pieces (bolt and tang). It may be made all of the same metal, but more often has a bolt of brass or bronze and a tang of sheet metal, as in Figure 43. When the lock is intended for high security use, the bolt may have been drilled from the back (or inside end of the bolt) nearly through to the outside face of

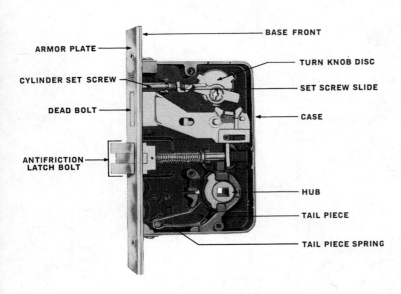

ARMOR PLATE

CYLINDER SET SCREW

DEAD BOLT

ANTIFRICTION LATCH BOLT

BASE FRONT

TURN KNOB DISC

SET SCREW SLIDE

CASE

HUB

TAIL PIECE

TAIL PIECE SPRING

Figure 41. A mortise deadbolt. Note that the parts terminology in this illustration is different in some instances from that used in the text. The variation in nomenclature of corresponding parts from one manufacturer to another is typical throughout the industry. Illustration courtesy of Eaton Yale & Towne, Inc.

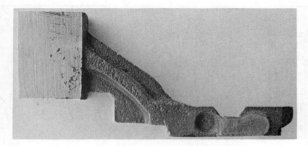

Figure 42. A one-piece cast deadbolt.

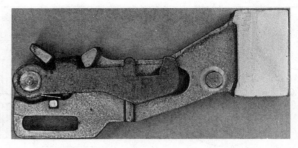

Figure 44. A cylinder cam-operated deadbolt. No deadbolt cam is provided with this type of deadbolt.

the bolt, usually in two places, and hardened steel roller bearings or pins are inserted loosely into these holes. The holes may then be crimped to retain the rollers in place, or the rollers may be retained by the attaching of the tang. Any attempt to saw through the bolt will be frustrated by these hardened rollers rolling under the saw blade.

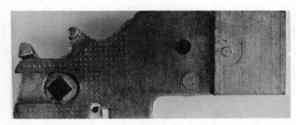

Figure 43. A sectional deadbolt. The bolt is made of brass; its tang is of steel. Attachment is caused by pinning. The insertion of hardened rollers in this type of deadbolt is very easy.

The Tang

The inside portion, or tang, of the deadbolt will include either straight edges or a slot to guide the deadbolt and keep its motion in a straight line throughout the length of its travel. Unless the deadbolt is of the cam-operated variety, it will include a Y-shaped notch in the upper side of the deadbolt tang as in Figure 44. In the mid-point between its limit of travel positions, this notch will be directly below the center of the lock cylinder. When the deadbolt is in either the locked or unlocked position, movement of the cylinder cam across the face of the sloping side of the Y serves to make any necessary small adjustment in the position of the deadbolt to permit disengagement of the *detent*, and holds it in position while this is being accomplished.

As the cylinder cam enters the base of the Y-shaped notch, it encounters a folded-over tang of the *deadbolt detent lever*, depressing and dis-

engaging it from the case detent (stump) and allowing it to travel with the deadbolt tang (on which it is pivoted) until it is in position to engage the opposite side of the case detent. At that position, the cylinder cam climbs up out of the base of the Y-shaped notch, gradually releasing the deadbolt detent lever while holding the deadbolt in position by its action on the opposite side slope of the Y from that which the cylinder cam first encountered. This allows full 360° rotation of the cylinder cam and permits withdrawal of the key with the deadbolt in either position.

The Deadbolt Detent Lever

The *deadbolt detent lever* of the type shown in Figure 45 is attached, in a pivot arrangement, directly to the deadbolt tang. Stops to limit its travel may be provided on the deadbolt tang, case, or front; or limits of travel may be provided which are inherent in its method of engaging the case detent; or limit of travel may be established in only one direction (also provided by the case detent). The deadbolt detent lever may have a second tang, further from the pivot, not necessarily folded over at right angles as is the tang which the cylinder cam engages, but projecting straight out from the body of the detent lever.

When the detent lever is depressed by the cylinder cam, it clears a square post (or stump) on the lock

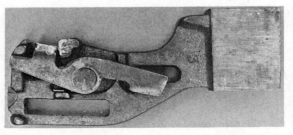

Figure 45. Deadbolt detent lever utilizing protrusion locking rather than notch locking.

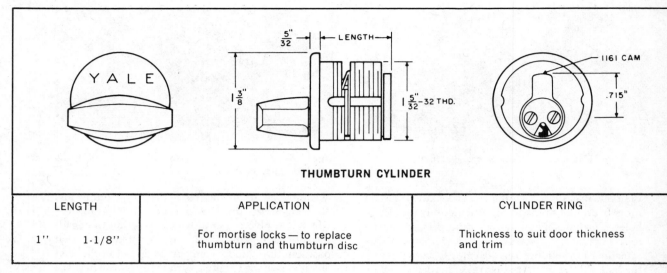

THUMBTURN CYLINDER

LENGTH	APPLICATION	CYLINDER RING
1'' 1-1/8''	For mortise locks — to replace thumbturn and thumbturn disc	Thickness to suit door thickness and trim

Figure 46. A thumbturn (or turn knob) cylinder. Illustration courtesy of Eaton Yale & Towne, Inc.

case, unlocking the *deadbolt assembly* from the case and permitting it to be moved to the opposite position. When the deadbolt assembly has reached the opposite position, the cylinder cam gradually releases the deadbolt detent lever which is raised by the action of the *deadbolt detent lever spring* to re-engage the case detent, but with the case detent on the opposite side of the tang from the side in which it was initially engaged. If a downward limit of travel is to be imposed on the deadbolt detent lever, the procedure is to enclose the detent tang in the body of the lever. The tang then becomes a fully enclosed notch in the lever, roughly in the shape of an inverted U.

The *deadbolt detent lever spring* of the older type is customarily a piece of flat spring steel, one end of which is firmly affixed in a notch which has been provided for that purpose in the detent lever. The other end rides on a protrusion which is part of the deadbolt tang. A more modern variation of this spring is made by the use of a torsional constricting coil type of spring which is mounted on the detent lever's pivot pin, with one arm of the spring thrusting against the deadbolt detent lever and the other thrusting against the deadbolt tang, as in Figure 44.

Turn Knob Cylinders and Discs

This mode of operation will work very well if lock cylinder operation from both inside and outside of the door is intended. If, however, the lock is intended for operation by a cylinder outside and a turn knob inside, as is common in this function,

another feature must be added. It may take the form of a *turn knob cylinder*, on the inside of the door, which is similar to a regular keyed lock cylinder, but is provided with a turn knob where the key would ordinarily be inserted and a spring arrangement for holding the turn knob centered in the "at rest" position when not in use.

More commonly, this cylinder is shortened to about the thickness of the lock case, as in Figure 47, with the turn knob attached to the inside trim, rather than to the cylinder itself. The cylinder cam is driven by a square or flat piece attached to the turn knob and protruding through (or into) an appropriately shaped hole in the cylinder cam. This type of turn knob cylinder is known as a *turn knob disc*.

The Turn Knob Cam

Alternatively, a *turn knob cam* may be added to the deadbolt mechanism to provide an inside turn knob operation. The case bearing for such a turn knob mechanism is usually centered on the *backset line** of the lock, irrespective of the shape of the cam or its mode of operation. Although the shape of the turn knob cam may vary widely, it is confined to one of two basic modes of operation. One of these modes of operation utilizes the turn knob cam to actuate the deadbolt detent lever, while in the

---

*An imaginary line along the lock case, equidistant at all points from the high edge of the bevel of the outside of the door front — usually 2-3/8'' or 2-3/4'' — along which the cam bearing, the lock cylinder hole, and the turn knob cam bearing are centered.

other, the turn knob cam replaces the deadbolt detent lever.

The mode of operation in which the turn knob cam operates the deadbolt detent lever usually places the cam in a hole in the detent lever as shown in Figure 48. The hole in the detent lever is usually of such a size that it comes against the body of the turn knob cam to provide limits of travel for the deadbolt. If the limits of travel so imposed leave the turn knob cam in a position to provide adequate resistance to end pressure on the deadbolt, the case stump may be eliminated in the process of manufacture. If protection against end pressure is not provided, or if additional resistance to end pressure is desired for extra security, the case stump should be retained. Here the mode of operation may be achieved in two manners, although the specific shapes of the parts used can and do vary widely.

One method uses a two-armed cam, one arm being used to depress the detent lever to throw the bolt. Another method uses a cam with one arm which is notched on its end to engage a protrusion or pin set at right angles to the body of the detent

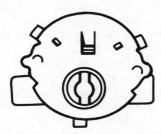

Figure 47. A thumbturn disc (or turn knob disc). The spring which maintains the dual cam in a neutral position is on the opposite side of the disc. A short section of it is visible just above the connecting bar hole. Illustration courtesy of Eaton Yale & Towne, Inc.

lever. The depth of the notch is such that the pin, while always engaged in the notch, comes against the bottom of the notch as the cam is turned, moving the detent lever out of the locked position during the initial phase of throwing the deadbolt. Either of these methods retains the direct operation of the deadbolt detent lever by the cylinder cam in addition to operation by the turn knob cam, with the cam being turned by the operation of throwing the deadbolt.

The mode of operation in which the turn knob cam replaces the deadbolt detent lever makes use of a check spring, usually of flat spring steel, functioning on an irregularly-shaped portion of the cam in order to keep the cam (and deadbolt) in the desired

position, either locked or unlocked, and incapable of being accidentally moved by vibration, or deliberately moved by an attempt to force the bolt back. In some applications, this *cam check spring* serves a purpose in overcoming thrust on the end of the deadbolt; in others, no thrust is absorbed by the action of the spring, and it serves merely to provide a means of keeping the cam properly in line to fulfill its function.

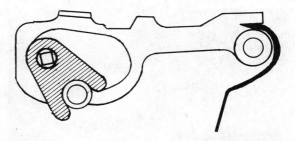

Figure 48. Turn knob cam operation of the deadbolt and its detent lever.

One of the design principles used in this mode of operation utilizes the turn knob cam as a sort of floating lever, with the portion below the cam bearing acting in a Y-shaped notch in the deadbolt tang in much the manner of a substitute cylinder cam, and with the upper portion being operated by the cylinder cam, as illustrated in Figure 49. Deadlocking is achieved by the fact that the bolt-throwing cam arm is never completely disengaged from the Y-shaped notch in the deadbolt, but comes to a stop with the end of the cam arm against the sloping side of the Y. With the cam check spring maintaining the position of the deadbolt cam, the deadbolt is made secure against end pressure.

The specific application of this principle shown in Figure 49 makes use of an interesting relationship between the upper portion of the turn knob cam and the cylinder cam which engages it. The cylinder cam is of the cloverleaf type. The middle leaf first

Figure 49. Turn knob cam replacing the deadbolt detent lever.

engages in a roughly U-shaped notch in the upper end of the cam, moving the deadbolt to the opposite position without deadlocking it. As rotation of the cylinder cam is continued, one of the side "leaves" of the cloverleaf slides along the outside of the U, moves the lower arm of the turn knob cam onto the sloping side of the Y in the deadbolt tang, and completes the deadlocking operation as the side arm of the cloverleaf slides on past the turn knob cam.

Figure 50. Turn knob cam replacing the deadbolt detent lever in a piston type of operation. The presence of the latch unlocking lever behind the deadbolt tang is an alignment factor. Placement of a cylinder lever in the hole of the tang just to the left of the floating lever would make the latch unlocking lever operational when the bolt is in the unlocked position.

In the application shown in Figure 50, the turn knob cam (or deadbolt cam) replaces the deadbolt detent lever, utilizing much the same principle as the crankshaft in an automobile engine. A three-armed cam is used in this application, one arm being notched to receive and be operated by the cylinder cam, and another at about right angles to it and somewhat pointed on the tip to slide back and forth over a hump on the cam check spring. Approximately opposite this arm of the turn knob cam (or deadbolt cam) is the third arm, which has a pivot pin near its tip. To this pivot pin is attached a floating lever whose other end is pivoted to the deadbolt tang.

If one visualizes the turn knob cam as the crankshaft in an automobile engine, and the floating lever as the connecting rod, it is easy to see how the turning of the turn knob cam causes the deadbolt (or piston) to move back and forth in a straight line. The bolt is maintained in alignment by the lock front and the deadbolt tang is kept in alignment vertically by a notch in the tang which straddles the hub of the deadbolt cam. This same notch bottoms against the hub bearing to provide a

limit-of-travel in the retracted position. Horizontal alignment of the deadbolt tang is provided on one side by the presence of the cam arms against the tang, and on the other side by the latch unlocking lever which fits between the case and the deadbolt tang. Limit-of-travel in the extended position is provided by the length of the floating lever. Protection against end pressure on the bolt is afforded by permitting the deadbolt cam to turn to a position just past the fully extended position, at which point the cam arm to which the floating lever is attached comes up against a stop on the deadbolt tang. End pressure on the bolt will then force the cam arm more tightly against its stop on the tang. Action of the cam check spring in this instance serves as a protection against vibration and to hold the bolt in the extended or retracted position, rather than making an active contribution to the deadlocking function. The presence of this spring in position in the lock case also serves as a retainer to keep the latch unlocking lever in position on its pivot pin.

It should be noted that this example of the latch unlocking lever does not have a tang for operation by the cylinder cam. Instead, it is operated by a *cylinder lever* which is pivoted on the deadbolt tang. The cylinder lever, as shown in Figure 51, has the tang for operation by the cylinder cam. But the tang, being attached to the deadbolt, is out of reach of the cylinder cam at such times as the deadbolt is in the extended position. The arm of the cylinder lever which operates the latch unlocking lever slides back and forth along a matching arm of the latch unlocking lever as the deadbolt is thrown. Many deadbolt tangs have provision for accepting a cylinder lever whether or not that cylinder lever is used, in order to achieve maximum utilization of parts.

The cylinder lever is always provided in those cases where push buttons are added to combine the mortise deadbolt function with a mortise springlatch function in the same case. Guarding of the push buttons is optional in this instance, since the latch bolt is not guarded. One method of guarding the stops provides for a shoulder on the deadbolt tang to come behind the floating stop, securing it against end pressure only when the deadbolt is extended. This system also requires that extending the deadbolt, either by key or by turn knob, automatically sets the push buttons as well. The push buttons may be set without throwing the deadbolt, in which case the stops are not secured against end pressure.

Another method uses a spring operated rocking

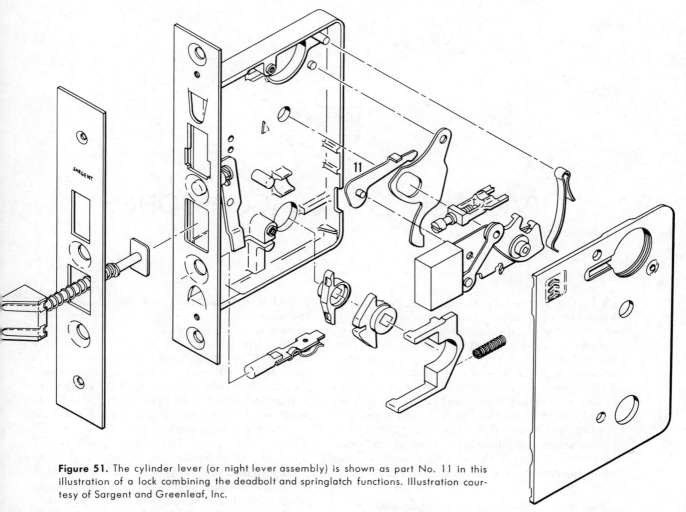

Figure 51. The cylinder lever (or night lever assembly) is shown as part No. 11 in this illustration of a lock combining the deadbolt and springlatch functions. Illustration courtesy of Sargent and Greenleaf, Inc.

lever, one end of which rises behind the floating stop, while the other rests on the toe of the latch lever, as in Figure 39. This type of push-button stop is always secured against end pressure whenever set in the locked position, irrespective of any operation of a deadbolt. Unlocking of the stop is accomplished by turning the inside knob sufficiently to tip the latch lever to a degree whereby the latch lever toe will rise enough to cause the opposite arm of the *stop deadlocking lever* to clear the floating stop.

Other variations of the mortise deadbolt include the use of latch lever lift blocks in an entry lock mode of operation using thumb levers, or a thumb lever on one side and a knob on the other.

Yet others include a split, or dual, deadbolt for use in mutual consent access type of situations, as for example, for communicating doors between offices or hotel rooms. Typically, this consists of a pair of deadbolts, each of which occupies half of the hole in the lock front which would ordinarily be occupied by a single deadbolt, with the division between them made horizontally. The tangs of these two deadbolts lie side by side in the lock case, being only about half the thickness of an ordinary deadbolt tang, but in other respects similar in both appearance and operation.

LESSON 5

Mortise Lock Accessories

Mortise lock accessories, that is, those parts which are not contained inside of the lock case, are often considered (with the exception of the lock cylinder) as nuisance items, a necessary evil, or of minor importance. This is far from true. The workmanship, the accuracy of manufacture and design, and the positioning of these parts in relation to the corresponding case-housed parts are of no less importance than are these same factors when applied to parts inside of the lock case. Mortise lock accessories provide alignment and bearing surfaces, as well as a support function for those parts by which operating forces are exerted on the internal parts.

Improperly made, misaligned, or damaged accessories can cause malfunctions at any time; and can cause the rate of wear of both accessories and internal parts to be almost unbelievably accelerated. Although relatively little space will be devoted to these accessory-type items, it is not because they are lacking in importance, but because their design is simple and readily understood once their purpose becomes clear. Actual construction details may vary considerably, even among locks of a single manufacture, and space prohibits a comprehensive examination of these variations except in principle.

Knob Spindles

Mortise lock spindles, or *knob spindles*, come in a number of variations; the most common of these fitting into one of three classes, each class of which is intended to fulfill a specific purpose relative to the mode of knob operation. The three major types

of knob spindles are shown in Figure 52 and include the *plain spindle* (or *straight spindle*), the *swivel spindle*, and the *hook spindle*.

The *plain spindle*, like all other mortise lock spindles, is essentially square in cross section and of such size as will readily permit its being slipped through the square hole in the hub, and yet be without excessive torsional play inside the lock hub, which would cause both hub and spindle to wear rapidly. It is solid throughout its length and usually 5/16 or 3/8 in. square. This type of spindle is used with locks whose function permits both knobs to be solidly coupled and always to turn at the same time, as in the passage latch and deadbolt functions where neither knob is locked at any time. To lock one knob when the other is solidly coupled to it by the spindle would result in both being locked at the same time. In a lock intended for one cylinder operation, this would mean that there would be no means of unlocking the door from the side which does not have a cylinder.

Since the spindle fits closely in the hole in the hub, its alignment with the hub is very critical if binding of the cam arms or cam bearings (of the hub) against the lock case is to be avoided and the knob bearings are not to bind.

Three different types of ends are used on these spindles to provide for attachment of knobs. The *pin end spindle* has one or both ends drilled, usually in a number of places, to provide a degree of adjustment and to accept a pin which passes through the knob bearing and the hole in the spindle. The pin is retained in position either by the tightness of its fit

⁵⁄₁₆″ TRIPLEX SPINDLES

5⅛″ Plain Spindle—pin end x triplex end
For D35, G35 and D134 knobs in pairs—1⅜″ to 2½″ doors inclusive.

5⅛″ Swivel Spindle—pin end x triplex end
For D35, G35 and D134 knobs in pairs—1⅜″ to 2½″ doors inclusive.

⁵⁄₁₆″ THREADED SPINDLES

4″ Plain Spindle—pin end x threaded end
For D133 and D134 knobs in pairs (808 and 802 locks)—1⅜″ and 1¾″ doors.

5⅛″ Plain Spindle—pin end x threaded end
For D35S, G35S, Advance and Deluxe knobs in pairs—1⅜″ to 2¼″ doors inclusive.

5⅛″ Swivel Spindle—pin end x threaded end
For D35S, G35S, Advance and Deluxe knobs in pairs—1⅜″ to 1¾″ doors inclusive.

5⅜″ Swivel Spindle—pin end x threaded end
For D35S, G35S, Advance and Deluxe knobs in pairs—2″ to 2¼″ doors inclusive.

⁵⁄₁₆″ THREADED HOOK SPINDLE

3″ Spindle—threaded end x hook end
For D35, G35 and D134 single knobs—1⅜″ to 2½″ doors inclusive.

For D35S, G35S, Advance and Deluxe single knobs—1⅜″ to 1¾″ doors inclusive.

3¼″ Spindle—threaded end x hook end
For D35S, G35S, Advance and Deluxe single knobs—2″ to 2¼″ doors inclusive.

Figure 52. Knob spindles. Courtesy of Eaton Yale & Towne, Inc.

or by being covered by the rose bearing which fits over the knob bearing, or both. Knobs intended for this method of attachment usually have a plain, square hole in the knob shank, which merely slips onto the spindle to be retained in position by the pin.

A variation of the pin end has these holes in the spindle drilled and tapped to accept a machine screw which is passed through the knob shank and screwed into the spindle. If both ends of the spindle are provided with pin ends and the holes are not tapped, the first knob is used with a plain rose and the second (inner) with an *adjustable rose,* which permits the pin to be inserted in the second knob after the spindle has passed through the door. Then the bearing surface of the adjustable rose is unscrewed until the bearing surface is in its proper position on the knob bearing, covering the second pin and retaining it in position.

The *threaded end spindle* may have one or both ends threaded, in which case the knob shank is also threaded. A set screw is provided through the side of the knob shank in a position to bear on a flat side of the spindle, thus keeping the knob shank from coming unscrewed from the spindle. A variation in the mode of attachment of the knob to the threaded end spindle utilizes what is called a *screwless shank knob.*

The screwless shank knob, as seen in Figure 53, is essentially a three-piece knob. A *thimble,* carrying the knob bearing and having internal threads which mate with external threads on the knob shank, is inserted over the spindle. A nut which is specially shaped to lock to the knob shank is screwed onto the spindle and into the cup of the thimble. A knob with a shank having a square hole and externally threaded is slipped onto the spindle and made fast by screwing the thimble onto the shank threads. The specially shaped nut is locked to the knob shank at the same time, forming a unit which cannot come out of adjustment until the thimble is completely unscrewed and the knob removed.

The *triplex end spindle,* featured on a number of Yale locks, is divided into three portions along its length. Each of these three sections is roughly triangular in shape, leaving one quadrant of the spindle open to pass by an alignment protrusion inside the otherwise square hole in the knob shank. A set screw in the side of the knob shank which is opposite this alignment protrusion thrusts against the center section of these triangular sections, wedging them apart and tightly against the knob shank, thus securing the knob tightly to the spindle,

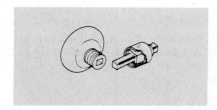

Figure 53. A screwless shank nob. Illustration courtesy of Eaton Yale & Towne, Inc.

without threads or pins. Only one end of a triplex spindle is attached in this fashion, the other end commonly being a pin end. In a plain spindle, the pin is drilled through in such a way as to catch all three segments of the triplex. In a swivel spindle only one end is triplexed.

The *hook spindle* is a variation of the plain spindle, resembling half of a plain spindle. It is used where a knob is desired only on one side of a door and the spindle is divided lengthwise into two halves. The knob end of these halves is threaded for knob attachment, with the opposite end of one of the halves bent (or hooked) outward to engage behind the hub. The hooked half (usually) has a pin protruding from the mating surface near the threaded end which enters a corresponding hole in the straight half to provide for proper alignment of the threads as they enter the knob shank. The hooked half is inserted through the hub first to such depth as will allow the hook to engage the far side of the hub, and then the straight half is pushed straight into the hub with the mating surfaces together, until the alignment pin engages its hole. The knob may then be attached to the threaded end in the usual manner, although it may be necessary to be careful to avoid pushing the spindle too far into the door.

The *swivel spindle* is a two-part spindle, coupled in the center so that either half is free to turn in either direction, independently of the other. The swivel spindle is used primarily in those functions in which one knob may be locked while the other is always free, allowing independent operation of the knobs without regard to the locked or unlocked condition of the other.

Two methods are commonly used to achieve this coupling. One method utilizes a projecting threaded stud at the inner end of one half of the spindle which screws into a drilled and tapped hole in the inner end of the other half of the spindle. In operation, this stud is unscrewed about three-fourths of a turn from the snug position to provide ample free swiveling action before the spindle is

OUTSIDE KNOB—PIN END
All knobs in pairs are supplied with outside knob pinned to spindle. Knob pin is concealed by rose or thimble. Multiple holes are drilled in spindle to provide adjustment for door thickness.

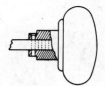

INSIDE KNOB—TRIPLEX TYPE
D35, G35 and D134* knobs in pairs are supplied with triplex spindles.

Set screw pressure on central wedge piece expands the spindle to a tight fit in the shank and effectively prevents loosening of the set screw.

*D134 Knob is also supplied with a threaded spindle for lever tumbler locksets.

STANDARD, ADVANCE AND DELUXE KNOBS
(Typical Construction)
.035", .050" or .081" thick top as specified in trim section, closed into a groove in knob base. Solid rod shank screwed into knob base.

INSIDE KNOB—SCREWLESS SHANK
D35S, G35S, Advance and Deluxe knobs are supplied with threaded spindles.

Screwless shank prevents removal of knob without the use of a special wrench. Sleeve is adjusted for end play by positioning the nut on threaded spindle. When sleeve is screwed onto shank, nut is retained in position by a recess in knob shank.

INSIDE KNOB—THREADED TYPE
D35, G35 and D134 single knobs are supplied with threaded hook spindles.

D134* knobs in pairs are supplied with a threaded spindle for lever tumbler locksets.

*D134 knob is also supplied with triplex spindles.

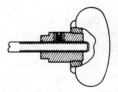

WROUGHT STUFFED SHANK KNOB
(D133 and D134 Knob Construction)
Wrought Knob Top closed into a groove in the stuffed shank.

Shank has a closely fitting rustless insert with a long spindle bearing.

Figure 54. Knob construction. Illustration courtesy of Eaton Yale & Towne, Inc.

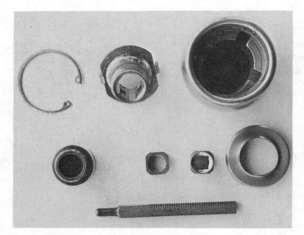

Figure 55. Sectional knob construction. This knob is of the screwless shank type.

inserted into the lock. Another method utilizes a similar stud which is not threaded but has a groove around it, and which fits into a smoothly drilled hole in the opposite half of the spindle. The drilled end of the spindle is crimped or swaged to lock into the groove.

An alternative to this method is the use of a snap ring in the groove of the stud which will expand into a corresponding groove inside of the drilled hole to such an extent as to have the ring partially in both grooves when it expands. The non-threaded coupling does not require adjustment of the spindle coupling before it is inserted into the lock, since it allows 360° rotation of the two halves. An essential feature of spindles in this class of service is a provision to keep the outside knob from being removed, and the inside knob and the complete spindle from being pushed through the hubs and out of the door. Otherwise, a tool could be used to reach through the outside hub to turn the inside hub, thus bypassing the requirement for a key from the outside. For this reason, the spindle half which is intended for use on the locked side of the door may have a pin through it in close proximity to the hub, or be crimped or otherwise distorted at that point to such an extent that it will not pass completely through the hubs.

An alternative solution to this problem utilizes hubs with the locking side having a larger spindle hole than the non-locking side. The locking side of the spindle is correspondingly larger at the point where it enters the hub. Alternatively, protection against this may be obtained by the use of a pinned attachment for the knob on the locked side of the door, together with a fixed or rigid type of rose.

Ends on the swivel spindle may be both threaded or both pinned (if an adjustable rose is used); or the outside (or locking) end may be a pin end and the inside threaded or triplex.

KNOBS

Mortise lock knobs may be solid, wrought, or sectional. Solid knobs are usually made by casting, and as the process implies, are of a single, uniform material, with top and shank in one piece. Wrought knobs have a hollow top which is usually attached to the shank by closing the top into a groove near the outside end of the shank. Knob tops of this type may be formed from sheet metal, while the shanks may be cast, forged, machined, or pressure cast with a thin sheet metal shell covering for the pressure cast. (The latter type is known as a stuffed shank knob.)

The sectional knob is made up of a number of parts, or sections, which vary from one manufacturer to another. The sectional knobs are usually made by several methods, depending on the nature of the part, and may include casting or die casting, stamping, forging, or machining of metallic parts. Sectional parts may be assembled by use of threaded parts, by pinning, by screws, by snap rings, or by a combination of these methods.

The knob bearing, located at the inner end of the shank, fits into a complementary bearing surface in the rose or trim plate (escutcheon). The knob bearing should be sufficiently snug in the rose bearing to avoid excessive wobble of the knob and spindle, and yet be free enough to allow unimpeded turning and spring return of the knob. It is this knob bearing, in association with the bearing provided by the trim, which keeps the spindle running true and squarely through the cam, thus minimizing stresses and consequent wear in the internal lock mecha-

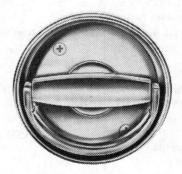

Figure 56. Cup (ring) handle. Illustration courtesy of Eaton Yale & Towne, Inc.

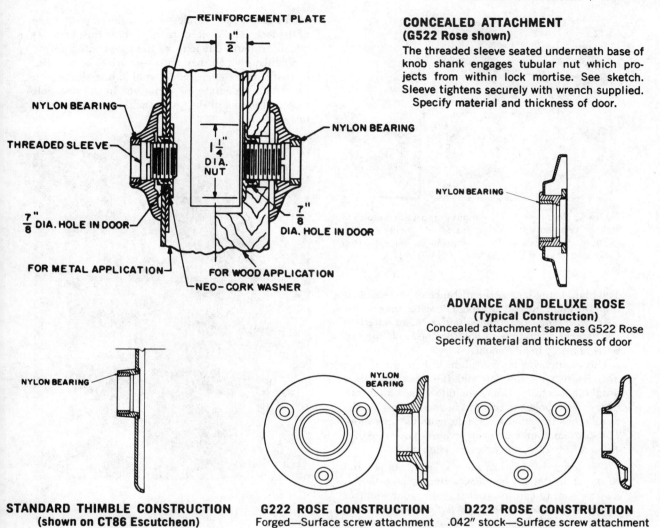

CONCEALED ATTACHMENT
(G522 Rose shown)
The threaded sleeve seated underneath base of knob shank engages tubular nut which projects from within lock mortise. See sketch. Sleeve tightens securely with wrench supplied. Specify material and thickness of door.

ADVANCE AND DELUXE ROSE
(Typical Construction)
Concealed attachment same as G522 Rose
Specify material and thickness of door

STANDARD THIMBLE CONSTRUCTION
(shown on CT86 Escutcheon)

G222 ROSE CONSTRUCTION
Forged—Surface screw attachment

D222 ROSE CONSTRUCTION
.042″ stock—Surface screw attachment

Figure 57. Rose construction and knob bearing attachment to escutcheon plate. Illustration courtesy of Eaton Yale & Towne.

nism. For this reason, it is very important that the spindle, knob bearing, trim bearing, and cam all be concentric with one another.

Variations of the knob top may include the use of materials such as plastics, ceramics, glass, or wood. Design variations may include such items as ring handles, thumb turns, and lever handles.

ROSES

A *rose*, or *rosette*, as it is often called, is essentially a small, but thick, metal plate which is attached to the surface of the door either by screws or by a threaded tube and nut arrangement which is part of the rose itself. The rose contains a mating bearing for the knob bearing which serves to correctly align the spindle with the lock cam. The rose may be of an all-metal construction, or it may include a plastic insert for its bearing to reduce the friction of the knob bearing with a resultant increase in wear resistance of both of these bearings.

ESCUTCHEONS

As an alternative to the use of roses, the knob support bearing may be attached to a larger metal plate called an *escutcheon*. It serves the same purpose with respect to the knobs as does the rose. It is, however, dimensionally larger than the rose, and is usually attached to the door by four screws.

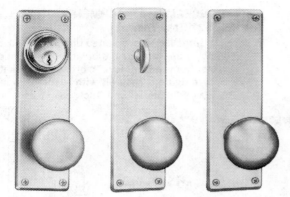

Figure 58. Escutcheons for various purposes in a matched styling. Left, outside; center, inside for deadbolt; right, multiple use—passage latch or inside for springlatch or deadlatch. Illustration courtesy of Eaton Yale & Towne, Inc.

The plate being larger, the distance of these screws from the knob support bearing is greater than in the case with the rose, providing a more effective mechanical advantage in holding the knob support bearing firmly in alignment.

The escutcheon may include a hole for a lock cylinder to pass through and into the lock. In a deadbolt function, it may include a turn knob but no cylinder hole on the inside escutcheon, if the deadbolt is intended for single cylinder operation. This turn knob may be designed and positioned for either a turn knob cam or a turn knob disc.

The design of the turn knob is quite simple and easily understood when encountered; however, it may be well to caution that the turn knob shaft or tang may need to be cut somewhat shorter than that suppled by the manufacturer to suit various thicknesses of doors.

In those used with a deadbolt cam, the turn knob shaft may bind against the opposite side of a thin door; or, if used with a turn knob disc, the tang may

bind against the cylinder cam on the opposite side of the lock. The shaft, or tang, of these turn knobs is usually square, but may be flat, depending on the mechanism which they must operate. The spindles or shafts are usually free to wobble just slightly, in order to compensate for variations in the deadbolt cam or turn knob disc and small eccentricities of installation.

Figure 59. Deadbolt escutcheons, for use with deadbolts having the turn-knob cam type of construction. Note the difference in positioning of the turn-knobs in this illustration compared with that of the corresponding escutcheon in Figure 58.

The Mortise Panic Exit Device

Simply stated, the *mortise panic exit device* consists of a mortise deadlatch which has the inside knob or thumb lever replaced by a panic bar mechanism. The panic bar mechanism, when depressed, will retract the latch bolt and open the door. While it is true that the deadlatch itself is usually an especially rugged and heavily built model and avoids the use of the push button stopworks in favor of the key-operated stopworks, there is no deviation of principle involved in its construction. Outside operation remains the same as that for other applications of the mortise deadlatch, either by knob or by thumb lever unless the stopworks is set, in which case operation from the outside is by key only.

A panic exit device must always open out; that is, away from the side of the door on which the panic bar is mounted, in order that the same pressure which retracts the latch bolt will also open the door. This is essential since in panic situations, if people are pressed tightly against a door, it may be impossible to open it if it opens inward. Thus, it is this difference of the door *always* opening outward, together with the addition of the panic bar mechanism, which creates the essential difference between the mortise deadlatch and the mortise panic exit functions.

The panic bar mechanism consists of three basic assemblies and their attendant mechanisms: the bar assembly, the inactive end case, and the active end case. Both of the end cases are surface-mounted on the door and each of them is equipped with a pivot pin for attaching the ends of the panic bar to the case. In a few instances, this completes the mechanism at the inactive end of the panic bar; however, in most applications, the *panic bar end* extends into the case in order that it may be acted upon by a coiled return spring, usually of the compression type, but sometimes of the torsional type. The torsional type, where used, is commonly mounted in a recess in the panic bar end and around the pivot pin, with one free end thrusting against the panic bar end and the other end thrusting against the inactive end case. In many instances where the inactive end is sprung, the limit-of-travel stops are provided by arranging for a portion of the panic bar end to come into firm contact with some portion of the case.

Since most of the mortise lock units used with panic exit devices are of the thumb lever operated type, it would appear that panic bar operation could be readily achieved merely by extending the active arm tang of the panic bar end deeply enough into the door to pick up the latch lever lift block and operate the lock mechanism in this fashion. This is not the usual situation, even though it is true that it would operate the lock unit for a time.

Because of a combination of two factors, failures of such a mechanism would be both frequent and severe. The fact that the motion of a tang of this type would be essentially a rotary motion would mean that the moment of thrust at various points in the process of raising the latch lever lift block would not be in a straight line with the motion of the latch lever lift block. This would result in a sidewise strain on the lock case in one or both directions with

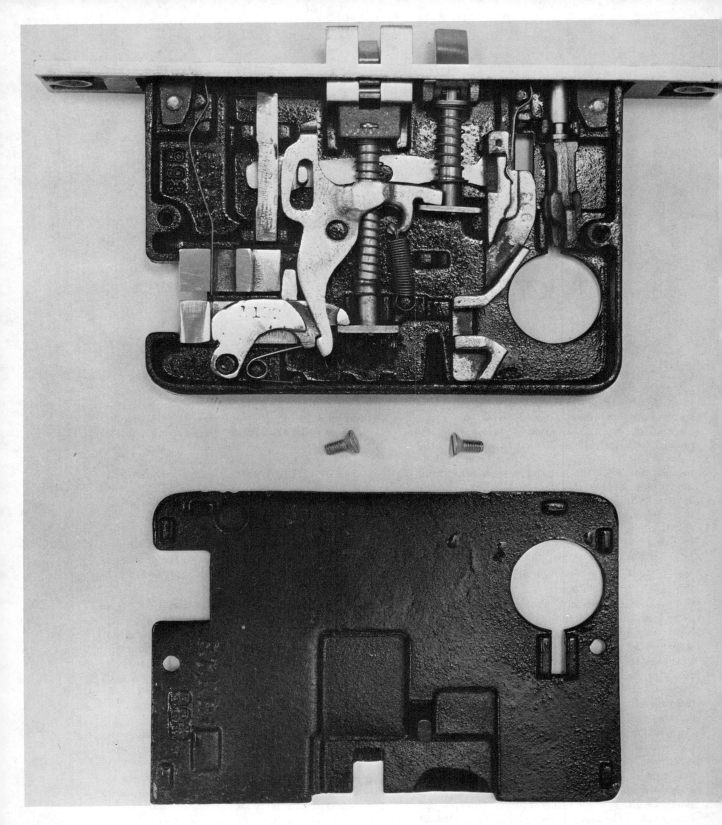

Figure 60. Mortise panic exit lock. Note the use of the
latch lever lift blocks and the stopworks operating lever.

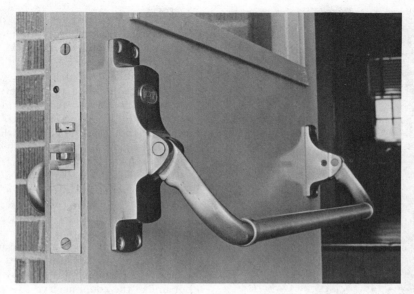

Figure 61. Mortise panic exit device installed on a door.

greater potential for damage to the lock case or lock front, and an accelerated rate of wear on the latch lever lift block. This condition, when added to the fact that panic bar operation may be of a considerably more violent nature than is thumb lever operation, could quickly result in an inoperative lock unit.

An *intermediate lift lever* is provided in the active end case to change the rotary motion of the active panic bar arm into a straight line motion for lifting the latch lever lift block. The intermediate lift lever travels in a straight line inside the active end case and has a surface which extends across the top of the arm tang, providing a means by which it may be lifted by the active arm tang. Spring return of the intermediate lift lever is usually provided by a coil spring of either the compression or expansion type. A retaining cover or cap in the back of the case holds the lever in place and has a hole, or slot, through which the tang of the intermediate lift lever protrudes to engage the latch lever lift block of the deadlatch unit. This completes the *single acting panic bar* mechanism.

The *double acting panic device* is one in which the panic bar may be either depressed or lifted in order to retract the latch bolt. This mode of operation is achieved by the addition of a *secondary lever* which has one surface extended across the lower side of the panic bar arm tang and is connected to the intermediate lift lever by means of a *coupling lever*. The coupling lever is pivoted in the center on a pivot post. It has notches in both

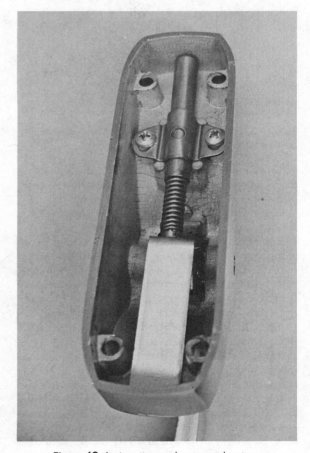

Figure 62. An inactive end case mechanism.

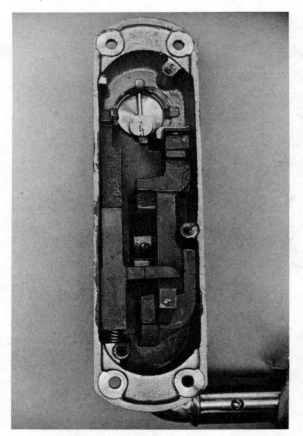

Figure 63. An active end case mechanism for a double acting panic device.

ends to fit over pegs which are part of the respective levers whose operation is coordinated by the coupling lever in a teeter-totter sort of action.

The installed height of the intermediate lever's tang in relation to the latch lever lift block of the lock is of crucial importance in panic exist devices of this type. Too low a tang position would result in the latch bolt being only partly drawn into the lock case, resulting in failure of the panic bar to open the door. Too high a tang position would partially raise the latch lever lift block without depressing the panic bar, causing the deadlocking mechanism to malfunction. For this reason, some units of this type are made self-aligning; that is, the screws used to mount the panic device's active end case are screwed to lugs on the back of the front escutcheon; in the process of attaching, they pass through holes in the lock case, thus maintaining proper alignment.

Another variation of this mechanism has the lift tang pivoted to the intermediate lever, with the position of the tang on the pivot adjusted by means

of a set screw, thereby providing a means of establishing the proper positioning of the tang relative to the latch lever lift block.

Some panic exit devices operate in conjunction with a mortise deadlatch which would ordinarily use a knob rather than a thumb lever on the inside. A requirement is that the panic device provide a rotary motion to operate the latch bolt, just as in turning a knob. This is accomplished by the addition of a *spindle lever* in the active end case, in a position below the active arm pivot.

The spindle lever is activated by either the secondary lever or the intermediate lever and moves with a rotary motion which is at right angles to the motion of the panic bar arm. In appearance, it resembles a hybridization of the coupling lever with the lock cam. It fits over a short stump in the active end case and has an arm with a notched end to accept a short stump on the lever which actuates it, but unlike the coupling lever, it has only the one arm. That portion of the spindle lever which is above the pivot post is round and has a distinct

Figure 64. Active end case mechanism with spindle lever.

resemblance to a knob cam in that it is round, and has a square spindle hole through it, with a bearing much like a knob cam at the point where it passes through the cap of the active end case.

If the outside operation of the lock is intended to be by knob, the knob spindle is usually inserted into the spindle lever in a free floating arrangement. If the outside operation of the lock is by thumb piece, the spindle is usually pinned into the spindle lever to maintain its position in a permanent and rigid mount. Because of the depth to which the spindle is inserted in both the knob cam and the spindle lever,

alignment of the active case on the inside surface of the door is often more critical than if a simple knob were used rather than a panic exit device.

Panic exit devices which utilize the spindle lever type of operation share an unique advantage with those which are self-aligning in that both of them provide for a firm stop in the active end case. They prevent the transmission of excessive motion to the latchbolt operating mechanism (as when the panic bar is hit too hard), with a resultant force which, when applied to the latch tail, could either rip it off the latch bolt assembly or break the lock case.

Figure 65. Mortise panic exit lock for the knob spindle type of operation. Compare this unit with the mortise deadlatch of Figure 29.

Bit Key Mortise Locks

THE bit key mortise lock is the only type of mortise lock in which the keying of the lock is integral with the lock case. The old-fashioned skeleton key is a good example of the type of key which this lock requires, although not all of the keys are of this simple a design or manufacture. The lock itself differs from the more modern cylinder type lock in that the key actually operates the mechanism inside the lock in direct contact with the mechanism, rather than indirectly as in the case of a cylinder lock. The bit key mortise lock has been produced in most functions, although the author cannot recall having seen the mortise deadlatch function. Current production of these locks in the United States has been largely supplanted by the various types of pin tumbler units, although they are still being produced in quantity in some other countries.

WARDED LOCKS

Locks of this type fall into one of two general classes of keying. One is the simple *warded lock*, and the other utilizes the *lever tumbler* type of construction. Warded type locks produced in the United States are almost uniformly of very cheap construction and quite easy to make keys for. In principle, the warded lock is very simple: it utilizes various *wards*, or baffles, placed in such a way as to either bar the insertion of a wrong key into the lock, or to obstruct the turning of a wrong key in the lock after it has been inserted; or both types of wards may be used. The correct key is, of course, cut away so as to clear, or *pass*, all of these wards.

Warding in locks of American manufacture usually consists of small protrusions from the inside of the lock case and cap on both sides of the keyhole to obstruct the turning of a wrong key, but once a key of about the right dimensions has been notched on both ends of the *bit*, it will turn and operate the boltwork as soon as these notches have been cut out to the proper location and size to pass the wards. In some cases, the key bit must have a groove down its side on one or both sides in order to pass *keyway wards* and enter the lock.

In some instances, a thin-bitted key will pass these keyway wards, but at other times, the keyway wards extend at least halfway into the keyway from either side of the keyhole, making it absolutely essential that a thicker-bitted key be used and that it be grooved on each side of the bit. This has the effect of causing the key bit to zig-zag in order to pass both sets of keyway wards. Such keyway warding is known as the *paracentric keyway*, although the term is much more frequently used in connection with a similar keyway construction in a pin tumbler lock cylinder.

As the bit key is inserted into the lock and turned, the bit contacts and lifts a detent, much like that used on the deadbolt in some of the modern pin tumbler mortise locks. At the same time, it engages the bolt to move it into the locked or unlocked position, as the case may be. A detent type of construction is illustrated in Figure 66.

Although locks of American manufacture which are of warded construction are almost uniformly of this simple type of construction, locks of some of the

Figure 66. Warded lock using a deadlatch detent type of construction.

older European manufacture, made when locks were being produced by fine hand craftsmanship contained what was known as *fine wards* (when made of sheet metal) or *bridge wards*. The bridge wards (although a case ward rather than a keyway ward) served much the same purpose as the paracentric keyway ward in that the key could not be skeletonized without cutting the key bit completely off. It was necessary for such keys to be slowly and patiently turned out by hand workmanship or by casting. The production of an intricate pattern of this type by the casting method was comparatively simple when a sample key existed; but when no sample existed, casting of an original key was frequently facilitated by the use of wax on the side of the bit of a test key, thus giving rise to the term "wax impression." For the old type of warded locks, it was a valid term, as it was from this impression in wax on the test key that a mold was constructed from which an original key could be cast.

Since bridge warded locks did not lend themselves well to mass production techniques, they fell into disfavor with the advancement of the industrial revolution. Only the more crude models were produced as they were more suited to the criteria of mass production. The ease with which warded locks could be circumvented by the use of a skeletonized or *skeleton key* was largely responsible for the falling into disfavor of all the types of locks which required a bit key. Because of the similar appearance of the keys, most people thought that none of the bit key locks offered any more protection than did the cheap warded type. This skepticism of bit key locks extended to the bridge ward type and even to those utilizing the lever tumbler construction.

LEVER TUMBLER LOCKS

The lever tumbler locks offer protection ranging from moderate to excellent, utilizing anywhere from one to five lever tumblers in their construction. The protection offered varies with the number of tumblers and the precision with which they are constructed. The lever tumbler is made from flat sheet metal, may be pivoted on either end, and is provided with spring action, usually by means of a flat piece of spring stock which is affixed to the tumbler at a point near the pivot. The *belly*, or that portion of the tumbler which is in contact with the key as the key is rotated in the lock, is usually curved in a segment of arc described by the bottom of that particular cut on the key as the key is rotated. This maintains the height of the gating at a constant level during the rotation of the key in order to keep the stump from binding on the tumbler as it passes through the gating.

Gating of the tumbler comes in two classes, *open* or *closed*. In open gated tumblers, the rectangular stump which is to enter the gating of the tumbler is outside of the tumbler itself, at the end of the tumbler opposite the pivot. The end of the tumbler is customarily provided with a small protrusion, usually at the top corner, to prevent it from falling completely out of position. This protrusion on the end of the lever may rest (in the locked position) on the bolt stump, or it may rest on a special stump or peg which has the sole purpose of supporting the tumbler. The stump which enters the gating is usually attached to the bolt, whether deadbolt or latchbolt. The configuration of the gating will vary with the type of function with which it is to operate and the quality of lock that is desired. In locks utilizing closed gating, the stump is entirely enclosed within the gating of the tumbler at all times.

Figure 67 shows the corresponding open and closed gatings of lever tumblers in relationship to the function for which they are intended. Other than this factor of the keying being contained in the lock case itself rather than in a separate cylinder, the mechanism of the lock is very similar to that used in corresponding functions of mortise pin tumbler lock construction, and further explanations

Figure 67. Lever lock showing both single and double pocketed closed gating in a combined deadbolt-springlatch function. The lower keyhole operates the deadbolt; the upper actuates the latch unlocking lever. Note the use of paracentric keyways.

should not be required as to their mechanical features. These locks are not currently a major production item, and their use has been largely supplanted in the United States by locks of pin tumbler construction. The locksmith will encounter them with diminishing frequency, but he must be prepared to cope with them occasionally because of the vast quantities which have been produced in the past. It should be stressed that although these locks can offer excellent protection, their limitations restrict their desirability for certain applications; for example, the number of key changes are extremely limited in any type of master key operation, and are considerably more limited than are the pin tumbler locks in a situation where master keying is not required.

Rim and Cylinder Locks

II

Rim Locks

I$_N$ times past, the term "rim lock" applied merely to a lock much like the corresponding mortise unit which was applied to the inside surface of the door, rather than being mortised into the edge. This type of rim lock has no front like the separate front of a similar mortise unit, but the latchbolt and deadbolt, if any, simply protrude through the edge of the lock case. The lock case completely surrounds the edges of the lock, obviating the need for a separate front.

These locks are universally of the bit key type, and most commonly of warded, rather than of lever construction. At one time they were produced in quality units of lever construction, but more recently produced units have tended to be of the cheaper warded type and very few, even of these, are currently being produced. Because of the huge numbers which have been produced in the past, there are still many in existence on older installations. These locks were made principally in the springlatch and deadbolt functions. More modern rim locks are used in conjunction with the pin tumbler lock cylinder, which is mounted through a hole bored in the door to receive it, and held in place by means of a *cylinder clamp plate*.

The cylinder clamp plate is affixed to the rear of the cylinder by a pair of *cylinder clamp screws*, with the thickness of the door intervening between the cylinder clamp plate and the cylinder's rim (or cylinder ring, as the case may be). A usually flat metal strip called a *cylinder connecting bar* connects the *plug*, or turning portion of the lock cylinder, with the cam or operating portion of the rim lock.

For a number of years, the production of pin tumbler rim locks was practically confined to the springlatch function, but more recent production has diversified to include greater proportions of the so-called jimmy-proof lock, the deadbolt, and the deadlatch functions. Although rim locks are produced by many manufacturers, both in this country and abroad, they are consistently of a relatively simple design which utilizes many of the same principles used in the construction of mortise locks, although in a slightly modified application consistent with the more restrictive use requirements in a smaller and differently shaped case. Bearing in mind that it would be impractical to discuss all of the variations found in the construction of this type of lock, and that even these variations are simple to one who understands the principles of the other locking devices which we have considered, our discussion will be limited to a few specific types.

THE RIM SPRINGLATCH

The rim springlatch utilizes a latch bolt much like the latch bolt of a mortise passage latch in external appearance; however, the latch shaft is more in the form of a trident, the center line of which has been cut short to allow a compression type coil spring to be slipped over it in order to provide the necessary expelling tension for the latch bolt. The two outer tangs of this trident have been blunted at the tip into two squared knobs to form a bearing face for the two cams (a *knob cam* and a *cylinder's cam*).

The knob cam is operated by the turn knob from the inside of the door, and the other (cylinder's) cam

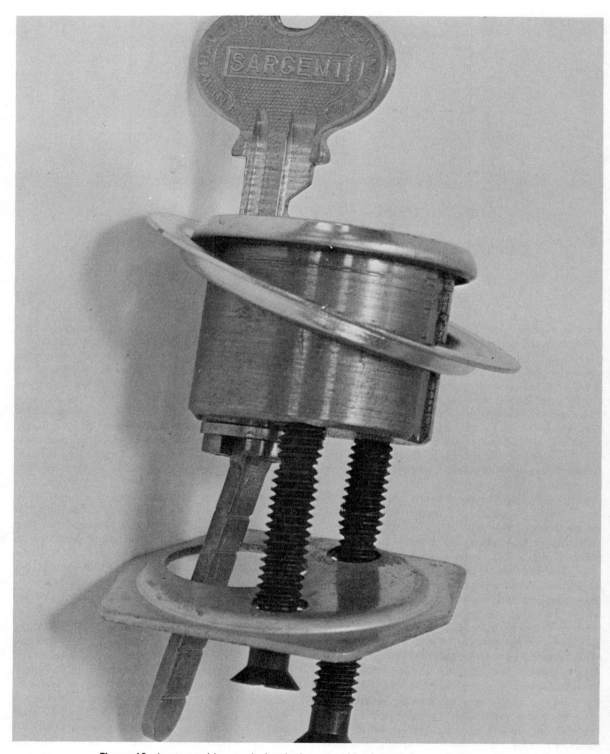

Figure 68. A pin tumbler rim lock cylinder assembly showing key, cylinder proper, cylinder ring, cylinder connecting bar, cylinder clamp plate, and cylinder clamp screws.

is operated by the pin tumbler lock cylinder through the use of the *cylinder connecting bar*.* The latch bolt is held into the retracted position, when it is not desired to lock the door, by a sliding spring which moves in a shallow channel in the inside of the lock case. This spring is actuated by a sliding button type control from the face of the lock.

There is a hump on the spring which moves into position to engage a peg which is attached to, or part of the trident shaped yoke, which, in turn is part of the latch bolt. This holds the latch bolt in the retracted position until the hump of the latch retainer spring is slid into position to free it by clearing the obstruction. These parts, together with the case, cap, and a simple strike, comprise the rim springlatch.

A variation of the rim springlatch has a lock cylinder on the inside of the door, usually keyed to match the outside cylinder. It requires a key from either side of the door to unlock it. This type is rather limited in its usefulness, since the lock itself is mounted on the inside surface of the door, with a consequent vulnerability to the latch bolt from that side even though the screws may be concealed and not subject to attack.

Other rim lock functions may include an inside lock cylinder with somewhat better success, providing that the mounting screws are concealed inside the lock case. Access to them can be obtained only by removing two screws on the front edge and removing the case. Mounting screws in this type of construction hold the cap, rather than the lock case to the door. That is one way whereby screws make an acceptable substitute for concealed mounting screws, since they are extremely difficult to remove when once installed.

Cases for these locks, as for most rim locks, are made from cast iron or steel, cast bronze, or more recently, from pressure cast metals such as aluminum alloys, brass, or bronze. The latch bolt is frequently of brass or bronze, but may well be of a ferrous metal (an iron-based metal); it may be made of two dissimilar metals if the latch bolt and yoke are made in two pieces. Other parts are customarily made from steel, with the exception of such items as the turn knob and the latch retainer button, which are often made of some more ornamental material such as brass. Where the trident shaped yoke is used, the lock case includes a protrusion or a depression on the front side of the post (which is

---

*Lock cylinders used with rim locks, although predominantly of the pin tumbler type, may also be of the disc tumbler type.

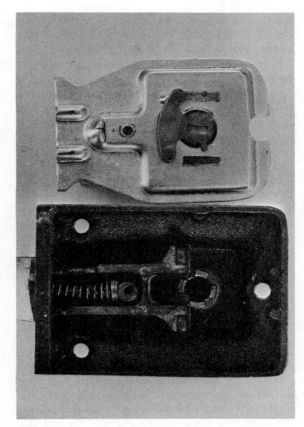

Figure 69. A rim springlatch. A crossbar connects the two outer tangs of the trident. The notch just left of the center of the upper tang would be the point engaged by the latch retaining stop spring if the latchbolt were turned over.

used to attach the cap) to maintain the latch bolt spring in position. Other types which utilize a rudimentary latch shaft may have this post in the form of a yoke, similar to that used to accommodate the latch shaft of a mortise unit. No latch tail is used with this type of construction, as forward motion of the latch bolt is limited by other means, and the shaft is not used to draw the bolt.

THE RIM DEADLATCH

The usual construction of the rim deadlatch is different in its deadlocking principle from that of the mortise deadlatch and is more of an adaptation from the rim springlatch. The essential differences between the rim springlatch and the rim deadlatch are in the latch tail and the fact that a pin, rather than a cam, is used to operate it. This *operating pin* is mounted eccentrically on the knob cam and protrudes through a specially shaped slot in the rearmost position of the latch tail. Although the cylinder's cam is of the more conventional two-armed type, it does not directly engage the latch tail, but is used to move the operating pin, which in turn operates the latch bolt assembly.

The latch tail may be formed integrally with the latch bolt, but frequently is made separately from sheet metal and riveted to the latch bolt, or the bolt is formed around it by casting or pressure casting. Refer to Figure 70 and note the three indentations in the rearmost surface of the slot in which the operating pin functions; the center one is directly behind the rudimentary latch shaft. The lock serves as an ordinary springlatch until the operating pin is deliberately forced over the hump from the outer depression into the center depression. The center depression holds the operating pin in a position which is centered behind the end of the latch shaft; thus, any end pressure on the latch bolt merely presses the end of the "latch shaft" against the operating pin, and thereby the lock is deadlocked when the operating pin is in the central position.

This type of deadlatch, because it is equipped with an extra depression on the opposite end of the operating slot in the latch tail, may be turned over and used for doors which open out, although there are certain limitations to this ability. Two of these limitations are the availability of a proper strike or the suitability of an existing one, and the fact that the opening of the door demands that the lock be mounted somewhat further from the edge of the door in order for the lock case to clear the door frame. Latch retention may be by the sliding button method described under the rim springlatch, or by any of several other methods. These may take the form of a turn lever on the outside of the case which simply uses a rotary motion, rather than a sliding one to place an obstruction in the path of travel of some part of the latch bolt. Another latch retaining method (used in the lock shown in Figure 70) passes the shank of the turn knob through a tube set into the lock case, and a sliding dog in one side of the turn knob enters a slot in the tube to hold the turn knob in the retract position.

Yet another method utilizes a spring-loaded plunger in the center of the turn knob to interpose an obstruction in the path of travel of the latch tail. The obstruction is slightly undercut at the point of contact with the latch tail so that the spring tension of the latch bolt assembly against it will maintain the obstruction in place until the spring tension is relieved by manual pressure on the end of the latch bolt, or by turning the key or turn knob enough to relieve the spring tension supplied by the latch bolt spring.

THE JIMMY-PROOF RIM LOCK

The so-called jimmy-proof rim lock offers as great a degree of protection as a rim lock, and while it can usually be forced, the effort requires considerably more time and skill than for other rim locks. This type of lock features a locking arrangement whereby two sets of fingers, one on the edge of the lock case and one on the striker plate, are interleaved and locked together by means of a round locking bolt. In the unlocked position, the locking bolt is wholly contained in the fingers of the lock case, which are considerably thicker than are those on the strike. As the mechanism is locked, the round locking bolt comes out from the fingers of the lock case, passing through the holes in the fingers of the strike plate, and on into the next hole in the adjoining finger of the lock case. Because of the mechanisms which must be housed in the lock case, the hole in the fingers of the case assumes a U shape rather than being round as are the holes in the fingers of the strike.

The round locking bolt is actually two bolts and both of them are connected to a single deadbolt by an extension of the deadbolt to two fingers of the lock case. (One finger of the lock case contains no locking bolt of its own, but merely receives the one adjacent to it as the bolts are moved into the locked position.) The deadbolt mechanism strongly resembles the classical mortise deadbolt which utilizes the deadbolt locking lever, with the exception of the extensions leading to the round locking bolts. It is thrown, not by the cylinder cam, but by a cylinder's cam, mounted inside the cap of a rim lock and actuated by the cylinder connecting bar.

The cylinder's cam acts the same as, and in lieu of, a cylinder cam, in disengaging the deadbolt locking lever and moving the deadbolt itself from one position to the other. The motion of the deadbolt being vertical (as the lock is mounted on the door) makes it necessary for the jimmy-proof lock case to be somewhat wider than other rim lock cases in order to contain not only the length of the deadbolt, but also the length of its throw.

The turn knob on the face of the lock, by means of which the lock is thrown manually, terminates inside the lock case in a double-armed cam which is very much like the turn knob disc used with mortise deadbolts. The double-armed cam makes it possible to throw the mechanism to its limit of travel in either direction with a 180° rotation of the turnknob

Figure 70. A rim deadlatch. When the operating pin is turned into the center notch in the latch tail, the latch bolt extends a little further and is deadlocked by the blocking action of the operating pin on the center protrusion of the opposite cross member. Latch holdback on this particular model is accomplished by a slide which is integral with the inside knob, and the knob is locked in much the same fashion as a mortise lock hub.

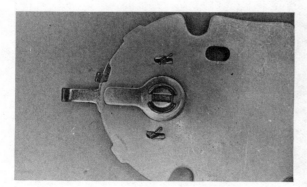

Figure 71. Jimmy-proof rim lock. Note the similarity of the boltwork to that of a mortise deadbolt. The shutter over the hole which accepts the cylinder connecting bar is spring loaded and closes off access to the hole if the cylinder is removed.

instead of requiring a complete 360° rotation as would be the case if the turn knob cam had only one arm.

An additional feature commonly, but not always, found on the jimmy-proof lock is a shutter arrangement used to protect the cylinder's cam when the cylinder connecting bar is not in place in the lock. The shutter arrangement makes entry through a locked door by means of forcible removal of the lock cylinder considerably more difficult. The shutter consists of a plate of heavy gauge sheet metal with a hole in it through which the cylinder connecting bar may pass in order to reach the cylinder's cam. The plate is spring loaded so that the hole passes to one side of the cylinder's cam, covering it with a metal plate when the cylinder connecting bar is not in place.

Some shutters include a notched or stepped construction which is designed in such a manner that pressure against the shutter, as in trying to force it back from the outside after the cylinder has been removed, will engage the notched or stepped portions of the shutter with the case or with a shoulder on the cap, holding it even more firmly in place. The shutter is housed in an extremely shallow auxiliary case attached to the cap on the side next to the door, and the sides of the lock case are extended enough to seat firmly on the door without binding the auxiliary case or shutter. The auxiliary case has a hole through it which is directly opposite the cylinder's cam and is just large enough to permit the cylinder connecting bar to pass through.

The Rim Deadbolt

Customarily, the deadbolt of the rim deadbolt lock is actuated by a bolt throwing lever which also serves as a deadlocking lever in a piston-like manner somewhat similar to the mortise lock deadbolt mechanism of Figure 50. In both cases, the lever which moves the deadbolt back and forth in the lock case is thrown into a slight overtravel position in locking the bolt — that is, the lever moves slightly further than is necessary to fully extend the bolt. At that point, the lever comes against a stop, which may be either part of the bolt or part of the lock case, and is held against the stop by spring tension so that end pressure on the bolt presses the lever more firmly against the stop. In the rim lock application, the spring tension is provided by a torsional type of coil spring.

Figure 72. Shutter on a jimmy-proof rim lock. Note the protruding tang which simplifies installation and locks the shutter against force.

Actual throwing of the bolt of the lock shown in Figure 73 is inititated by a pin projecting from the turn knob cam and passing through *one side only* of the *bolt actuating lever,* which is divided into two halves before reaching this point and is separated by about the thickness of the bolt from the other half. As the turn knob is operated, the pin lifts the actuating lever off its stop and pushes or pulls the bolt along its line of travel, depending on whether the bolt is being retracted or extended.

Operation by key of the bolt lever is accomplished by means of a similar pin on the cylinder's cam which engages a notch, rather than a hole, in the bolt actuating lever, lifting it off the stop, moving it along its limit of travel, and bringing it to rest against the other stop. The open end of the notch in which this pin operates allows it to continue its rotation, permitting a complete revolution of the key in order that the key may be withdrawn from the lock cylinder. The deadbolt itself is not sprung; in fact the spring which supplies tension to the bolt actuating lever is the only spring used in this type of rim deadbolt.

Figure 73. A rim deadbolt. The locking principle is representative of locks of this type and a number of tubular deadbolts as well.

Variations in the rim deadbolt mechanism do exist and are encountered from time to time, but their similarity to mortise deadbolt mechanisms are quite apparent and should create few problems if one is sufficiently familiar with mortise lock construction.

The Rim Panic Exit Device

Bᴇᴄᴀᴜsᴇ of the wide divergence in current manufacture of rim exit devices, it has been necessary to select a specific device which illustrates the basic construction principles of a number of them and which utilizes mechanical principles not discussed in connection with other types of lock units. With this in mind, the Sargent & Greenleaf 5300 series of rim panic exit devices has been selected to illustrate many of these principles.

Tʜᴇ Bᴀsɪᴄ Exɪᴛ Dᴇᴠɪᴄᴇ

The basic unit, shown in Figure 74, is intended solely as an escape device, with no means provided for operation from outside of the door. The active panic bar end, or *operating handle,* has a tang which thrusts directly against a curved surface extending behind, and which is part of the rolling type of latch bolt with which this unit is equipped. Although some lateral slippage occurs between the panic bar end's tang and the latch bolt extension, it is dimensionally minor in relation to the angular motion of the latch bolt which occurs in the process of withdrawing the latch bolt into the case. As a result of this ratio of slippage, the amount of friction generated is relatively insignificant.

A comparatively lightweight spring (number 4 in Figure 74) is introduced, bearing on the latch bolt in order to produce easy action for the latch bolt in closing the door. A spring (number 5) is introduced, bearing on the active end tang (operating handle tang) in order to support the weight of the bar assembly (with an assist from the inactive end spring), and to keep the panic device from being too

readily tripped by accidental pressure on or by the weight of the bar itself. It is of necessity considerably stiffer than the latch bolt spring. The inactive end of this panic device is shown at the right of the illustration, and is very similar to those used with mortise panic exit devices.

Kᴇʏ Oᴘᴇʀᴀᴛɪᴏɴ ᴏғ ᴛʜᴇ Lᴀᴛᴄʜ Bᴏʟᴛ

The next level of operation of the rim panic exit device provides for key operation from the outside of the door. Necessarily, a pull handle is added to the outside to provide a means of pulling the door open from the outside once the latch bolt has been released by the key. The internal construction of the device is shown in Figure 75.

The essential change in the internal mechanism of this unit in providing for a rim lock cylinder type of operation involves the addition of a cylinder cam (part number 3), which supplies a true camming operation in that it is used to wedge the latch bolt extension from its normal "at rest" position into a position which retracts the latch bolt. As the cam is turned, the inclined ramp thrusts against the roller which has been provided on the latch bolt extension, causing the roller to roll up the ramp and rotate the latch bolt into a retracted position. At that time, the door may be opened from the outside by pulling on the pull handle provided with this type of exit device.

It readily will be seen from the example that this unit is intended for operation from the outside only by use of the key. This may be bypassed only by dogging down the panic bar ends from the inside. In

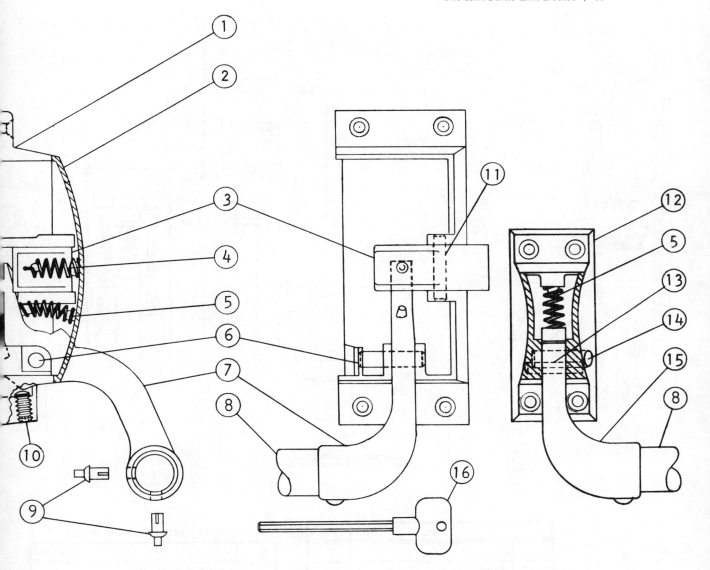

*DENOTES PARTS THAT ARE HANDED

No.	NAME	PER UNIT
1 *	Case	1
2 *	Cover	1
3	Bolt	1
4	Bolt Spring	1
5	Operating Handle Spring	2
6	Pivot Pin	1
7 *	Operating Handle	1
8	Handle Bar	1
9	Handle Bar Rivet	4

No.	NAME	PER UNIT
10	Dogging Screw 7/16-14	2
11	Bolt Pivot Pin	1
12	Outboard Bracket	1
13	Outboard Stop Pin	1
14	Outboard Pivot Pin	1
15 *	Outboard Handle	1
16	Dogging Key	1

Figure 74. A basic panic exit device. Note the "easy action" (separately sprung) latch bolt. Illustration courtesy of Sargent & Greenleaf, Inc.

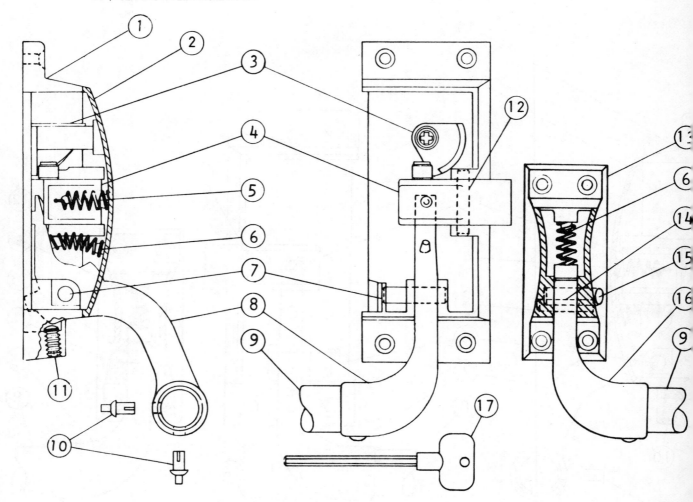

No.	NAME	PER UNIT
1 *	Case	1
2 *	Cover	1
3 *	Cylinder Cam	1
4 *	Bolt Ass'y	1
5	Bolt Spring	1
6	Operating Handle Spring	2
7	Pivot Pin	1
8 *	Operating Handle	1
9	Handle Bar	1

*DENOTES PARTS THAT ARE HANDED

No.	NAME	PER UNIT
10	Handle Bar Rivet	4
11	Dogging Screw 7/16—14	2
12	Bolt Pivot Pin	1
13	Outboard Bracket	1
14	Outboard Stop Pin	1
15	Outboard Pivot Pin	1
16 *	Outboard Handle	1
17	Dogging Key	1

Figure 75. Key operation of rim panic exit latch bolt. Outside operation of the latch bolt is by key only. Note the addition of cam and roller to the design of Figure 74. Illustration courtesy of Sargent & Greenleaf, Inc.

other words, once the door is latched, it cannot be left unlocked because no means of operating the latch bolt from outside of the door is provided except by use of the key to rotate the cam.

THUMBLATCH OPERATION

The next stage of operation of the rim panic exit device provides for operation by the panic bar from inside, as in the case of all panic devices, and also allows for operation from outside by means of a thumblatch — except when the thumblatch is locked by key from outside. In this instance, the key makes a full revolution in throwing the dogging mechanism from the locked to the unlocked position, or vice versa. The key may be withdrawn in either position as soon as the full revolution is completed.

Both the latch bolt and active panic bar end (operating handle) are identical to those used in the basic No. 5300 unit. In order to achieve this mode of operation, a number of parts must be added to the basic unit. One of these is the thumblatch lever (item 10, Figure 76). The thumblatch lever has one arm which extends upward behind the latchbolt extension immediately beside the operating handle tang and which operates the latchbolt in the same manner. The other arm of the latchbolt lever protrudes through the back of the case, into a cutout hole in the door where it is engaged by the thumblatch tang.

The thumblatch lever is equipped with a return spring of the constricting torsional coil type (item 11 in Figure 76), one end of which bears on the lock case and the other on the thumblatch lever. In order to prevent the thumblatch from being operative at those times when it is desired to lock the door, a slide plate (item 6 in Figure 76) is added to slide back and forth in the case. In the locked position, this plate comes in front of the tip of the thumblatch lever tang to prevent it from moving against the latch bolt extension; in the unlocked position, it moves out from the path of travel of the thumblatch lever, permitting it to operate freely.

The motion of the slide plate is permitted and limited by two slots through which two shoulder screws pass. These shoulder screws hold the slide plate in place in the case and permit it to move freely through the limits of travel imposed by the length of the slots. The shoulder screws are identical to item 4 in Figure 76, which is used to provide a pivot for a formed flat spring whose purpose is to maintain the slide plate in either the locked or unlocked position in which it has been placed by

the cylinder plug assembly, item 3 in Figure 76.

The cylinder plug assembly rotates as the key is turned in the rim lock cylinder, and moves the slide plate from side to side by means of two opposed pins, one in the cylinder plug assembly and one in the slide plate at a right angle to the first. As the slide plate is moved from one limit of travel position to the other, the slide plate spring rotates on its pivot screw to a sufficient degree to change the direction in which it tends to push the slide plate. This action maintains the slide plate in whichever position it has been placed. The retaining action is in addition to that provided by the cylinder plug assembly and permits more positive positioning when the lock cylinder is not mounted squarely in the door.

Knob Adaptation

Referring again to Figure 76, we find that when a knob type of operation is desired, parts numbered 23 through 25 are added to this mechanism. The basic intent of this is to provide a knob cam with a lug to thrust against the tang of the thumblatch lever in order to raise it, instead of using a thumblatch for that purpose. Spring return of the knob cam is made possible by the same return spring which returns the thumblatch lever to its normal position, with the energy of the spring being applied to the knob cam by the thumblatch lever.

VARIATIONS

A number of variations occur in the rim panic exit device; some utilize a mechanism which disengages the thumblatch lever so that it is free to move, but which does not operate the latchbolt. Some make use of an internal mechanism reminiscent of the compensating type of devices usually associated with mortise panic exit devices. Others (of recent production) include a deadlatching device for the latchbolt which, although different in configuration, is similar in principle to those incorporated in mortise deadlatch units. Still others are made reversible by turning over or by inverting certain of the internal parts in the unit.

At first glance, these units may appear quite complicated, but nearly all of them are based upon principles of leverage and motion which have already been discussed. This becomes apparent upon detailed examination of these units. In illustration of this, note in Figure 77 the use of an inclined ramp to draw the latch bolt. Although this inclined ramp is, in this instance, operated by the panic bar, the method by which it operates the latch-

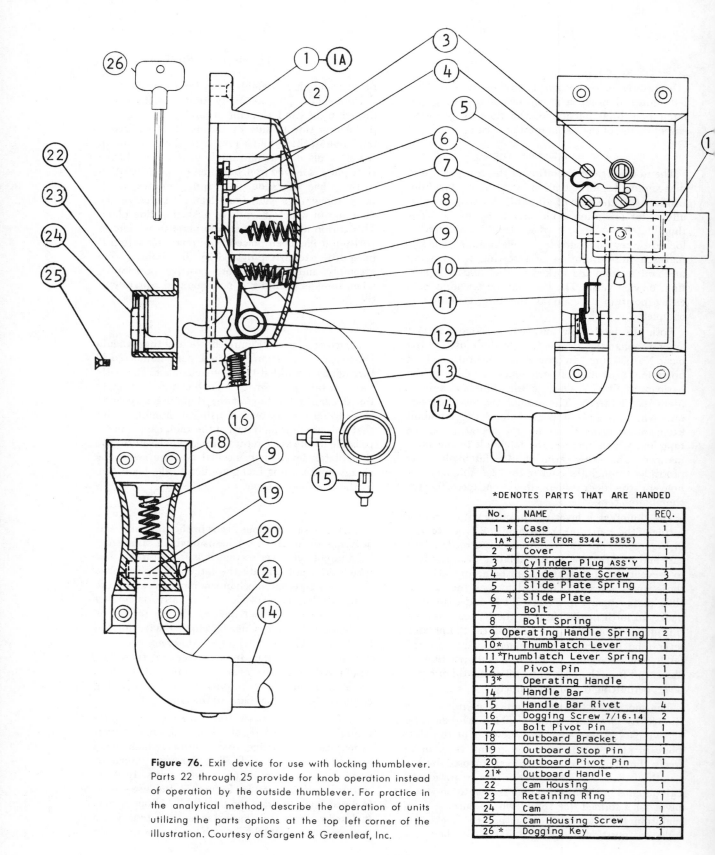

Figure 76. Exit device for use with locking thumblever. Parts 22 through 25 provide for knob operation instead of operation by the outside thumblever. For practice in the analytical method, describe the operation of units utilizing the parts options at the top left corner of the illustration. Courtesy of Sargent & Greenleaf, Inc.

*DENOTES PARTS THAT ARE HANDED

No.	NAME	REQ.
1 *	Case	1
1A*	CASE (FOR 5344, 5355)	1
2 *	Cover	1
3	Cylinder Plug ASS'Y	1
4	Slide Plate Screw	3
5	Slide Plate Spring	1
6 *	Slide Plate	1
7	Bolt	1
8	Bolt Spring	1
9	Operating Handle Spring	2
10*	Thumblatch Lever	1
11*	Thumblatch Lever Spring	1
12	Pivot Pin	1
13*	Operating Handle	1
14	Handle Bar	1
15	Handle Bar Rivet	4
16	Dogging Screw 7/16-14	2
17	Bolt Pivot Pin	1
18	Outboard Bracket	1
19	Outboard Stop Pin	1
20	Outboard Pivot Pin	1
21*	Outboard Handle	1
22	Cam Housing	1
23	Retaining Ring	1
24	Cam	1
25	Cam Housing Screw	3
26 *	Dogging Key	1

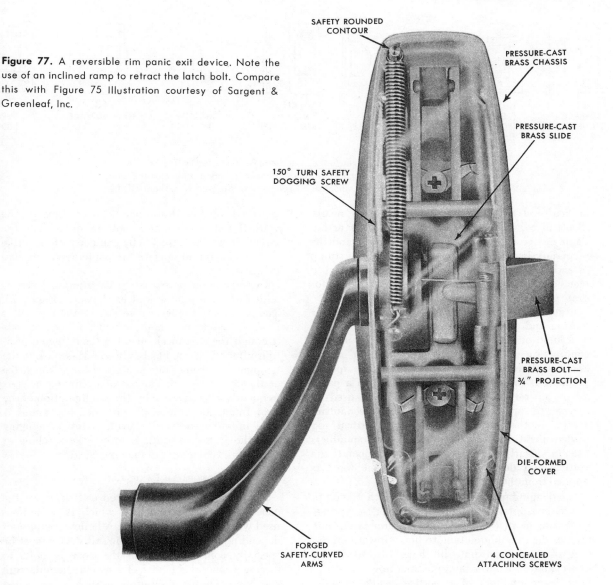

Figure 77. A reversible rim panic exit device. Note the use of an inclined ramp to retract the latch bolt. Compare this with Figure 75 Illustration courtesy of Sargent & Greenleaf, Inc.

SAFETY ROUNDED CONTOUR

PRESSURE-CAST BRASS CHASSIS

PRESSURE-CAST BRASS SLIDE

150° TURN SAFETY DOGGING SCREW

PRESSURE-CAST BRASS BOLT— ¾" PROJECTION

DIE-FORMED COVER

FORGED SAFETY-CURVED ARMS

4 CONCEALED ATTACHING SCREWS

bolt is quite similar to that used in Figure 75, showing Sargent & Greenleaf unit No. 5311 and operated by a key. The difference of the ramp having been straightened out, rather than embodied in a curved form, and moving in a straight line motion rather than in a rotary motion, makes no significant change in principle with respect to its operation of the latch bolt.

Strikes

A considerable number of different styles of strike plates are provided for use with rim panic exit devices. In most cases, the choice of these strikes

depends upon the type of door frame in which the door is set. Other styles are available for one of a pair of doors. Some of the various strikes used are shown in Figure 78.

Vertical Rod Panic Exit Devices

Vertical rod panic exit devices consist of a top and a bottom latch unit, with operating rods connecting them to a basic panic device unit. This basic panic device does not have a latch bolt directly associated with the mechanism itself, since latching of the door is accomplished by the top and bottom latch units. The internal mechanism in the active end case of a

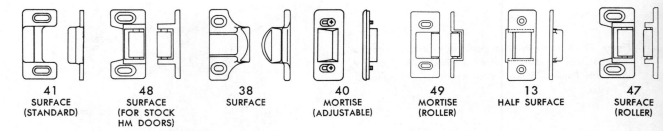

41	48	38	40	49	13	47
SURFACE (STANDARD)	SURFACE (FOR STOCK HM DOORS)	SURFACE	MORTISE (ADJUSTABLE)	MORTISE (ROLLER)	HALF SURFACE	SURFACE (ROLLER)

Figure 78. Strikes for the rim panic exit devices illustrated in Figures 74 and 76. This variation in strikes considerably enhances the adaptability of the units to a variety of situations. Illustration courtesy of Sargent & Greenleaf, Inc.

vertical rod device (short term) is often nearly identical to, or reminiscent of, the double acting panic device mechanism. The intermediate and the secondary levers have been equipped with a stump extending through the lock case, which is tapped to receive the threaded end of the *vertical rods* (or *draw bars*).

A dogging mechanism may have been added for locking and unlocking any outside functions, such as a knob or a thumblatch, which may be used in conjunction with the device. The addition of this knob or thumblatch mechanism poses no special problem as the same mechanism by which a panic device operates a mortise latch bolt is used in reverse to operate the vertical rod mechanism. Stated another way, instead of the vertical rod device operating a knob spindle, the knob spindle is used in a nearly identical application to operate the vertical rod device instead because the force is applied from the opposite direction.

The dogging mechanism by which such a knob is locked is similar to that used in mortise spring-latches or mortise deadlatches. A thumblatch unit utilizes the thumblatch tang to press upward on the tang of the intermediate lift lever instead of the intermediate lift lever tang pressing upward on the latch lever lift block of a mortise deadlatch unit. Somewhat different shaping of these parts is often required by virtue of this change, but such changes are not so drastic as to alter their principle of operation.

An interesting variation in the compensation lever principle is shown in Figure 79, and utilizes two pivoted levers in a compensating arrangement. The bottom latch unit shown in the illustration may also be used as a top latch; it secures the door to the door frame or threshold, depending on whether it is used as a top or bottom latch. It should be noted that the latch bolts are of the roll-back type, and are separately sprung in their own case. A slot is provided in the draw bar which connects the vertical rod to the latch bolt in order that the latchbolt may be retracted by contact with its strike without having to operate the panic device mechanism also.

In some cases a *trip lever,* or *tripping lever,* is added to the top latch unit as shown in Figure 79. The purpose of the lever is to engage the top rod or its draw bar so as to hold both of the vertical rods (through the case mechanism) in the retracted position after the panic bar has been released until the tripping lever comes into contact with the top strike as the door is closed. The use of a tripping lever is frequently necessitated by the configuration of the door frame and the wall with which the unit is used in order to avoid scarring paint, or gouging into the frame or wall, or otherwise catching or binding in the process of opening the door.

CONCEALED PANIC EXIT DEVICE

A variation of the vertical rod exit device is the *concealed panic exit device* which is sometimes used with hollow metal doors. A hollow metal door depending on this type of panic exit device must be specially constructed when the door is built to receive a concealed device of a specific manufacture and type. The configuration of such a concealed panic exit device differs from that of the surface mounted vertical rod unit, since the rods, both upper and lower, are concealed inside of the door.

Sometimes part of the case mechanism of such a device also protrudes into the door, but the principle of its operation and construction is very similar to that of surface mounted vertical rod devices, even though the dimensions of some of the parts and even of their configuration may be quite different from that of the surface mounted installations. Aside from the means by which they are mounted in and on the door, the difference in principle is negligible. See Figure 80.

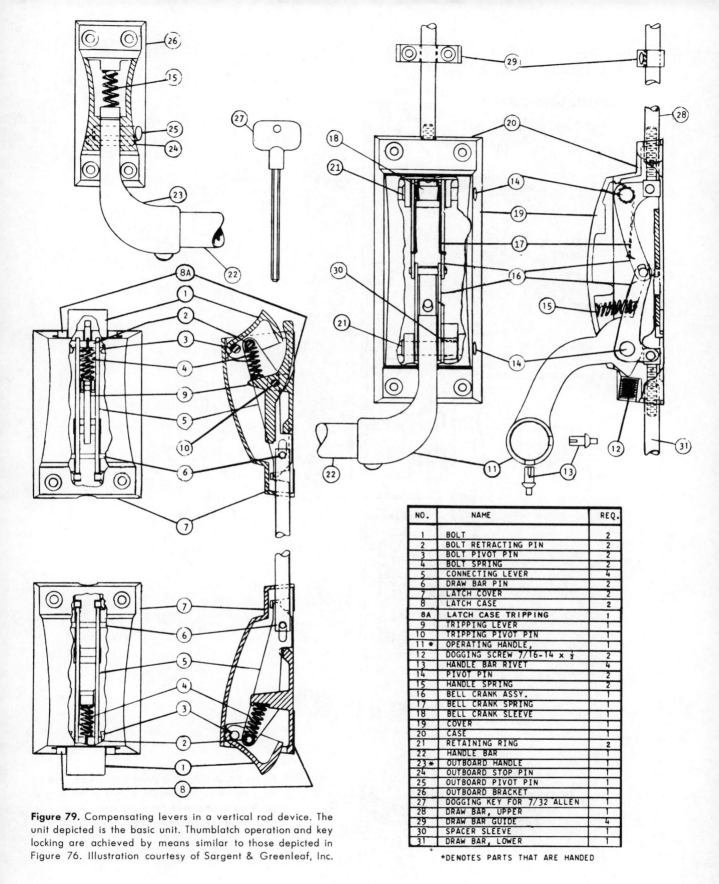

Figure 79. Compensating levers in a vertical rod device. The unit depicted is the basic unit. Thumblatch operation and key locking are achieved by means similar to those depicted in Figure 76. Illustration courtesy of Sargent & Greenleaf, Inc.

NO.	NAME	REQ.
1	BOLT	2
2	BOLT RETRACTING PIN	2
3	BOLT PIVOT PIN	2
4	BOLT SPRING	2
5	CONNECTING LEVER	4
6	DRAW BAR PIN	2
7	LATCH COVER	2
8	LATCH CASE	2
8A	LATCH CASE TRIPPING	1
9	TRIPPING LEVER	1
10	TRIPPING PIVOT PIN	1
11 *	OPERATING HANDLE,	1
12	DOGGING SCREW 7/16-14 x ½	2
13	HANDLE BAR RIVET	4
14	PIVOT PIN	2
15	HANDLE SPRING	2
16	BELL CRANK ASSY.	1
17	BELL CRANK SPRING	1
18	BELL CRANK SLEEVE	1
19	COVER	1
20	CASE	1
21	RETAINING RING	2
22	HANDLE BAR	1
23 *	OUTBOARD HANDLE	1
24	OUTBOARD STOP PIN	1
25	OUTBOARD PIVOT PIN	1
26	OUTBOARD BRACKET	1
27	DOGGING KEY FOR 7/32 ALLEN	1
28	DRAW BAR, UPPER	1
29	DRAW BAR GUIDE	4
30	SPACER SLEEVE	1
31	DRAW BAR, LOWER	1

*DENOTES PARTS THAT ARE HANDED

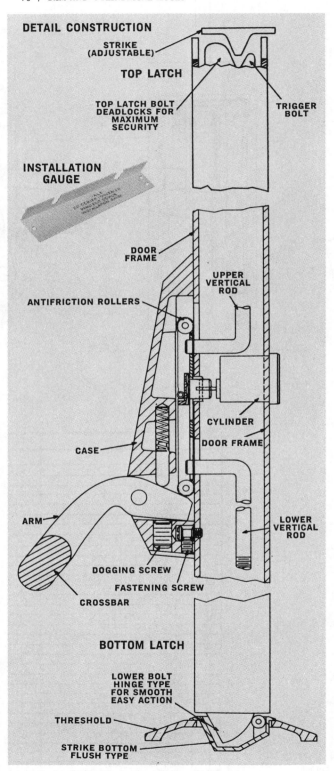

DETAIL CONSTRUCTION

STRIKE (ADJUSTABLE)

TOP LATCH

TOP LATCH BOLT DEADLOCKS FOR MAXIMUM SECURITY

TRIGGER BOLT

INSTALLATION GAUGE

DOOR FRAME

UPPER VERTICAL ROD

ANTIFRICTION ROLLERS

CYLINDER

DOOR FRAME

CASE

ARM

LOWER VERTICAL ROD

DOGGING SCREW

FASTENING SCREW

CROSSBAR

BOTTOM LATCH

LOWER BOLT HINGE TYPE FOR SMOOTH EASY ACTION

THRESHOLD

STRIKE BOTTOM FLUSH TYPE

Figure 80. A concealed panic exit device. Look for the similarities in principle to that of the mortise panic bar's intermediate and floating lever. Illustration courtesy of Eaton Yale & Towne, Inc.

Cylindrical Locks

CYLINDRICAL locks, though distinctively a part of the twentieth century and derived from its technology, have accrued sufficient popularity and versatility to constitute a major factor in the sales of locks today in the comparatively short course of their existence. Perhaps the most eloquent evidence of their acceptance by the public is the variety of names by which they have become known: they are variously referred to as cylindrical, bored-in, or key-in-the-knob locks. They are available in a wide range of functions capable of fulfilling a great many modes of operation, all of which are based on a variation of either the springlatch or deadlatch function.

Since most cylindrical locks are made with either a considerable number of sheet metal or pressure cast parts, or both, the techniques necessary to their construction on a mass basis did not exist until the

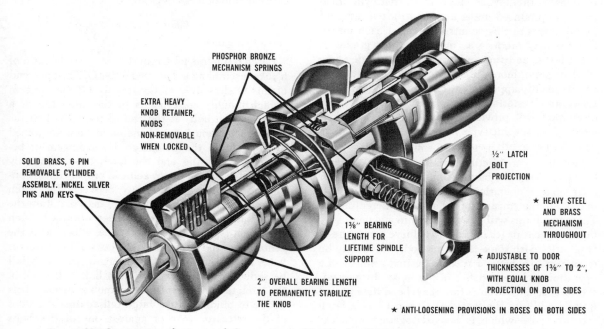

PHOSPHOR BRONZE
MECHANISM SPRINGS

EXTRA HEAVY
KNOB RETAINER,
KNOBS
NON-REMOVABLE
WHEN LOCKED

SOLID BRASS, 6 PIN
REMOVABLE CYLINDER
ASSEMBLY. NICKEL SILVER
PINS AND KEYS

1⅜″ BEARING
LENGTH FOR
LIFETIME SPINDLE
SUPPORT

2″ OVERALL BEARING LENGTH
TO PERMANENTLY STABILIZE
THE KNOB

½″ LATCH
BOLT
PROJECTION

★ HEAVY STEEL
AND BRASS
MECHANISM
THROUGHOUT

★ ADJUSTABLE TO DOOR
THICKNESSES OF 1⅜″ TO 2″,
WITH EQUAL KNOB
PROJECTION ON BOTH SIDES

★ ANTI-LOOSENING PROVISIONS IN ROSES ON BOTH SIDES

Figure 81. Construction features of Group I cylindrical locks. Illustration courtesy of Falcon Lock Co.

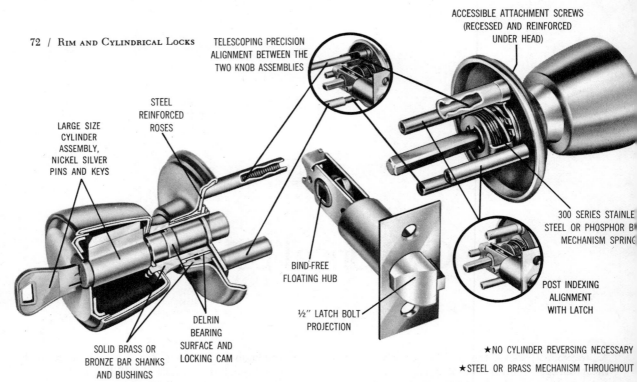

ACCESSIBLE ATTACHMENT SCREWS
(RECESSED AND REINFORCED
UNDER HEAD)

TELESCOPING PRECISION
ALIGNMENT BETWEEN THE
TWO KNOB ASSEMBLIES

STEEL
REINFORCED
ROSES

LARGE SIZE
CYLINDER
ASSEMBLY,
NICKEL SILVER
PINS AND KEYS

300 SERIES STAINLE
STEEL OR PHOSPHOR B
MECHANISM SPRING

BIND-FREE
FLOATING HUB

POST INDEXING
ALIGNMENT
WITH LATCH

½" LATCH BOLT
PROJECTION

DELRIN
BEARING
SURFACE AND
LOCKING CAM

SOLID BRASS OR
BRONZE BAR SHANKS
AND BUSHINGS

★NO CYLINDER REVERSING NECESSARY
★STEEL OR BRASS MECHANISM THROUGHOUT

Figure 82. Group II lock construction. Group II cylindrical locks have the cam or cams in the latch tube. Note the plunger locking of the outside knob. In some functions, locking the outside knob also disconnects it from the spindle to permit independent function of the inside knob. Illustration courtesy of Falcon Lock Co.

twentieth century. The advent of the pin tumbler, and later of the disc tumbler locking mechanism, also did much to make the cylindrical lock set a practical reality. When one realizes that cylindrical locks are produced under at least a dozen different brand names in this country, and that each manufacturer may produce as many as three grades of these locks, and that one grade may include some three dozen or more different modes of operation, it can be readily appreciated that discussing all these types and variations is impractical. Rather, our discussion will provide a basis for an individual study of the specific type, manufacture, and mode of operation currently popular in a given area.

Two basic construction forms account for a number of specific units having similar characteristics in their mode of operation, and in many cases employing a similar means of function differentiation. These two forms will be considered in a general way in subsequent pages in order to achieve a measure of familiarity with some of the methods which may be employed to achieve a specific desired result in the operational characteristics of the locks. In general, the two construction forms may be differentiated quite readily by the fact that Group I has a latch retractor yoke which is housed together with the knobs, rather than being contained in the latch tube

as are locks in Group II. Having convenient, if arbitrary, labels for discussion purposes, we are ready to proceed with the essential features of the construction of locks in Group I.

Group I Locks

Latch Retraction

Latch retractors for Group I locks take the form of a yoke resembling a distorted letter C. The open end of this C slips over and mates to a T bar section (which roughly corresponds to the latch tail of a mortise latch bolt) at the end of the latch bolt, as the retractor housing and the latch tube are joined in the door. A tongue-and-groove arrangement between the housing and the latch tube is usually provided in order to help achieve proper dimensional and angular relationship between the latch retractor and the latch tail and to help maintain the proper alignment once it has been obtained. Such alignment is important in this type of lock, as is the centering of the latch tail in the jaws of the retractor yoke. Should misalignment occur, or the latch tail be seriously off center, the latch bolt will stick in the latch tube the first time the knob is turned. This is due to binding of the latch tail against the sides of the retractor yoke or against one of the cams which operate the yoke (see Figure 83).

The cams which operate the latch retractor yoke are an integral part of the "eared spindles." The spindles serve a number of purposes in most cylindrical locks. Besides the ears which work against shoulders on the latch retractor to fulfill the function of cams, the spindles support and retain the knobs by means of a knob catch which is spring loaded and operates through holes in both the spindle and the knob to lock them together into a firm unit until the knob catch is once again depressed by a pin or punch. The knob catch on the knob carrying the cylinder is guarded when the lock is in the locked condition, either by requiring that the knob is turned slightly before the access hole is aligned with the knob catch, or by guarding the knob catch by interposing part of the internal lock cylinder mechanism in its path of travel until the key is turned slightly.

Locking of the one ear spindle, and consequently of the cylinder carrying knob, is accomplished in conjunction with a key spindle assembly. The assembly carries the second ear for the one ear spindle, but instead of being attached to the spindle, it is joined with the cylinder tailpiece in a floating attachment. The key spindle is carried along with the knob at times when the lock is unlocked, but is also capable of being separately operated by the lock cylinder when the knob itself is immobilized by the plunger, which is also part of the key spindle assembly. The plunger is spring loaded and has a locking lug at the side. It is operated by the main button assembly, which presses the locking lug further into the slot in the hub and housing assembly. This immobilizes the spindle within the hub, and the lock is locked.

As can be seen from Figure 83 and the accompanying parts list showing the usage of these parts in various locks, a great many different modes of operation may be achieved by building from the foundation of this basic mechanism, simply by the addition or substitution of various parts. Such practice is by no means unique, being representative of the usual manufacturing process.

Latch Tubes

Latch tubes for this type of lock may consist simply of an ordinary springbolt (latchbolt assembly) adapted to house in the round tube, or a deadlatching mechanism may be added. In the event that a deadlatch mechanism is incorporated in the latch tube, there is an auxiliary plunger (or deadlocking plunger) incorporated at the flat face of the latchbolt. This plunger does not enter the

pocket of the strike plate, but merely rides the surface of it, as does the auxiliary latch of a mortise deadlatch. It fills much the same function in that a deadlocking lever is incorporated into the latch tube in order to deadlock the latchbolt against end pressure. The deadlocking lever is controlled by the deadlocking plunger.

The construction of the latch tube is primarily of academic interest, since the difficulty of disassembling the unit (which is usually sealed) is disproportionate to the value of doing so; as a consequence, repairs are usually confined to cleaning and relubricating. Figures 84 through 92 show the approach of another manufacturer to this type of construction and the changes in function obtained by parts substitution from the basic unit.

Group II Locks

Another approach to tubular lock construction utilizes a latch tube, with a cam or a pair of cams inside, which operates directly against the latch tail, which in this instance is in the form of a yoke, permitting both arms of the cam to operate the latch bolt. Locking of one or both knobs in this type of construction is achieved by locking the spindle to the rosette of the lock. This is commonly accomplished by the small spindle within the larger, hollow, outer spindle which is attached to the knob, with the smaller inner spindle being operated by the key. The inner spindle moves a sliding part into a position astride the knob spindle and since the sliding part is fixed so that it cannot rotate on the rosette, this will effectively lock the knob spindle to the rosette.

Alternative methods include the use of a camming arrangement to bring lugs into a slot in a plate attached to the rosette, accomplishing substantially the same purpose. Locks utilizing the same principle have the locking device on the rosette of the inside knob; consequently, both knobs are locked when either is locked. Variations of this arrangement may utilize a hollow knob spindle that is partially cut away so that the inner spindle, which is key operated, may throw the locking plate or locking lug into the locked position on the outside rosette, thus locking the outside knob to its rosette. This also locks the inside knob unless a swivel arrangement is provided at a point midway between the two knob cams in order to provide for independent operation. Other locks of this type use a locking plunger much like the arrangement in Group I. (See Figure 82.) Normally the latch tubes for most of the cylindrical type locks are sealed

(continued on page 80)

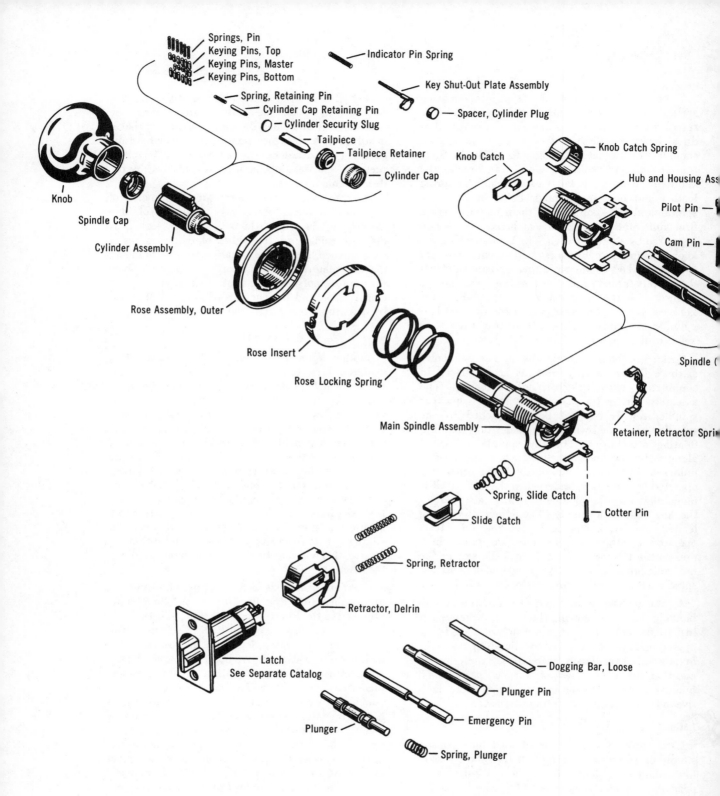

Springs, Pin
Keying Pins, Top
Keying Pins, Master
Keying Pins, Bottom

Indicator Pin Spring

Spring, Retaining Pin
Cylinder Cap Retaining Pin
Cylinder Security Slug
Tailpiece
Tailpiece Retainer
Cylinder Cap

Key Shut-Out Plate Assembly
Spacer, Cylinder Plug

Knob Catch
Knob Catch Spring
Hub and Housing Ass
Pilot Pin
Cam Pin

Knob
Spindle Cap
Cylinder Assembly

Rose Assembly, Outer
Rose Insert
Rose Locking Spring

Main Spindle Assembly

Spindle (

Retainer, Retractor Spri

Spring, Slide Catch
Slide Catch
Cotter Pin

Spring, Retractor

Retractor, Delrin

Latch
See Separate Catalog

Dogging Bar, Loose
Plunger Pin
Emergency Pin
Plunger
Spring, Plunger

Figure 83. Exploded view showing parts utilization. Not all of the parts are used in any one function. See parts utilization in the following lists. Illustration and parts lists courtesy of Falcon Lock Co.

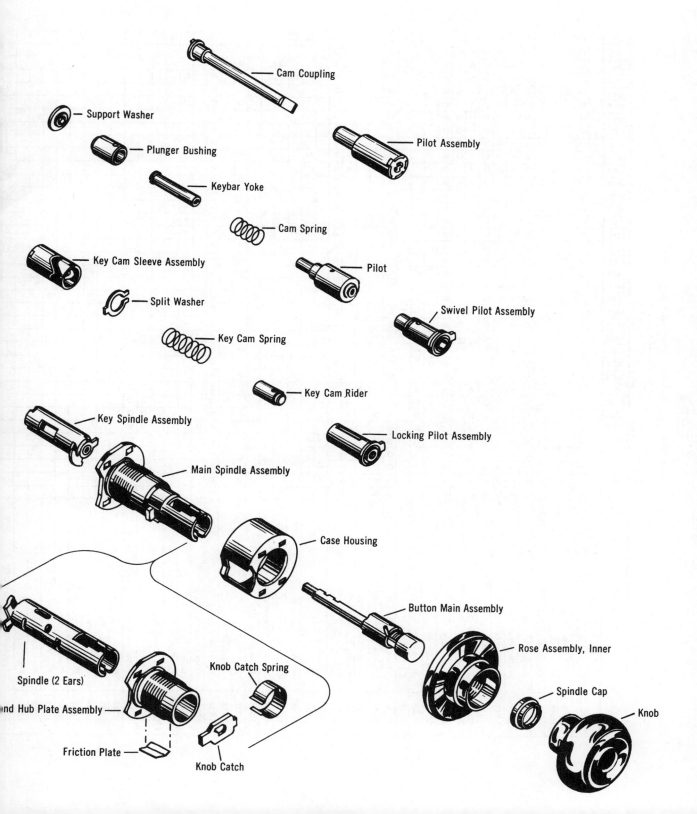

Cam Coupling

Support Washer

Plunger Bushing

Pilot Assembly

Keybar Yoke

Cam Spring

Key Cam Sleeve Assembly

Pilot

Split Washer

Swivel Pilot Assembly

Key Cam Spring

Key Cam Rider

Key Spindle Assembly

Locking Pilot Assembly

Main Spindle Assembly

Case Housing

Button Main Assembly

Rose Assembly, Inner

Spindle (2 Ears)

Knob Catch Spring

nd Hub Plate Assembly

Spindle Cap

Knob

Friction Plate

Knob Catch

CODE

I = INNER TRIM   B = BOTH INNER AND OUTER TRIM
O = OUTER TRIM   C = RETRACTOR CHASSIS

PART NO.	DESCRIPTION	X591	X581	X571	X561	X551	X541	X531	X521	X501	X491	X481	X461	X451	X441	X431	X421	X411	X401	X391	X381	X371	X361	X351	X341	X331	X321	X311	X301	X261	X221	X211	X201	X191	X151	X141	X101
	X SERIES LOCK PARTS																																				
12101N	Retractor, Delrin	C	C	C	C	C	C	C	C	C	C	C	C	C	C	C	C	C	C	C	C	C	C	C	C	C	C	C	C	C	C	C	C	C	C	C	C
12104	Slide Catch, Long					C			C																												
12105	Slide Catch, Short			C			C					C	C	C	C																						
12107	Spring, Slide Catch			C	C	C	C	C	C	C	C	C	C	C	C														C	C							
12108	Spring, Retractor	C	C	C	C	C	C	C	C	C	C	C	C	C	C	C	C	C	C	C	C	C	C	C	C	C	C	C	C	C	C	C	C	C	C	C	C
12109	Friction Plate, Delrin	I	I	I	I	I	I	I	I	I	I	I	I	I	I	I	I	I	I	I	I	I	I	I	I	I	I	I	I	I	I	I	I	I	I	I	I
12112	Retainer, Retractor Spring	C	C	C	C	C	C	C	C	C	C	C	C	C	C	C	C	C	C	C	C	C	C	C	C	C	C	C	C	C	C	C	C	C	C	C	C
12114	Case Housing	C	C	C	C	C	C	C	C	C	C	C	C	C	C	C	C	C	C	C	C	C	C	C	C	C	C	C	C	C	C	C	C	C	C	C	C
12116	Spindle (2 Ears)		I	B	B		B						B						B	B							B	O		I	B	B	B				
12117	Spindle (1 Ear)	I																										B	I	I	B						
12118	Knob Catch	B	B	B	B	B	B	B	B	B	B	B		B	B	B	B	B	B	B	B	B	B	B	B	B	B	B	B	B	B	B	B				
12119	Key Spindle	B	B		O					O	O	O	B	B			B	B						B	O	O	B	B	B								
12126	Spring Cam		O	O														B		B			O		O		O	O	O	O	O	O	O	O			
12131	Cam Spring			O									O	O							B																
12134	Rose Locking Spring	O	O	O	O	O	O	O	O	O	O	O	O	O	O	O	O	O	O	O	O	O	O	O	O	O	O	O	O	O	O	O	O	O	O	O	O
12140	Spindle Cap, Passage	I	I				O																						O	O	O	O	O				
12141	Spindle Cap, Emergency			O																		O							I	I							
12142	Spindle Cap, Button						B																						B		I	I						
12143	Spindle Cap, Cylinder	O	O	O	O	O	O	O	O	O	O	O	O	O	B	B	B	B	B	B	B	B	B	B	I	I	O	O				B	I				
12146	Thumbturn																													O			I				
12207	Plunger	C	C							C	C					C	C		C	C	C	C	C	C			C							O			B
12208	Holdback Spindle, Double	O	O										O										O	O													
12210	Keybar Yoke				O		O												C													C			O		
12213	Holdback Spindle, Single										O								B			O			I		O	O									
12214	Cut-Away Spindle	O	O				B					O	O	O					B	B	B	B	B	B	I												
12215	Split Washer										I				I	I	I	I	B																		
12216	Short Spindle						B						O								I																
12220	Knob Catch Spring	B	B	B	B	B	B	B	B	B	B	B	B	B	B	B	B	B	B	B	B	B	B	B	B	B	B	B	B	B	B	B	B	B			
12225	Dogging Bar, Loose																		C			C							C			C					
12232	Cam Pin																		B	B		B							B			B					
12238	Key Cam Rider				O		O						O																						O	C	
12239	Key Cam Spring																		B																		
12249	Pilot Pin	O	O			O	O			O		I	O	O	I	I	I	I		I		I	I	I										I	I	I	
12251	Support Washer																											O	O	O	O	O					
12252	Pilot																												C	C		O					
12257	Emergency Pin										C						C		C					C				C	C	C	C	C					
12262	Cam Coupling																				O		O														
12269	Spring - Plunger				C										C																						
12273	Plunger														C																						
12279	Pilot Pin						B						O									O											O				
12280	Plunger Bushing	O									B								B	I		I		I										I			I
12281	Plunger Pin					C										C		C	C	C	C													C		C	
12300	Turn Button Main Assembly																																	I			

Figure 83a

PART NO.	DESCRIPTION	X101	X141	X151	X161	X201	X211	X221	X241	X301	X311	X321	X331	X341	X351	X361	X371	X381	X391	X401	X411	X421	X431	X441	X451	X461	X481	X491	X501	X521	X531	X541	X551	X561	X571	X581	X591
12306	Hub and Housing Assembly													o																							
12310	Hub and Hub Plate Assembly	-											o	o																							
12311	Hub and Housing Assembly	o	o	o		o	o	o		o	o				o		o	o	o	o		o							o	o	o	o	o	o	o	-	o
12312	Locking Pilot Assembly								-						-						-	-		-				-									o
12318	Main Spindle Assembly, Inside		o																		o																
12320	Main Spindle Assembly							o					o			o							o	o	o	o	o									o	
12328	Key Spindle Assembly																																		o		-
12339	Locking Pilot Assembly																							o		o											
12341	Swivel Pilot Assembly																															o		o			
12360	Locking Pilot Assembly																-					-							-	-	-						
12361	Pilot Assembly																o														o						
12362	Pilot Assembly																-														-						
12364	Swivel Pilot Assembly									o																											
12365	Turn Button Main Assembly						B	o	o	o																											
12370	Key Cam Sleeve Assembly																o																				
12371	Key Cam Sleeve Assembly																		B							-											
12372	Key Cam Sleeve Assembly						o										-		B	B															o		
12373	Key Cam Sleeve Assembly						o						-			-											o					o	o				o
12374	Swivel Pilot Assembly															-		-	B	B			-			o											
12375	Key Cam Sleeve Assembly																						-														o
12376	Key Cam Sleeve Assembly										-																									o	
12377	Key Cam Sleeve Assembly						-			-															o	o	o					o			o	o	
12406	Main Spindle Assembly			-	-	-	-								-		-										-	o	-	-	-		-	-	-	-	-
12407	Main Spindle Assembly			o		o	o	o	o	o	o						o	o	o	o	o	o	o	o	o	o	o	o	o	o	o	o	o	o	o		
12408	Main Spindle Assembly																																				-
12411	Main Spindle Assembly																-												-								
12412	Main Spindle Assembly								o																												
12413	Main Spindle Assembly									-																											
12414	Main Spindle Assembly	-	-			-	o	o	-	o	-				-														o	o	o	o	o	-	-	-	-
12415	Main Spindle Assembly	o	o		o	o	o			o					o		o	o			o							o									
12419	Key Spindle Assembly																																				
12422	Universal Button Main Assembly					-			-																												-
12423	Push Button Main Assembly																																				
12424	Retract Mechanism Assembly																												C								
12425	Retract Mechanism Assembly																		C														C				
12426	Retract Mechanism Assembly																													C							
12427	Retract Mechanism Assembly																														C						
12428	Retract Mechanism Assembly																																C		C		
12429	Retract Mechanism Assembly						C																								C						
12430	Retract Mechanism Assembly																C			C																	
12432	Retract Mechanism Assembly																							C													
12433	Retract Mechanism Assembly																								C												
12434	Retract Mechanism Assembly																																				
12435	Retract Mechanism Assembly	o																																			

Figure 83b

CODE   I = INNER TRIM   O = OUTER TRIM   B = BOTH INNER AND OUTER TRIM   C = RETRACTOR CHASSIS

PART NO.	DESCRIPTION	X101	X141	X151	X161	X201	X211	X221	X241	X301	X311	X321	X331	X341	X351	X361	X371	X381	X391	X401	X411	X421	X431	X441	X451	X461	X481	X491	X501	X521	X531	X541	X551	X561	X571	X581	X591
12436	Push Button Main Assembly								I																	I						I			I		
12437	Turn Button Main Assembly							I	I																					I							
12438	Push Button Main Assembly																										I		I								
12439	Push Button Main Assembly (Spanner)					I																			I												
12440	Main Spindle Assembly																			I																	
12441	Main Spindle Assembly																									O						O			O		
12442	Main Spindle Assembly							I																I													
12443	Main Spindle Assembly																											I									
12444	Main Spindle Assembly										O								O															O			
12445	Main Spindle Assembly														O			O				O						O			O		O				
12446	Main Spindle Assembly																			O											O					O	
12447	Main Spindle Assembly																																				O
12449	Main Spindle Assembly	O		O													C																				
12462	Main Spindle Assembly						I																														
12466	Main Spindle Assembly																																				
12469	Retract Mechanism Assembly	C																																			
12470	Retract Mechanism Assembly		C																																		
12471	Retract Mechanism Assembly					C																															
12472	Retract Mechanism Assembly							C																													
12473	Retract Mechanism Assembly															O																		C			
12474	Main Spindle Assembly												I						I													I					
12476	Main Spindle Assembly																																				
12477	Retract Mechanism Assembly																																				
12478	Retract Mechanism Assembly																		O													C			C		
12479	Retract Mechanism Assembly																																				
12480	Main Spindle Assembly																	I					I														
12481	Retract Mechanism Assembly			C																																	
12482	Retract Mechanism Assembly				C																																
12483	Retract Mechanism Assembly								C																												
12484	Retract Mechanism Assembly										C																										
12485	Retract Mechanism Assembly											C																									
12486	Retract Mechanism Assembly												C																								
12487	Retract Mechanism Assembly													C																							
12488	Retract Mechanism Assembly																										C										
12490	Retract Mechanism Assembly																				C																
12491	Retract Mechanism Assembly																						C														
12492	Retract Mechanism Assembly															C																					
12493	Retract Mechanism Assembly																											C									
12495	Retract Mechanism Assembly														C																						
12496	Retract Mechanism Assembly																					C															
12497	Retract Mechanism Assembly																					I	I	I													
12750	Cylinder Assembly (5 pin)									O	C	O	I	I	I	O	B	O	B	B	B	O															C
12752	Cylinder Assembly (5 pin)															I	B	B										O	O	O	O		O	O		O	O
12752-2A	Cylinder Assembly (5 pin)																																				
12754	Cylinder Assembly (5 pin)																						O			O										O	
12756-1	Cylinder Assembly (5 pin)																						O	O								O		O	O		

Figure 83c

12760	Cylinder Assembly (6 pin)											O		I	O	O		O			B	O						O	O	O	O		O			O	
12762	Cylinder Assembly (6 pin)										O		I		I	I	B	I	B	B		I	I	I										O			O
12762-2A	Cylinder Assembly (6 pin)																															O			O		
12764	Cylinder Assembly (6 pin)																						O				O										
12766-1	Cylinder Assembly (6 pin)																							O	O												
12766-2	Cylinder Assembly (6 pin)																									O											

	CYLINDER ASSEMBLY PARTS																																				
12162	Cylinder Cap										O	O	I	I	B	B	B	B	B	B	B	B	B	B	O	O	O	O	O	O	O	O	O	O	O	O	O
12163	Tailpiece Retainer										O	O	I	I	B	B	B	B	B	B	B	B	B	B	O	O	O	O	O	O	O	O	O	O	O	O	O
·12164	Tailpiece										O		I		I	I	B	I	B	B		I	I	I										O			O
12164-1	Tailpiece																							O	O												
12164-2	Tailpiece																									O											
12164-2A	Tailpiece																															O			O		
12165	Cylinder Cap Retaining Pin										O	O	I	I	B	B	B	B	B	B	B	B	B	B	O	O	O	O	O	O	O	O	O	O	O	O	O
12166	Spring, Cylinder Cap Retaining Pin										O	O	I	I	B	B	B	B	B	B	B	B	B	B	O	O	O	O	O	O	O	O	O	O	O	O	O
12167	Cylinder Security Slug										O	O	I	I	B	B	B	B	B	B	B	B	I	I				O	O	O	O	O	O	O	O	O	O
12169	Tailpiece											O		I	O	O		O			B	O	O				O	O	O	O	O		O			O	
12169-1	Tailpiece																																				
12174	Spacer, Cylinder Plug												**Use with 5 Pin Assemblies →**										O	O	O	O	O										
12266	Indicator Pin Spring												**(5 or 6 Pin) →**										O	O	O	O	O										
12367	Key Shut-Out Plate Assembly												**Use with 5 Pin Assemblies →**										O	O	O	O	O										
12367A	Key Shut-Out Plate Assembly												**Use with 6 Pin Assemblies →**										O	O	O	O	O										

	KNOBS																																				
12296	Beverly, Kim, Soto and Pico Knob .060″	B	B	O	I	B	B	B	I	B	O	O	I	I	B	B	B	B	B	B	B	B	B	B	B	B	B	B	B	B	B	B	B	B	B	B	B
12147	Beverly, Kim, Soto and Pico Knob .080″	B	B	O	I	B	B	B	I	B	O	O	I	I	B	B	B	B	B	B	B	B	B	B	B	B	B	B	B	B	B	B	B	B	B	B	B
12290	Crown Royal, Hana, Vega and Gala Knob .060″	B	B	O	I	B	B	B	I	B	O	O	I	I	B	B	B	B	B	B	B	B	B	B	B	B	B	B	B	B	B	B	B	B	B	B	B
12149	Crown Royal, Hana, Vega and Gala Knob .080″	B	B	O	I	B	B	B	I	B	O	O	I	I	B	B	B	B	B	B	B	B	B	B	B	B	B	B	B	B	B	B	B	B	B	B	B

	ROSE ASSEMBLIES, OUTER																																				
12301	Beverly and Crown Royal	O	O	O		O	O	O	O	O	O	O			O	O	O	O	O	O	O	O	O	O	O	O	O	O	O	O	O	O	O	O	O	O	O
12329	Kim and Hana	O	O	O		O	O	O	O	O	O	O			O	O	O	O	O	O	O	O	O	O	O	O	O	O	O	O	O	O	O	O	O	O	O
12342	Soto and Vega	O	O	O		O	O	O	O	O	O	O			O	O	O	O	O	O	O	O	O	O	O	O	O	O	O	O	O	O	O	O	O	O	O
12352	Gala and Pico	O	O	O		O	O	O	O	O	O	O			O	O	O	O	O	O	O	O	O	O	O	O	O	O	O	O	O	O	O	O	O	O	O

	ROSE ASSEMBLIES, INNER																																				
12302	Beverly and Crown Royal	I	I	I	I	I	I	I	I	I	I	I	I	I	I	I	I	I	I	I	I	I	I	I	I	I	I	I	I	I	I	I	I	I	I	I	I
12330	Kim and Hana	I	I	I	I	I	I	I	I	I	I	I	I	I	I	I	I	I	I	I	I	I	I	I	I	I	I	I	I	I	I	I	I	I	I	I	I
12343	Soto and Vega	I	I	I	I	I	I	I	I	I	I	I	I	I	I	I	I	I	I	I	I	I	I	I	I	I	I	I	I	I	I	I	I	I	I	I	I
12353	Gala and Pico	I	I	I	I	I	I	I	I	I	I	I	I	I	I	I	I	I	I	I	I	I	I	I	I	I	I	I	I	I	I	I	I	I	I	I	I

Figure 83d

together at the time of construction at the factory and it is rarely practical to make repairs in the latch tube itself. Usually it is more practical to simply replace the latch tube assembly on a unit basis, in the event of a breakdown.

The majority of parts in these locks are usually formed from sheet metal, with a few being forged or cast. Some plastics are beginning to be used in the construction of cylindrical locks. Pressure castings are being used successfully for some of the more massive parts which are not subject to a great deal of strain or wear; however, die castings in operational parts, where stresses are high, fracture easily in service, particularly if the older, so-called "pot metal" materials are used. Die castings of brass or bronze, or higher strength alloys, do not pose such great problems.

A great many variations of these locks are possible, and do exist, both with respect to design and materials used; however, once basic principles are established for any given line of locks, they can usually be followed through from one mode of operation to another by simply observing the similarities and differences of the parts used in their construction.

Because of the multiplicity of these locks and the fact that some are used much more frequently in a given area than others, it is recommended that each individual study the popular locks with particular care, physically disassembling and reassembling specimens until he is thoroughly familiar with their construction. In the process of this study, one should learn precisely why they function as they do; thus, small discrepancies which are causing trouble and might otherwise be overlooked will become readily apparent.

THE PROGRESSIVE NATURE OF CHANGES IN FUNCTION

The illustrations and charts beginning with Figure 83 show the requirements of two manufacturers of locks that utilize the techniques of addition, removal, or substitution of parts. Locks designed and built to this principle have orderly transition from one function to another, effecting economies in both production and parts stocking by the locksmith.

While the following illustrations by no means represent all of the functions of Yale cylindrical locks, they are representative of the means by which functions are changed. A great many combinations

of the variations used to produce the illustrated functions may be employed to produce a multiplicity of functions. The order in which these illustrations appear has been chosen as carefully as possible to show the nature and extent of the changes required to produce a specific operating characteristic.

Although hotel functions with occupancy indicators and "shut-out" keying are shown in both the Falcon and Yale cylindrical lock illustrations, no detailed discussion will be made of this type of lock because the specific means of accomplishing this function vary too widely from one manufacturer to another. It is sufficient to say that mechanically induced restrictions which operate on the tip of the key serve to do one of three things:

1. Exclude a key through the introduction of an obstruction into the rear of the keyhole.

2. Exclude a key through the pin of the occupancy indicator operating against one of the stops of the key to keep the key from going all the way into the lock.

3. Disconnect some part of the mechanism until a key with an especially long tip is inserted to nullify the disconnect mechanism while the key is being turned.

Application of the analytical method which has been followed in this text will quickly resolve the means by which this operation is accomplished in any specific lock.

RE-KEYING INEXPENSIVE CYLINDRICAL LOCKS

Although the less expensive cylindrical locks seldom warrant the expenditure of much time in repairing, it is frequently necessary to disassemble the lock cylinder for re-keying or key making. With this in mind, the following re-keying instructions for various locks of their manufacture is presented.

THE MONO-LOCK

The Mono-Lock may be considered a lock transitional between mortise and cylindrical locks. It has a number of features of both of these types in various of its functions. Unlike the cylindrical locks, the Mono-Lock is available with a deadbolt function; yet, like the cylindrical lock, it utilizes a wobbling plate to create the deadlocking function (part P in Figure 93). Other similarities and differences will become apparent upon examination of Figure 93.

PASSAGE LATCH
EXPLODED VIEW

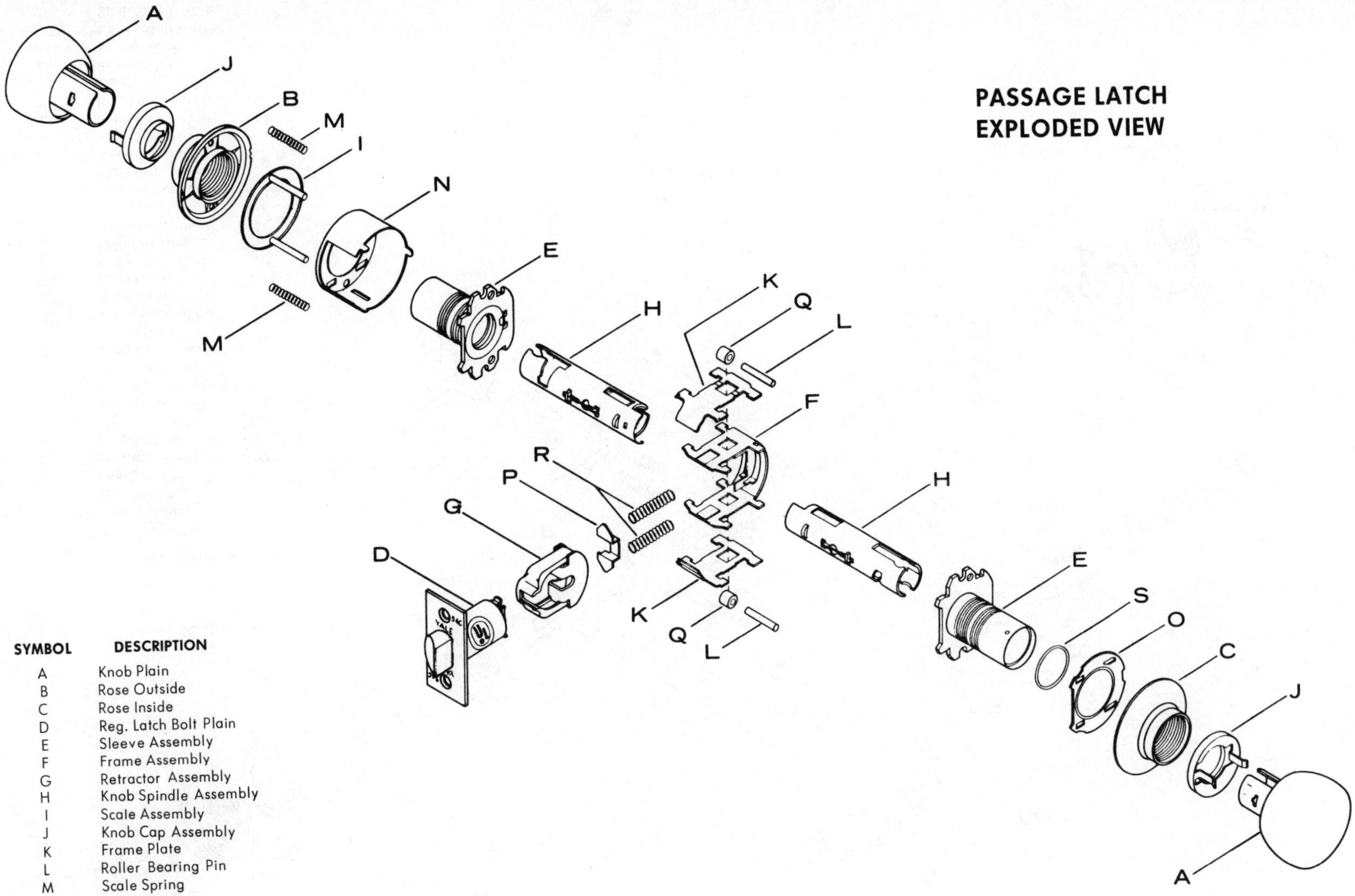

SYMBOL	DESCRIPTION
A	Knob Plain
B	Rose Outside
C	Rose Inside
D	Reg. Latch Bolt Plain
E	Sleeve Assembly
F	Frame Assembly
G	Retractor Assembly
H	Knob Spindle Assembly
I	Scale Assembly
J	Knob Cap Assembly
K	Frame Plate
L	Roller Bearing Pin
M	Scale Spring
N	Case Cover
O	Case Cover Plate
P	Retractor Spring Locator
Q	Roller Bearing
R	Retractor Spring
S	"O" Ring

Figure 84

PRIVACY LOCK
EXPLODED VIEW

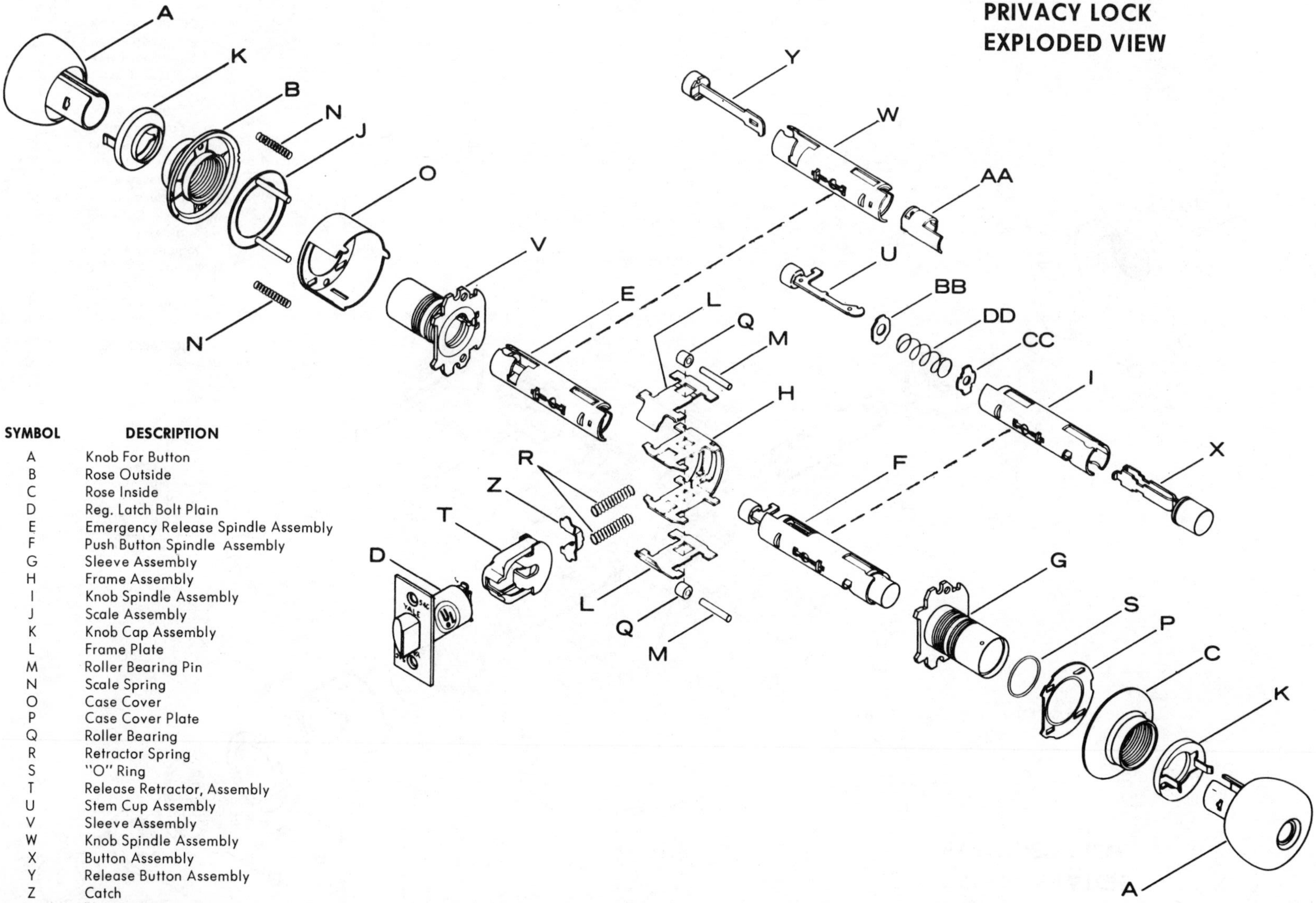

SYMBOL	DESCRIPTION
A	Knob For Button
B	Rose Outside
C	Rose Inside
D	Reg. Latch Bolt Plain
E	Emergency Release Spindle Assembly
F	Push Button Spindle Assembly
G	Sleeve Assembly
H	Frame Assembly
I	Knob Spindle Assembly
J	Scale Assembly
K	Knob Cap Assembly
L	Frame Plate
M	Roller Bearing Pin
N	Scale Spring
O	Case Cover
P	Case Cover Plate
Q	Roller Bearing
R	Retractor Spring
S	"O" Ring
T	Release Retractor, Assembly
U	Stem Cup Assembly
V	Sleeve Assembly
W	Knob Spindle Assembly
X	Button Assembly
Y	Release Button Assembly
Z	Catch
AA	Release Tube
BB	Spring Blank
CC	Connecting Washer
DD	Stem Spring

Figure 85

COMMUNICATING LOCK EXPLODED VIEW

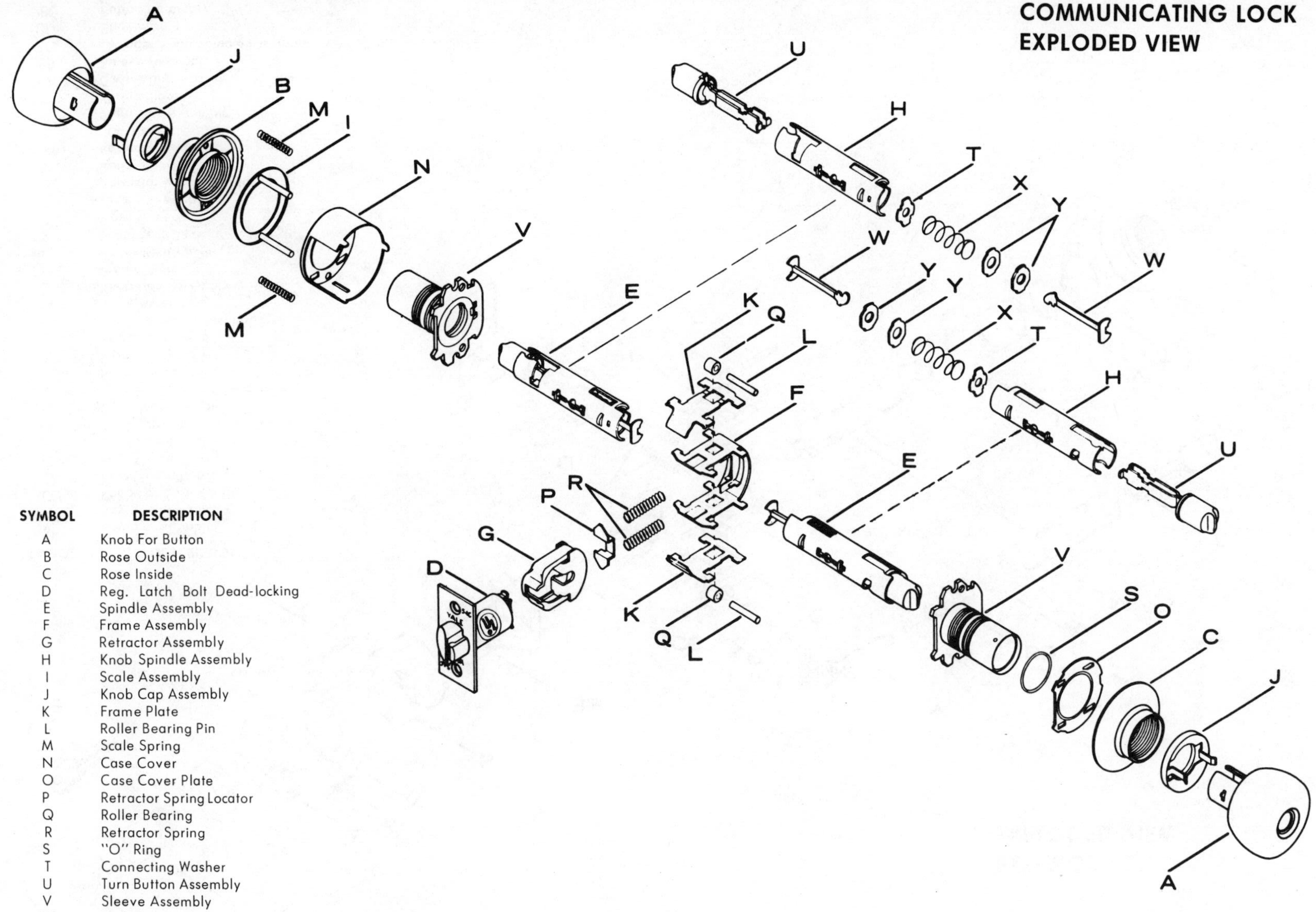

SYMBOL	DESCRIPTION
A	Knob For Button
B	Rose Outside
C	Rose Inside
D	Reg. Latch Bolt Dead-locking
E	Spindle Assembly
F	Frame Assembly
G	Retractor Assembly
H	Knob Spindle Assembly
I	Scale Assembly
J	Knob Cap Assembly
K	Frame Plate
L	Roller Bearing Pin
M	Scale Spring
N	Case Cover
O	Case Cover Plate
P	Retractor Spring Locator
Q	Roller Bearing
R	Retractor Spring
S	"O" Ring
T	Connecting Washer
U	Turn Button Assembly
V	Sleeve Assembly
W	Push Bar
X	Push Bar Spring
Y	Locking Washer

Figure 86

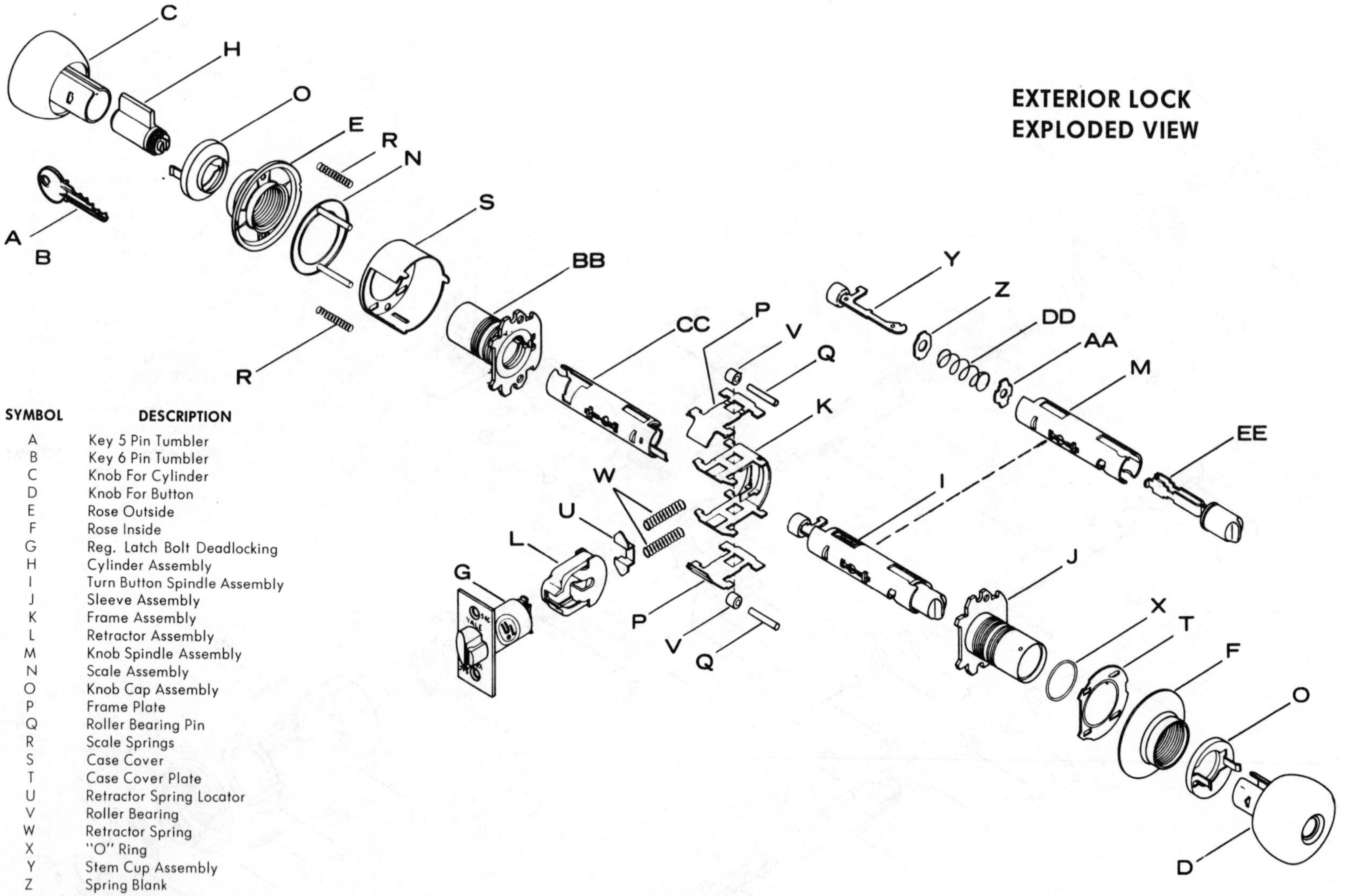

**EXTERIOR LOCK
EXPLODED VIEW**

SYMBOL	DESCRIPTION
A	Key 5 Pin Tumbler
B	Key 6 Pin Tumbler
C	Knob For Cylinder
D	Knob For Button
E	Rose Outside
F	Rose Inside
G	Reg. Latch Bolt Deadlocking
H	Cylinder Assembly
I	Turn Button Spindle Assembly
J	Sleeve Assembly
K	Frame Assembly
L	Retractor Assembly
M	Knob Spindle Assembly
N	Scale Assembly
O	Knob Cap Assembly
P	Frame Plate
Q	Roller Bearing Pin
R	Scale Springs
S	Case Cover
T	Case Cover Plate
U	Retractor Spring Locator
V	Roller Bearing
W	Retractor Spring
X	"O" Ring
Y	Stem Cup Assembly
Z	Spring Blank
AA	Connecting Washer
BB	Sleeve Assembly
CC	Cylinder Knob Spindle Assembly
DD	Stem Spring
EE	Button Assembly

Figure 87

ALL PURPOSE LOCK
EXPLODED VIEW

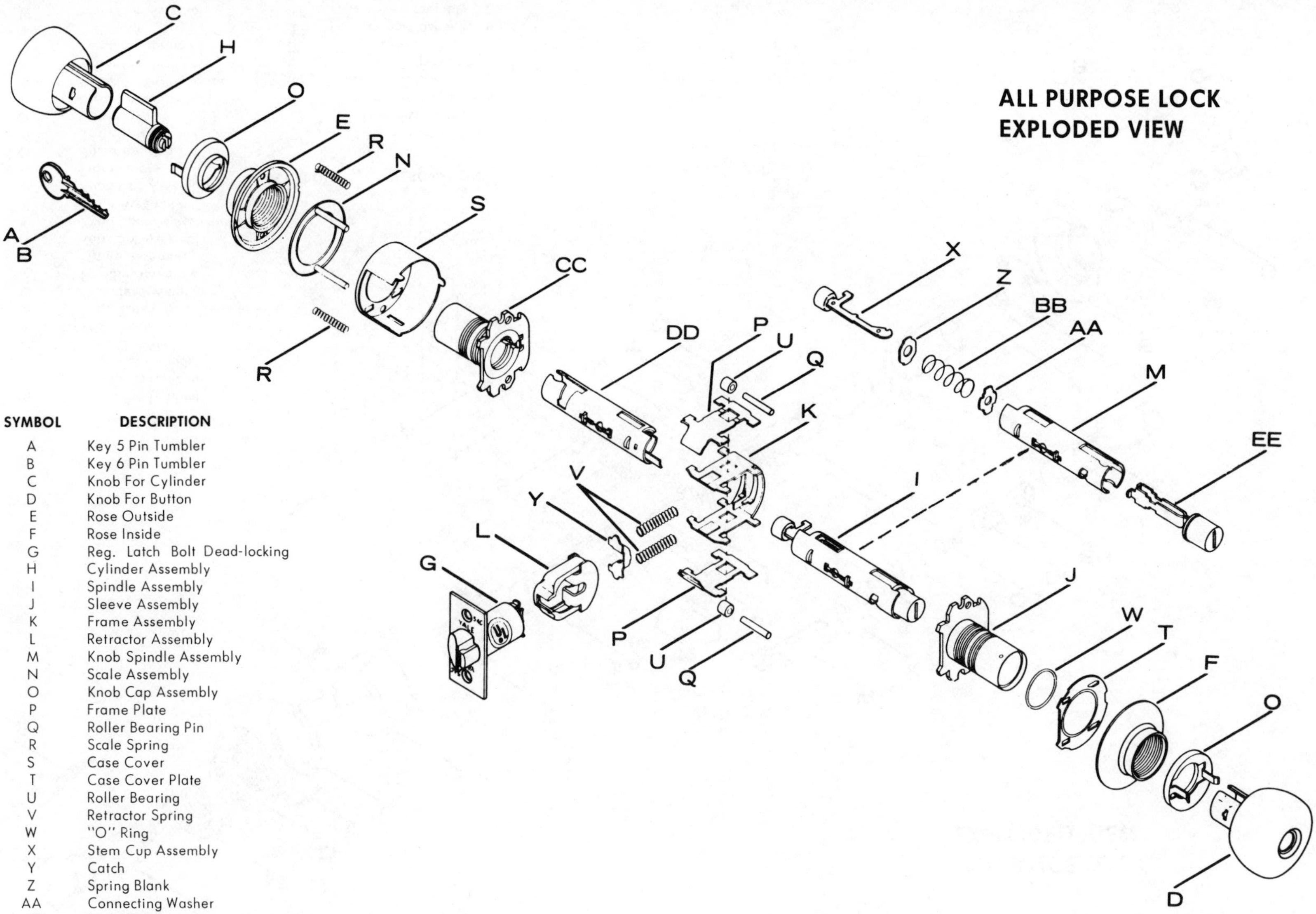

SYMBOL	DESCRIPTION
A	Key 5 Pin Tumbler
B	Key 6 Pin Tumbler
C	Knob For Cylinder
D	Knob For Button
E	Rose Outside
F	Rose Inside
G	Reg. Latch Bolt Dead-locking
H	Cylinder Assembly
I	Spindle Assembly
J	Sleeve Assembly
K	Frame Assembly
L	Retractor Assembly
M	Knob Spindle Assembly
N	Scale Assembly
O	Knob Cap Assembly
P	Frame Plate
Q	Roller Bearing Pin
R	Scale Spring
S	Case Cover
T	Case Cover Plate
U	Roller Bearing
V	Retractor Spring
W	"O" Ring
X	Stem Cup Assembly
Y	Catch
Z	Spring Blank
AA	Connecting Washer
BB	Stem Spring
CC	Sleeve Assembly
DD	Knob Spindle Assembly
EE	Universal Button Assembly

Figure 88

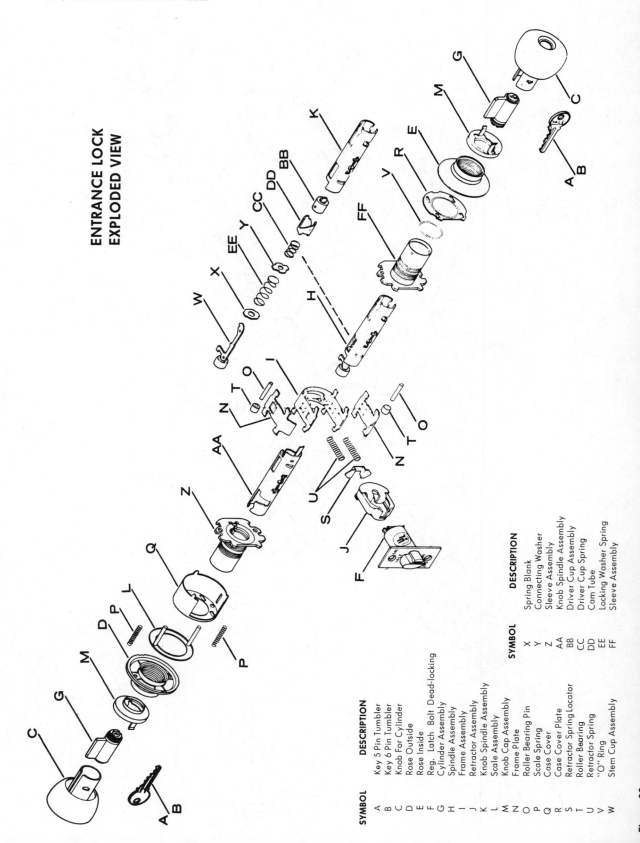

ENTRANCE LOCK
EXPLODED VIEW

SYMBOL	DESCRIPTION
A	Key 5 Pin Tumbler
B	Key 6 Pin Tumbler
C	Knob For Cylinder
D	Rose Outside
E	Rose Inside
F	Reg. Latch Bolt Dead-locking
G	Cylinder Assembly
H	Spindle Assembly
I	Frame Assembly
J	Retractor Assembly
K	Knob Spindle Assembly
L	Scale Assembly
M	Knob Cap Assembly
N	Frame Plate
O	Roller Bearing Pin
P	Scale Spring
Q	Case Cover
R	Case Cover Plate
S	Retractor Spring Locator
T	Roller Bearing
U	Retractor Spring
V	"O" Ring
W	Stem Cup Assembly

SYMBOL	DESCRIPTION
X	Spring Blank
Y	Connecting Washer
Z	Sleeve Assembly
AA	Knob Spindle Assembly
BB	Driver Cup Assembly
CC	Driver Cup Spring
DD	Cam Tube
EE	Locking Washer Spring
FF	Sleeve Assembly

STORE DOOR LOCK
EXPLODED VIEW

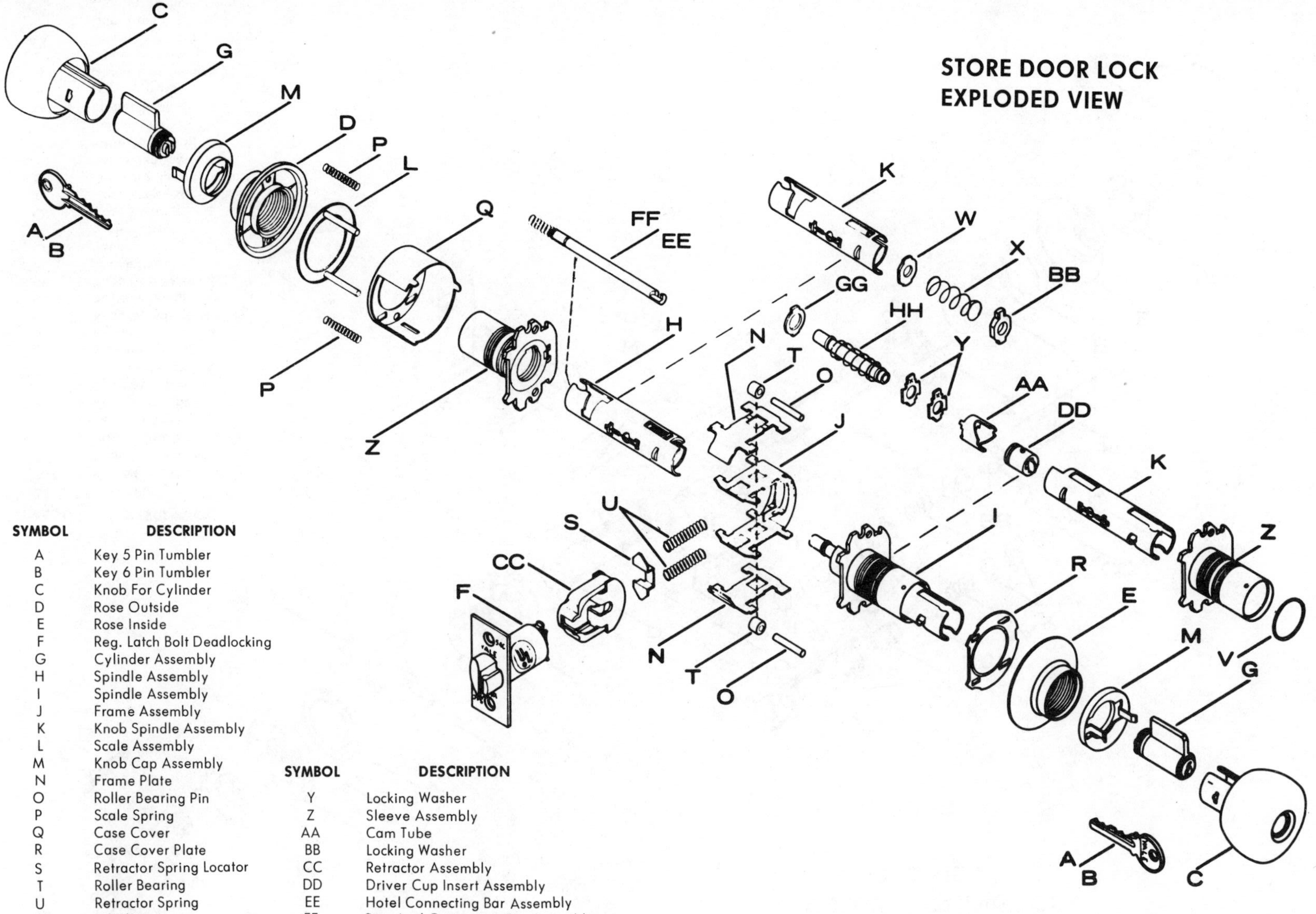

SYMBOL	DESCRIPTION
A	Key 5 Pin Tumbler
B	Key 6 Pin Tumbler
C	Knob For Cylinder
D	Rose Outside
E	Rose Inside
F	Reg. Latch Bolt Deadlocking
G	Cylinder Assembly
H	Spindle Assembly
I	Spindle Assembly
J	Frame Assembly
K	Knob Spindle Assembly
L	Scale Assembly
M	Knob Cap Assembly
N	Frame Plate
O	Roller Bearing Pin
P	Scale Spring
Q	Case Cover
R	Case Cover Plate
S	Retractor Spring Locator
T	Roller Bearing
U	Retractor Spring
V	"O" Ring
W	Spring Blank
X	Push Bar Spring

SYMBOL	DESCRIPTION
Y	Locking Washer
Z	Sleeve Assembly
AA	Cam Tube
BB	Locking Washer
CC	Retractor Assembly
DD	Driver Cup Insert Assembly
EE	Hotel Connecting Bar Assembly
FF	Standard Connecting Bar Assembly
GG	Spring Blank
HH	Connecting Bar Guide

Figure 90

COMMUNICATING LOCK EXPLODED VIEW

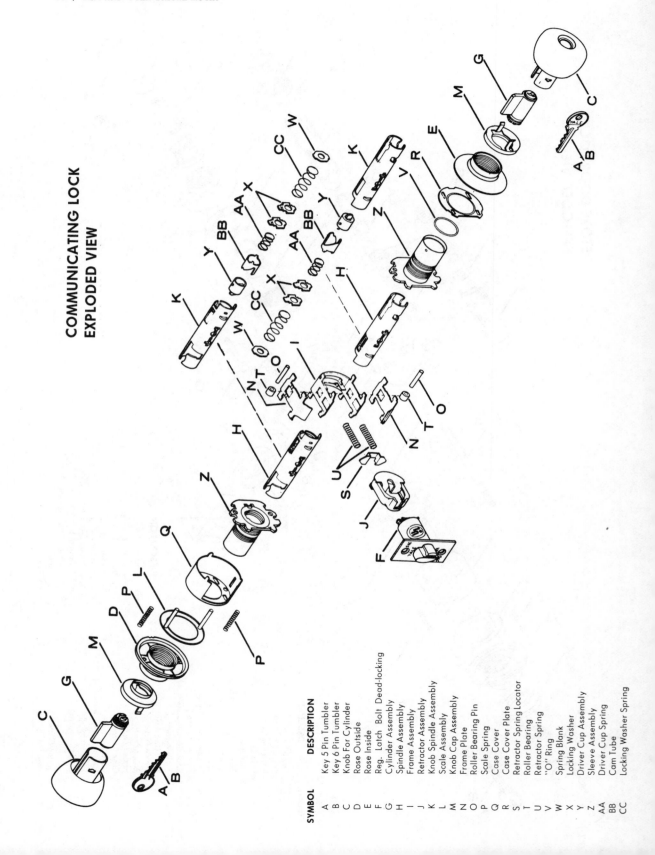

SYMBOL	DESCRIPTION
A	Key 5 Pin Tumbler
B	Key 6 Pin Tumbler
C	Knob For Cylinder
D	Rose Outside
E	Rose Inside
F	Reg. Latch Bolt Dead-locking
G	Cylinder Assembly
H	Spindle Assembly
I	Frame Assembly
J	Retractor Assembly
K	Knob Spindle Assembly
L	Scale Assembly
M	Knob Cap Assembly
N	Frame Plate
O	Roller Bearing Pin
P	Scale Spring
Q	Case Cover
R	Case Cover Plate
S	Retractor Spring Locator
T	Roller Bearing
U	Retractor Spring
V	"O" Ring
W	Spring Blank
X	Locking Washer
Y	Driver Cup Assembly
Z	Sleeve Assembly
AA	Driver Cup Spring
BB	Cam Tube
CC	Locking Washer Spring

HOTEL LOCK
EXPLODED VIEW

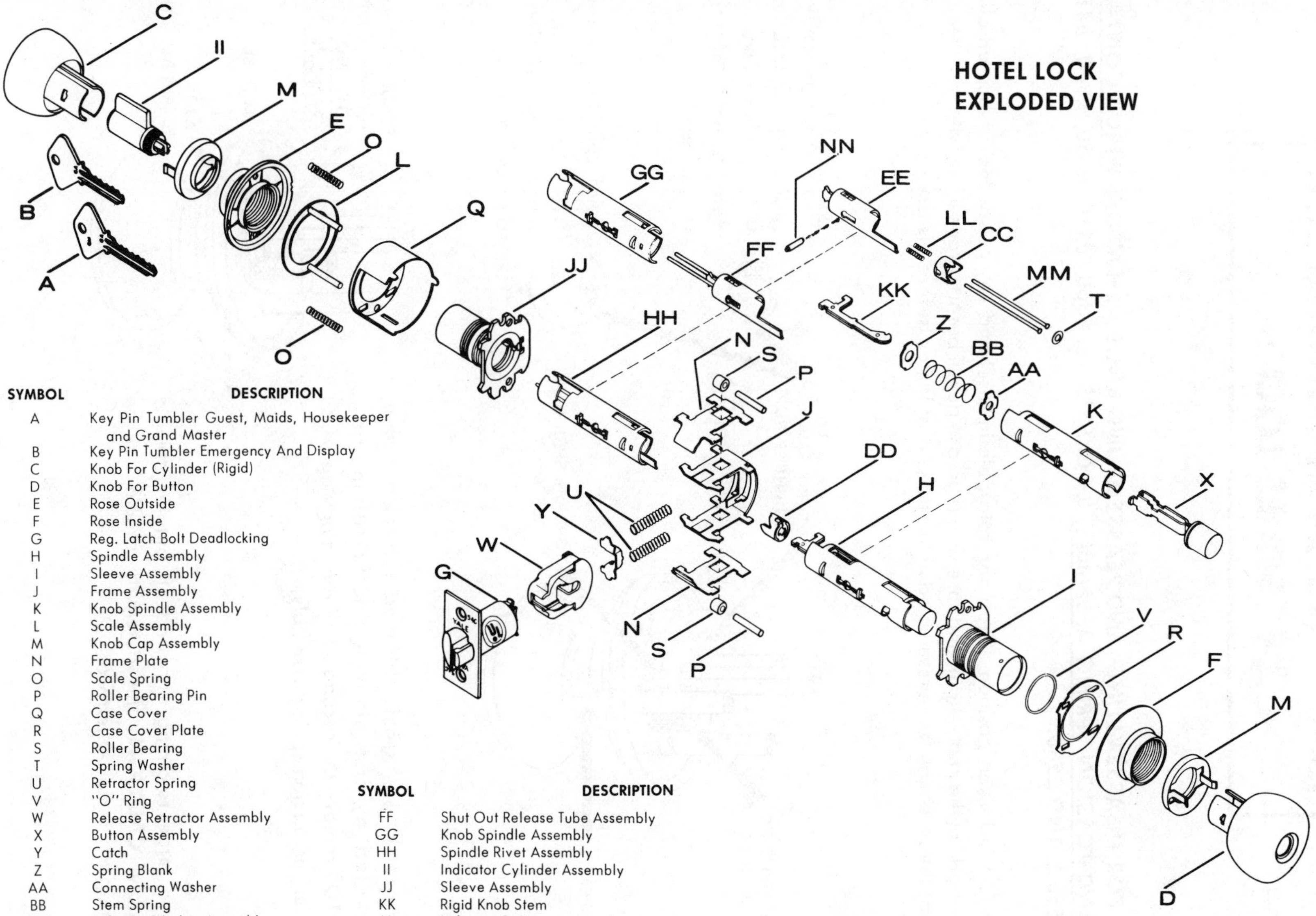

SYMBOL **DESCRIPTION**

A Key Pin Tumbler Guest, Maids, Housekeeper and Grand Master
B Key Pin Tumbler Emergency And Display
C Knob For Cylinder (Rigid)
D Knob For Button
E Rose Outside
F Rose Inside
G Reg. Latch Bolt Deadlocking
H Spindle Assembly
I Sleeve Assembly
J Frame Assembly
K Knob Spindle Assembly
L Scale Assembly
M Knob Cap Assembly
N Frame Plate
O Scale Spring
P Roller Bearing Pin
Q Case Cover
R Case Cover Plate
S Roller Bearing
T Spring Washer
U Retractor Spring
V "O" Ring
W Release Retractor Assembly
X Button Assembly
Y Catch
Z Spring Blank
AA Connecting Washer
BB Stem Spring
CC Indicator Washer Assembly
DD Cam Limit Washer Assembly
EE Shut Out Plug Bar Assembly

SYMBOL **DESCRIPTION**

FF Shut Out Release Tube Assembly
GG Knob Spindle Assembly
HH Spindle Rivet Assembly
II Indicator Cylinder Assembly
JJ Sleeve Assembly
KK Rigid Knob Stem
LL Indicator Spring
MM Indicator
NN Rivet

Figure 92

RE-KEYING INSTRUCTIONS FOR "E" SERIES LOCKS

PROCEDURE FOR DISASSEMBLING AND REASSEMBLING KNOB ASSEMBLIES WHICH CONTAIN A SPINDLE AND TORSION SPRING ASSEMBLY. (LOCKSETS No. E360, E370, E380, E390, E410, E520, E530, E560, AND E580.)

(1) With lock in locked position, (knob non rotatable in rose) and key out of lock, place an alignment mark on the knob shank and rose with a grease pencil before proceeding with disassembly, so that the rose may be reassembled in the same position as before disassembly.

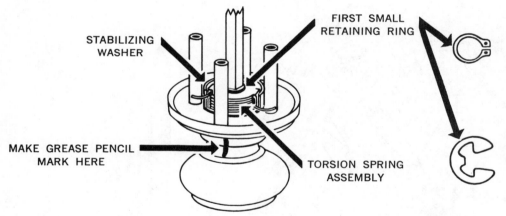

STABILIZING WASHER

FIRST SMALL RETAINING RING

MAKE GREASE PENCIL MARK HERE

TORSION SPRING ASSEMBLY

Tubular spindles with projecting tailpiece have this type retaining ring. Remove with retaining ring pliers.

Solid spindles have this type retaining ring. Remove with screwdriver or other suitable tool.

(2) Remove torsion spring assembly (and stabilizing washer, if included) by first removing small retaining ring. **DO NOT** *remove the second small retaining ring which is under the torsion spring assembly.*

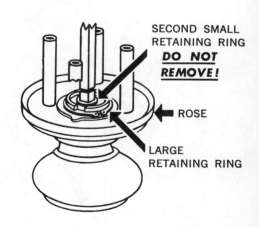

SECOND SMALL RETAINING RING *DO NOT REMOVE!*

ROSE

LARGE RETAINING RING

(3) Remove large retaining ring and lift out the spindle assembly. **CAUTION:** *this assembly should not be taken apart.* Remove the rose.

4 Remove cylinder cap by unscrewing it with cylinder cap tool (Part No. 1416). After removing cylinder cap, remove cylinder stop with tweezers.

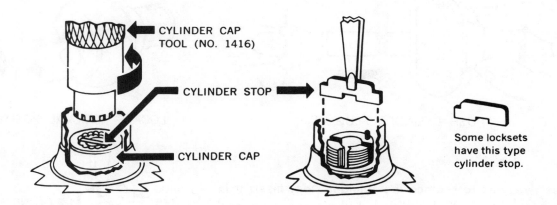

CYLINDER CAP
TOOL (NO. 1416)

CYLINDER STOP

CYLINDER CAP

Some locksets
have this type
cylinder stop.

"UNSCREW CYLINDER CAP"   "REMOVE CYLINDER STOP"

5 Insert key in cylinder plug, holding the knob so that the key-cuts are in an upward direction. Turn the key slightly clockwise to position the cylinder plug for removal. While holding the knob and key in this position push the cylinder plug out the key side by pushing a plug follower (Part No. 1421) in from the shank side. Continue to hold the key with the key-cuts in an upward direction to prevent the pins from falling out. **NOTE!** *Plug follower must be left in the cylinder until the cylinder plug is replaced to keep top pins and springs in position.*

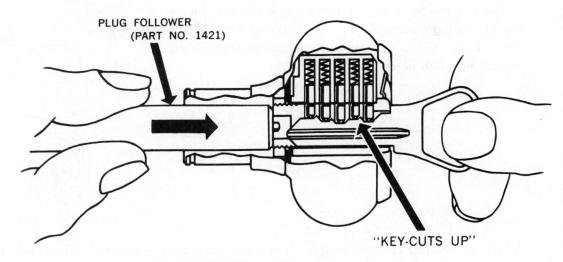

PLUG FOLLOWER
(PART NO. 1421)

"KEY-CUTS UP"

"PUSH CYLINDER PLUG
OUT WITH PLUG FOLLOWER"

6 After fitting the cylinder plug with pins to suit the new key, push the cylinder plug in from the key side by pushing the plug follower out the shank side. Then, holding the cylinder plug face inward with the thumb, and at the same time, pull and rotate the key to locate removal position. *Remove key* and continue with reassembly.

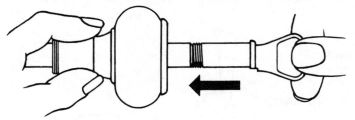

"PUSH CYLINDER PLUG IN"

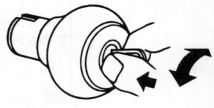

"LOCATE REMOVAL POSITION"

7 Use tweezers to replace the cylinder stop. Insert it in the slot of the cylinder plug with the notch down and the side tab to the left in relation to the retaining pin, at the top of the keyway. If the cylinder stop has two side tabs, either side tab may be placed to the left.

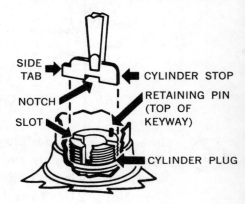

"REPLACE CYLINDER STOP"

8 Replace the cylinder cap by holding the cylinder plug in firmly from the front. Screw the cylinder cap on as far as it will go, using the cylinder cap tool (No. 1416). Then, back off two or three notches, allowing retaining pin to snap into notch.

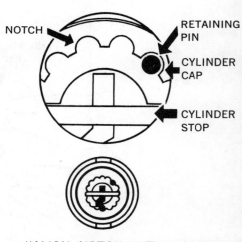

"ALIGN NOTCH WITH HOLE"

9 Insert key and test cylinder plug for freedom of action. If excessive bind or end play is evident, remove key and rotate cylinder cap with cylinder cap tool (No. 1416) as follows:

    a) **2** or **3** notches counterclockwise to eliminate binding.

    b) **2** or **3** notches clockwise to eliminate excessive end play.

Insert key and re-test action. *Remove key* and continue with reassembly.

Align slots of Delrin bearing with slots in knob shank, and *replace rose in its original position by re-aligning the grease pencil marks placed on the knob shank and rose prior to disassembly.*

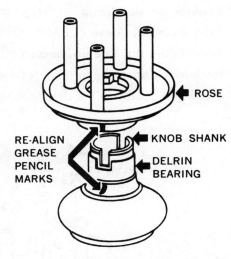

"REPLACE ROSE"

Align spindle assembly (except E410 and E580) so that the blocker and rose washer stop projections are parallel to each other and so that the cylinder tailpiece slot is positioned at right angles (90°) to them. (E410 and E580 locksets must be aligned with the cylinder tailpiece slot parallel with the lock plate notches.)

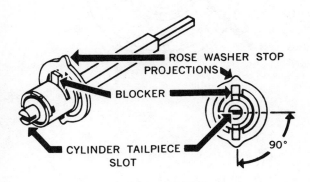

"ALIGN SPINDLE ASSEMBLY"

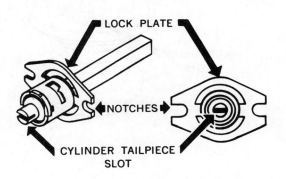

(E410 and E580)

Replace the spindle assembly so that the cylinder tailpiece slot will straddle the cylinder stop. (The beveled corner of the tailpiece projecting from the spindle of E380 locksets must be positioned to the bottom left, and of E370 locksets to the bottom right, in relation to the index sleeve at the top.) Secure with the large retaining ring making sure the ring is seated in the ring groove of the knob shank.

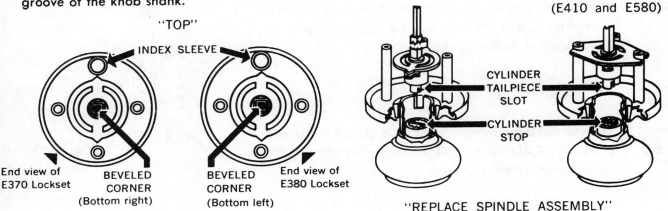

"REPLACE SPINDLE ASSEMBLY"

13 Replace the stabilizing washer (if included), aligning tabs with knob shank slots (and radial wall towards index sleeve of E360, E370, E380, E390 and E410). Replace the torsion spring assembly with the flat side up, and so that both of the spring legs are on the bottom side of the two latch index posts. Secure the torsion spring assembly with the small retaining ring, making sure the ring is seated in the ring groove of the spindle.

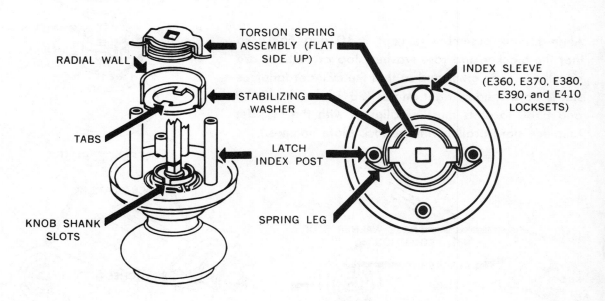

PROCEDURE FOR DISASSEMBLING AND REASSEMBLING KNOB ASSEMBLIES <u>WHICH DO NOT CONTAIN A SPINDLE AND TORSION SPRING ASSEMBLY</u>. (LOCKSETS No. E360, E370, E380, E390, AND E410.)

1 Place an alignment mark on the knob shank and the rose with a grease pencil before proceeding with disassembly, so that the rose may be reassembled in the same position as before disassembly.

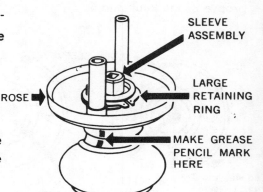

2 Remove the large retaining ring and lift out the sleeve assembly. **CAUTION:** *This assembly should not be taken apart.* Remove the rose.

Remove cylinder cap by unscrewing it with cylinder cap tool (Part No. 1416). After removing cylinder cap, remove cylinder stop with tweezers.

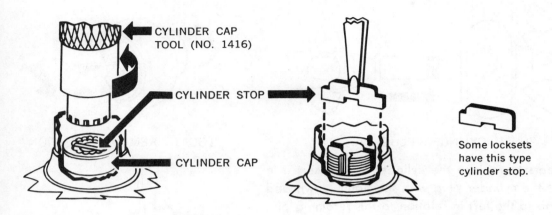

CYLINDER CAP TOOL (NO. 1416)

CYLINDER STOP

CYLINDER CAP

Some locksets have this type cylinder stop.

"UNSCREW CYLINDER CAP"   "REMOVE CYLINDER STOP"

Insert key in cylinder plug, holding the knob so that the key-cuts are in an upward direction. Turn the key slightly clockwise to position the cylinder plug for removal. While holding the knob and key in this position push the cylinder plug out the key side by pushing a plug follower (Part No. 1421) in from the shank side. Continue to hold the key with the key-cuts in an upward direction to prevent the pins from falling out. **NOTE!** *Plug follower must be left in the cylinder until the cylinder plug is replaced to keep top pins and springs in position.*

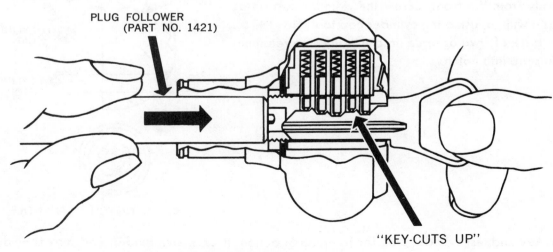

PLUG FOLLOWER (PART NO. 1421)

"KEY-CUTS UP"

"PUSH CYLINDER PLUG OUT WITH PLUG FOLLOWER"

5 After fitting the cylinder plug with pins to suit the new key, push the cylinder plug in from the key side by pushing the plug follower out the shank side. Then, holding the cylinder plug face inward with the thumb, and at the same time, pull and rotate the key to locate removal position. *Remove key* and continue with reassembly.

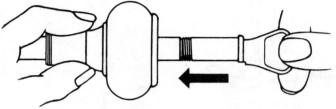

"PUSH CYLINDER PLUG IN"

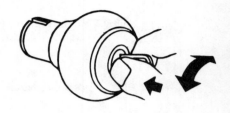

"LOCATE REMOVAL POSITION"

6 Use tweezers to replace the cylinder stop. Insert it in the slot of the cylinder plug with the notch down and the side tab to the left in relation to the retaining pin, at the top of the keyway. If the cylinder stop has two side tabs, either side tab may be placed to the left.

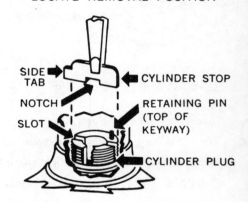

"REPLACE CYLINDER STOP"

7 Replace the cylinder cap by holding the cylinder plug in firmly from the front. Screw the cylinder cap on as far as it will go, using the cylinder cap tool (No. 1416). Then, back off two or three notches, allowing retaining pin to snap into notch.

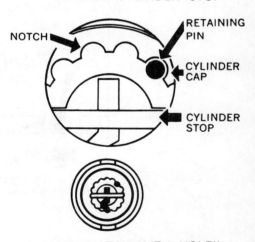

"ALIGN NOTCH WITH HOLE"

8 Insert key and test cylinder plug for freedom of action. If excessive bind or end play is evident, remove key and rotate cylinder cap with cylinder cap tool (No. 1416) as follows:

    a) **2** or **3** notches counterclockwise to eliminate binding.

    b) **2** or **3** notches clockwise to eliminate excessive end play.

Insert key and re-test action. *Remove key* and continue with reassembly.

Align slots of Delrin bearing with slots in knob shank, and replace the rose in its original position by realigning the grease pencil marks placed on the knob shank and rose prior to disassembly.

Align the sleeve assembly so that the cylinder tailpiece slot is positioned:
  a) (Locksets No. E360 and E410)
     parallel with the notches of the lockplate.
  b) (Locksets No. E370, E380 and E390)
     at right angles (90°) to the rose washer stop projections. (Blocker of No. E390 lockset to be positioned parallel with rose washer stop projections.)

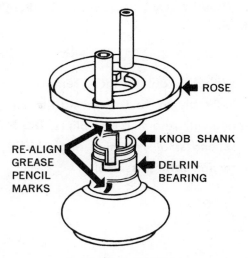

"REPLACE ROSE"

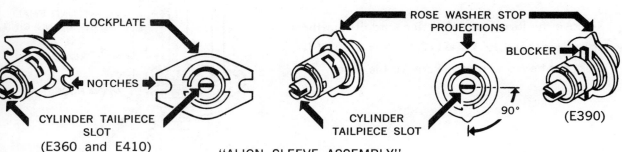

"ALIGN SLEEVE ASSEMBLY"

Replace the sleeve assembly so that the cylinder tailpiece slot will straddle the cylinder stop. (The beveled corner of the tailpiece opening within the sleeve, of E380 must be positioned to the bottom right, and of E370 to the bottom left, in relation to the index sleeve at the bottom.) Secure the sleeve assembly with the large retaining ring, making sure the ring is seated in the ring groove of the knob shank.

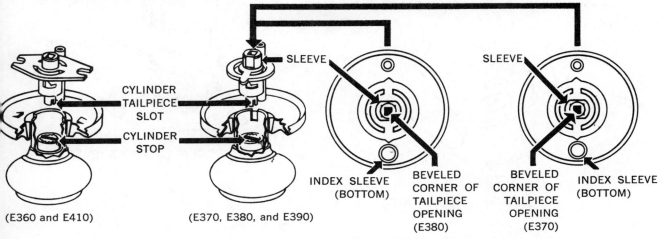

"REPLACE SLEEVE ASSEMBLY"

DISASSEMBLY AND REASSEMBLY PROCEDURE FOR RE-KEYING KNOB ASSEMBLY OF NO. E48 LOCKSET.

(1) With key out of lock, remove torsion spring assembly by first removing the small retaining ring with retaining ring pliers. **DO NOT** *remove second small retaining ring which is under the torsion spring assembly.*

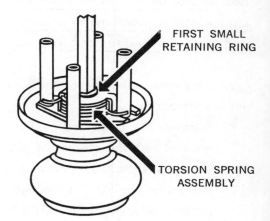

FIRST SMALL RETAINING RING

TORSION SPRING ASSEMBLY

(2) Remove the large retaining ring and lift out the spindle assembly. **CAUTION:** *this assembly must not be taken apart.* Remove the rose.

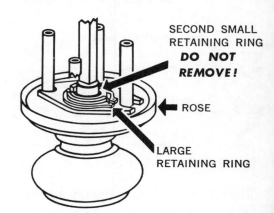

SECOND SMALL RETAINING RING *DO NOT REMOVE!*

ROSE

LARGE RETAINING RING

(3) Turn knob upside down and shake out the shut-out tailpiece. Remove the cylinder cap by unscrewing it with cylinder cap tool (Part No. 1416).

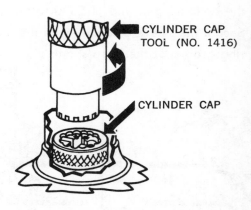

CYLINDER CAP TOOL (NO. 1416)

CYLINDER CAP

"UNSCREW CYLINDER CAP"

"SHAKE OUT"

SHUTOUT TAILPIECE

Insert key in cylinder plug, holding the knob so that the key-cuts are in an upward direction. Turn key slightly clockwise to position cylinder plug for removal. While holding the knob and key in this position push the cylinder plug out the key side by pushing a plug follower (Part No. 1421) in from the shank side. Continue to hold the key with the key-cuts in an upward position to prevent the pins from falling out. **NOTE!** *Plug follower must be left in the cylinder until the cylinder plug is replaced to keep top pins and springs in position.*

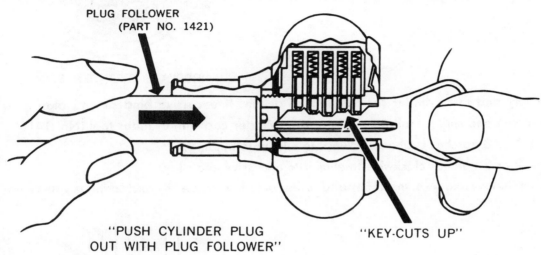

PLUG FOLLOWER
(PART NO. 1421)

"PUSH CYLINDER PLUG
OUT WITH PLUG FOLLOWER"

"KEY-CUTS UP"

After fitting the cylinder plug with pins to suit the new key, push the cylinder plug in from the key side by pushing the plug follower out the shank side. Then, hold the cylinder plug face inward with the thumb, and at the same time, pull and rotate the key to locate removal position. *Remove key* and continue with reassembly.

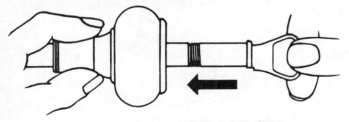

"PUSH CYLINDER PLUG IN"

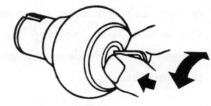

"LOCATE REMOVAL POSITION"

Replace cylinder cap by holding the cylinder plug in firmly from the front. Screw the cylinder cap on as far as it will go, using cylinder cap tool (No. 1416) then back off two or three notches, aligning two opposite notches parallel with the narrow slot in the cylinder plug.

CYLINDER CAP NOTCH

NARROW SLOT

(7) Replace the shutout tailpiece, using tweezers, in the narrow slot of the cylinder plug with the lanced tab down.

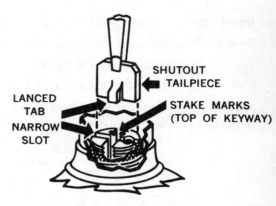

"REPLACE CYLINDER STOP"

(8) Insert key and test cylinder for freedom of action. If excessive bind or end play is evident, remove key and shutout tailpiece and rotate cylinder cap with cylinder tool (No. 1416).

    a) **2** or **3** notches counterclockwise to eliminate binding.

    b) **2** or **3** notches clockwise to eliminate excessive end play.

Replace shutout tailpiece. Insert key and re-test action. *Remove key* and continue with reassembly.

(9) Replace rose, aligning rose bushing slots at right angles with knob shank slots.

(10) Align the spindle assembly, first, by pushing in on the push rod to set the detent in the catch sleeve slot, then, position the two projecting tabs of the catch sleeve parallel with the lock plate notches.

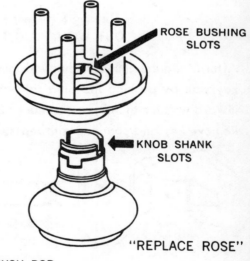

"REPLACE ROSE"

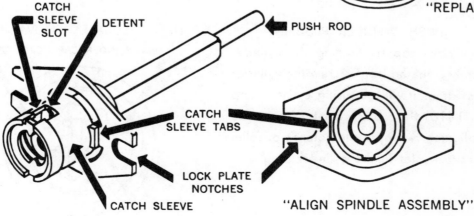

"ALIGN SPINDLE ASSEMBLY"

Replace the spindle assembly with the detent towards the bottom and so that the lazy cam washer will circumvent the shut out tailpiece. Secure the spindle assembly with the large retaining ring making sure the ring is seated in the ring groove of the knob shank.

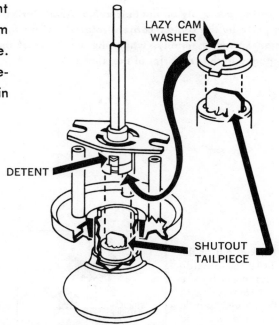

"REPLACE SPINDLE ASSEMBLY"

Replace the torsion spring assembly with the flat side up and so that both of the spring legs are on the bottom side of the two latch index posts. Secure the torsion spring assembly with the small retaining ring making sure the ring is seated in the ring groove of the spindle.

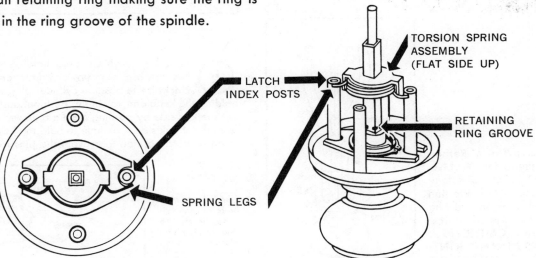

PROCEDURE FOR DISASSEMBLY AND REASSEMBLY OF FALCON NO. A530 TROY AND CROWN ROYAL LOCKS FOR RE-KEYING.

The torsion spring assembly on the Falcon No. A530 Series lock is removable for re-keying. It is held in place by the spindle corners rotated 45° to the square hole in the upper plate of the torsion spring assembly, while the two feet of the torsion spring assembly lower plate are engaged in the cross slot of the shank.

(1) With the key out of the lock, hold the outside knob assembly by the rose (around its diameter) with one hand, pull the tubular spindle outward with the other hand to disengage the two feet of the torsion assembly from the cross slots in the shank. While holding the tubular spindle in this outward position, rotate it to the right or left 45° to make the square of the spindle match the square of the torsion spring assembly. Torsion spring will then slip off the spindle.

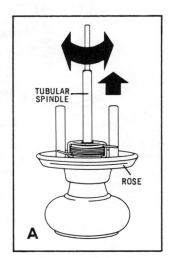

(2) Place points of Retaining Ring pliers (Part No. 1422) in holes of Retaining Ring as shown in Illustration B. Squeeze pliers to spread ring and permit removal of Hollow Spindle and rose. **CAUTION . . . SPREAD RETAINING RING ONLY FAR ENOUGH TO ALLOW REMOVAL. DO NOT OVERFORM RING.**

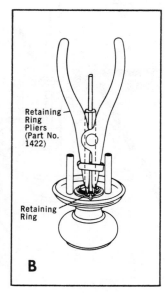

(3) Turn knob over and shake out tailpiece washer as shown in Illustration **D**.

(4) Remove tailpiece as follows (Note inset circled [illus]tration to see how the end of tailpiece must be [?] out of slot in order to permit unscrewing of the [cylin]der cap): Hold knob with left hand and using Retaining [Ring] pliers, pull tailpiece away from the knob about ¼" an[d] screw the cylinder cap by turning the tailpiece counterclock[wise]

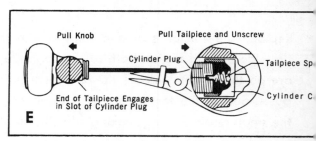

(5) Insert key in cylinder plug, holding the knob so the [?] key-cuts are in an upward direction. Turn key sli[ghtly] clockwise to position cylinder plug for removal. [While] holding the knob and key in this position, push the cylinder [?] out the key side by pushing a plug follower (Part No. 142[1]) from the shank side. Continue to hold the key with the key [cuts] in an upward position to prevent the pins from falling ou[t]

NOTE!

Plug follower must be left in the cylinder until the cylinder plug is replaced to keep top pins and springs in position.

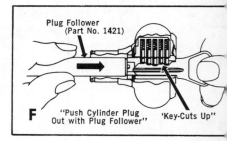

6 After fitting the cylinder plug with pins to suit the new key, push the cylinder plug in from the key side by pushing the plug follower out the shank side. Then, the cylinder plug face inward with the thumb, and at the time, pull and rotate the key to locate removal position. ...ve key and continue with reassembly.

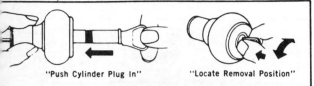

"Push Cylinder Plug In"   "Locate Removal Position"

7 Replace the tailpiece and brass cylinder cap as follows (In normal position the end of the tailpiece will hit the cylinder first and prevent the threads of the cylinder from starting on the cylinder):

To start threads, compress the tailpiece spring to permit ...rst threads of the cylinder cap to extend beyond the end ...e tailpiece. To compress the tailpiece spring, place the ...w Spindle (Illustration F) over the tailpiece and grip the ...ece with the loop of the Retaining Ring pliers so that the ...of pliers will overlap the Hollow Spindle about ¼". ...Hook end of the pliers behind the Hollow Spindle by pull-...e Hollow Spindle away from the pliers as shown in Illus-...n G.

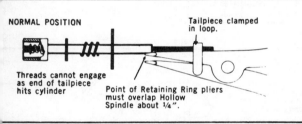

NORMAL POSITION

Tailpiece clamped in loop.

Threads cannot engage as end of tailpiece hits cylinder

Point of Retaining Ring pliers must overlap Hollow Spindle about ¼".

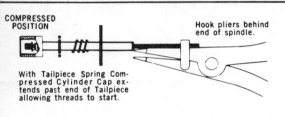

COMPRESSED POSITION

Hook pliers behind end of spindle.

With Tailpiece Spring Compressed Cylinder Cap extends past end of Tailpiece allowing threads to start.

8 Holding the knob assembly in one hand, thread the cylinder into the cylinder cap two or three turns. Release the Retaining Ring pliers and remove the Hollow ...dle. Again clamp the tailpiece with the Retaining Ring ...s (see Illustration H) and pull to compress the spring as ...re, so that the end of the tailpiece will not engage in the ...der plug slot. Then, screw the cylinder cap until snug. Now, ...se the tension and unscrew the cylinder cap approximately ...rn until the tailpiece snaps into the slot.

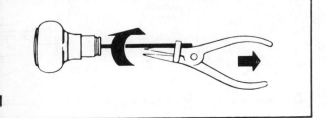

Insert key and test cylinder plug for freedom of action. For excessive bind or end play, *remove key*, pull tailpiece back with pliers to clear cylinder plug sector slots and turn tailpiece:
(a) ¼ turn *counterclockwise* to eliminate binding;
(b) ¼ turn clockwise to eliminate excessive end play.
Insert key and retest cylinder plug action.

9 Replace the tailpiece washer (Illustration I) as follows: Position the knob with the millcuts of the shank in horizontal plane. Then, place the washer on the tailpiece with the tabs in position as shown in Illustration J.

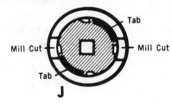

Tab

Mill Cut   Mill Cut

Tab

J

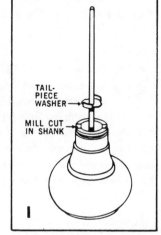

TAIL-PIECE WASHER

MILL CUT IN SHANK

I

Reassemble outer knob, Hollow Spindle assembly, and rose. Secure rose on shank with washer and retaining ring. Place entire assembly in upright position on flat surface. Slip torsion spring assembly over spindle with open end down and spring ends on the *same* side of posts exactly as shown. (Illustration K)

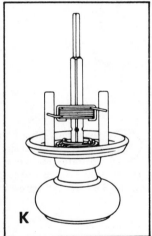

K

With the entire assembly in this upright position, push downward on the torsion spring assembly with one hand and with the other hand pull the tubular spindle outward. While holding the tubular spindle outward, rotate it in whichever direction is required (right or left) 45° to align the two feet of the torsion spring assembly lower plate with the cross slots in the shank. (This will make the corners of the square spindle 45° to the square hole in the torsion spring assembly upper plate.) Releasing the tubular spindle will lock the torsion spring in position and complete the assembly.

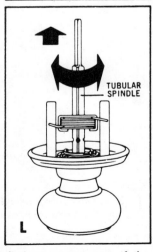

TUBULAR SPINDLE

L

10 Subsequent turning of the knob to full right or to full left should confirm that the torsion spring assembly is now securely in place. Key should now be tested for proper clockwise movement. If the key cannot be removed, recheck the tailpiece washer (point 9), as this washer may be improperly inserted.

PROCEDURE FOR DISASSEMBLY AND REASSEMBLY OF FALCON NO. A530 BEVERLY KEY-IN-KNOB LOCK FOR RE-KEYING.

TO REMOVE THE CYLINDER

1 In the No. A530 Beverly Key-In-Knob lock, the knob face and the pin tumbler cylinder are a sub-assembly which may be withdrawn from the outer end of the knob by turning the key in the cylinder approximately 60° in a counterclockwise direction. To permit key rotation in this counterclockwise direction, it is necessary that the tailpiece (shown and explained in ILLUSTRATION **A**) be disengaged from the inner end of the cylinder plug.

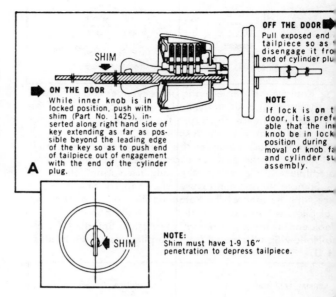

SHIM

OFF THE DOOR ▶
Pull exposed end tailpiece so as disengage it fro end of cylinder plu

▶ ON THE DOOR
While inner knob is in locked position, push with shim (Part No. 1425), inserted along right hand side of key extending as far as possible beyond the leading edge of the key so as to push end of tailpiece out of engagement with the end of the cylinder plug.

NOTE
If lock is on t door, it is pref able that the in knob be in lock position during moval of knob fa and cylinder su assembly.

A

SHIM

NOTE:
Shim must have 1-9 16" penetration to depress tailpiece.

2 To remove the knob face and cylinder sub-assembly from the knob shell, turn the key approximately 60° counterclockwise and oscillate the key very slightly to the right and to the left and at the same time *gently* pull on the key, as indicated in ILLUSTRATION **B**.

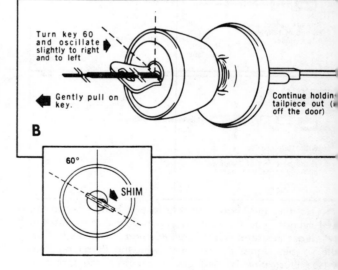

Turn key 60 and oscillate slightly to right and to left

Gently pull on key.

Continue holdin tailpiece out (off the door)

B

60°

SHIM

3 Knob face and cylinder sub-assembly will pull free of the knob shell, as indicated in ILLUSTRATION **C**.

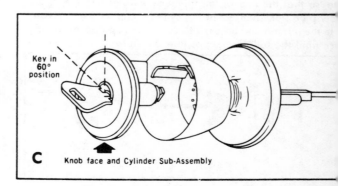

Key in 60° position

C Knob face and Cylinder Sub-Assembly

REPLACE THE CYLINDER

4 Insert the key in the keyway of the rekeyed or new knob face and cylinder sub-assembly and turn the key approximately 60° counterclockwise. Holding this assembly by the key, position it so the top of the cylinder is in upright position and in alignment with the cylinder guide.

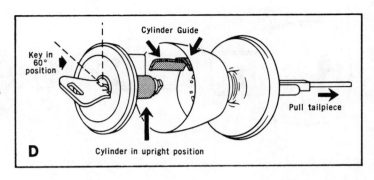

D

Key in 60° position

Cylinder Guide

Pull tailpiece

Cylinder in upright position

5 Place top of the cylinder into cylinder guide and press inward until the assembly is as far into the knob shell as possible. It is desirable that the tailpiece be prevented from engaging with the end of the cylinder plug until after the knob face and cylinder sub-assembly are in engagement within the knob. If the lock is off the door this may be accomplished by pulling the tailpiece as indicated in ILLUSTRATION E. Pressing inward, oscillate the key slightly to the right and to the left, as necessary from its 60° counterclockwise position to permit the assembly to fully seat into the knob.

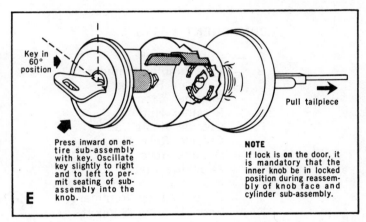

E

Key in 60° position

Pull tailpiece

Press inward on entire sub-assembly with key. Oscillate key slightly to right and to left to permit seating of sub-assembly into the knob.

NOTE
If lock is on the door, it is mandatory that the inner knob be in locked position during reassembly of knob face and cylinder sub-assembly.

A If lock is on the door with inner knob in locked position, rotate key clockwise beyond vertical position and oscillate key slightly between vertical and 45° clockwise position until tailpiece engages with the end of the cylinder plug.

B If lock is off the door, hold the tailpiece outward and rotate key clockwise beyond vertical position. Release tailpiece and oscillate key slightly between vertical and 45° clockwise position until tailpiece engages with end of cylinder plug. When engaged, the tailpiece will rotate in unison with the rotation of the cylinder plug by the key.

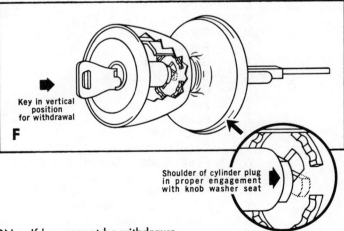

F

Key in vertical position for withdrawal

Shoulder of cylinder plug in proper engagement with knob washer seat

6 Return the key to vertical position and withdraw. (CAUTION — If key cannot be withdrawn in vertical position, tailpiece has incorrectly engaged the cylinder plug and must be disengaged per ILLUSTRATION **A** and repositioned per this step 5.)

OUTSIDE

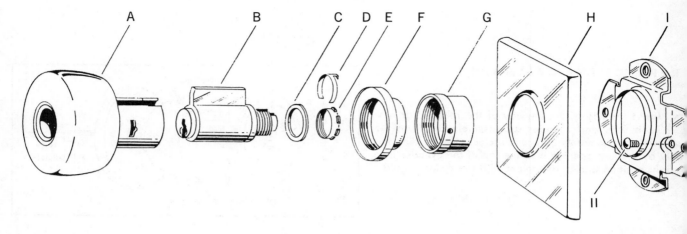

COMPONENT PARTS

A — Outside Knob
B — Pin Tumbler Cylinder
C — Retainer Bearing
D — Retainer Stop
E — Retainer Nut
F — Outside Knob Cap
G — Outside Rose Nut
H — Outside Rose
I — Attaching Plate
II — Rose Anchor Screw
J — Knob Bearing Retainer
K — Knob Bearing Support
KK — Knob Bearing
L — Tube Spindle
M — Retractor Frame Plate
N — Retractor Spring
O — Retractor
P — Guide Plate
Q — Stabilizer Spring
R — Deadlock
S — Pin
T — Latch Bolt
U — Pivot Pin
V — Lock Frame
W — Push Button Sleeve and Button Assembly
X — Knob Bearing Support
XX — Knob Bearing
Y — Knob Bearing Retainer
Z — Attaching Plate
AA — Attaching Screws
BB — Inside Rose
CC — Inside Rose Nut
DD — Inside Knob Cap
EE — Inside Knob

MONO-LOCK

OUTSIDE

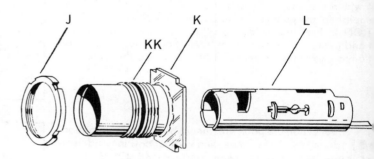

MONO-LOCK

INSIDE

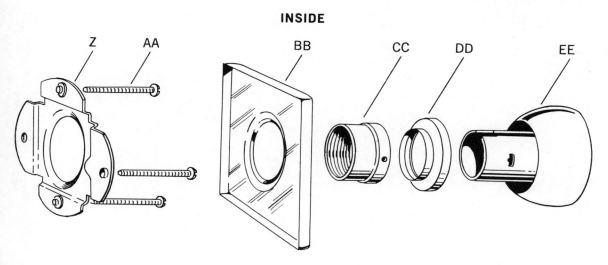

Trim and Fastening Plates

INSIDE

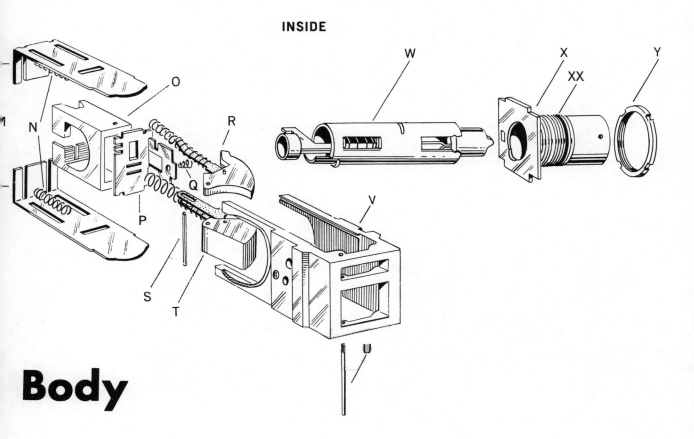

Body

Figure 93. Exploded view of Mono-Lock. Illustration courtesy of Eaton Yale & Towne, Inc.

III Key Operated Mechanisms

Locks of Warded Construction

It is the key operated mechanisms, whether in the form of a cylinder or as an integral part of the lock, which make one lock differ from another in the respect that only the key intended to operate a particular lock will serve to operate it. Other keys, even though similar, will not operate the lock.

Four methods are commonly used to achieve this goal: warded construction, lever tumbler construction, disc tumbler construction, and pin tumbler construction. These are all broad classifications and entail a number of variations, both minor and major. The degree of protection furnished by them is commensurate with the degree of skill with which they were designed and manufactured. Simplest of these is the warded construction, and it is to some degree common to all of the other types; therefore, the warded construction will be considered first.

When one says that a lock is of warded construction, he simply means that certain obstructions have been placed in either the keyhole (to prevent entry of the wrong key), or that certain fixed obstructions have been placed in the path of the turning of the bit (or throwing) portion of the key. A lock of such construction utilizes a key which has been cut away or notched in such a fashion as to clear or pass the wards, whether *keyway wards* or *internal wards*. This must be done without cutting away an excessive amount of the key and thus impairing its operation of the bolt or other mechanism with which it must function. Keyway wards may consist of simply a single protrusion which extends into the keyway to prevent an unauthorized key from being inserted into the lock.

Keyway Wards

This single keyway ward necessitates a lengthwise groove cut along the bit of the key in order to allow it to enter the lock. When two or more such keyway wards are used on opposing sides of the keyway, and extend far enough so that the wards overlap one another, then such wards constitute a *paracentric keyway,* since a key cannot be made thin enough to clear both of the wards without grooving the key bit in the proper places.

An attempt to make a key thin enough to clear the wards in a paracentric keyway would result in the key being cut in two. The grooves must be cut into the sides of a key bit, otherwise it will not enter the lock. Keyway wards may also take the form of a more elaborately shaped keyhole. It may involve several protrusions into, or the alternate narrowing and widening of, the keyhole; or the keyhole may have a wavy configuration.

Internal Wards

Internal wards consist of one or more fixed obstructions of almost any shape in the path of the turning of the key. Internal warding serves to obstruct the turning of the key in the lock after it has entered the keyway unless it has been suitably cut to clear these wards. Most internal wards in locks of recent construction are simple protrusions, more or less rectangular in cross section, and protruding sufficiently into the turning radius of the key bit as to require one or more cuts to be made in the key in order to permit the key to pass the wards.

Locks of certain quite old European construction

(notably British and Germanic) which were handmade, frequently utilized intricately shaped wards of a paracentric nature and were finely detailed. This type of construction necessitates the actual building of a key, frequently by casting it in a mold, or the construction of rather elaborate picks. Fortunately, such construction is extremely rare today, because the amount of time and work required to produce a key for these locks, either by casting or by careful hand work, would make the cost prohibitive. Even though these paracentric internal wards offered excellent protection, most of today's warded locks may be picked quite readily by any instrument having a tip suited to throw the particular type of mechanism used, by being sufficiently small or cut away to pass all wards, and yet retaining enough strength to move the locking mechanism in the required manner. Often a key may be cut down to bypass all of the wards, in which case it is said to be skeletonized or a *skeleton key.*

In order to make keys for modern locks of warded construction, it is necessary to select a *key blank** which will be of such a size and configuration to enter the keyhole, keeping the keyhole well filled. In order to allow some key blanks to enter the keyhole, it may be necessary to make one or two keyway cuts to bypass the keyway wards. Having a key blank which will enter the key hole, it is only necessary to turn it against the internal wards rather firmly. This causes the ward(s) to leave a faint mark or impression on the key blank. It is at this point that the blank must be cut away to pass the ward. Care should be taken in this process to avoid filing the ward cuts too deeply, as to do so may weaken the key excessively.

As one should avoid cutting the wards too deeply, so also should he avoid making the cuts too wide, as this may cause the key to operate locks it is not supposed to. It is generally considered unethical of the locksmith to take away more material than is required to make the key functional; doing so may cause the protection of someone else's lock to be compromised through the existence of an unauthorized key to fit it.

Warded locks are commonly found in association with one of four methods of operating boltwork. All of these methods involve the direct application of

*Key blank: A partially fabricated key from which a finished key may be made for any one of a number of specific locks, and which customarily will operate only one of these locks when completed. The points of similarity have been completed in the blank form, and the points of difference remain to be completed to change it from a key blank into a finished key.

the key to the part which moves to unlock.

Method 1. The key lifts a detent lever while throwing the bolt. This provides a deadbolt action.

Method 2. The key moves a bolt whose locked or unlocked position is maintained by the action of a humped flat spring in two notches in the bolt.

Method 3. The key moves directly against the latch tail of a latchbolt, or does so through the action of an intermediary floating lever.

Method 4. The key inserts between two springs and wedges them apart as it is turned. This is principally a padlock mode of operation and has changed little since its origin in ancient Rome.

PICKING

Picking a warded lock is an extremely simple matter, unless one of the very old locks with complex wards and a stout spring is encountered. That type of lock is likely to require either that a special pick be made to suit its requirements or that the proper key be made to unlock it. The more modern, cheap type of warded locks requires that the detent lever (if any is used) be lifted and sufficient pressure be exerted against the bolt to move it into the unlocked position.

If the lock is of the type which requires the spreading of a pair of springs (or a single spring in order that the spring may clear a notch in the bolt or shackle) the only requirements are that the position of the spring(s) be determined and reached with a tool which has been cut away sufficiently to pass the wards, while yet retaining sufficient strength to spread the springs as would the key.

Picks for warded locks are unique in that only a single tool is used. It must be shaped to avoid the wards while reaching the boltwork in such a shape, size, and strength as to be able to move it to the unlocked position. This sounds like a rather formidable requirement, but it actually is not. Most of the warded locks manufactured today fit into one of a very few types of key patterns. This is true to such a considerable extent that only three pick styles (see Figure 94) in two or three sizes will open about 90% of the warded locks. The principal variation which one is likely to encounter is in the spread-spring type of padlock. Occasionally one of these padlocks is found with two sets of springs, or even three sets, operating in corresponding notches in the shackle of the padlock.

There are two methods for dealing with the picking of such a lock. One is to have picks made on which two or three pairs of spreading lugs are left to operate all of the sets of springs simultaneously.

These picks must be made in at least two sizes and with several different spacings between the spreader lugs. This can become almost as complicated as carrying around a complete set of keys of all the combinations possible for this type of lock. The preferred method would be to pull rather firmly on the shackle while spreading the pairs of springs one pair at a time, repeating as often as necessary until the shackle opens.

The principle involved is that of maintaining a small enough pressure on the shackle to allow the pick to spread the locking springs, and yet enough pressure to keep them from snapping back to their full depth in the shackle notch as soon as the pressure of the pick is released. Tension must be maintained throughout the picking of the lock; otherwise, the springs will revert to full depth in the notch and the job must start over. The most important factor in picking this type of lock is the amount of pressure used on the shackle. Too much, and you either don't move the springs, or you break the pick trying; too little, and you just move the springs back and forth in the notch (unless it is a single spring) without accomplishing anything more than giving the spring its exercise and contributing to the frayed disposition of the would-be locksmith. That is not to say that even a master locksmith is going to pick every warded lock in 10 seconds flat, either.

Everyone has his difficulties occasionally, but three things are important in minimizing them so far as this or any other type of lock picking is concerned. They could be called three commandments for lock picking:

1. Ye shall *practice.*
2. Ye shall *not give up,* even though ye rest as ye must.
3. Ye shall *find out* why ye blew it before, so that ye *do it nevermore!*

Practice is the most important factor in being able to pick a lock with anything approaching efficiency. The author considers that a beginner's regimen should include a minimum of two hours of daily practice with all types of locks. By the end of his training, this should have tapered down to a quiet hour or two a week, and even the old timers don't get so good that this much practice won't help them, too. (Assuming that they practice on those locks which give them the most difficulty, instead of showing off before young apprentices, that is!)

Sometimes when a person tries to pick a lock, the difficulties seem insurmountable. The longer one tries, the worse matters seem to get. That is the time

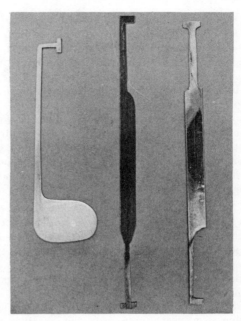

Figure 94. Picks for warded locks. Left to right: for bit keys; for flat steel keys; for locks using wedge-apart spring construction.

to back away and reconsider. Go sit down and see how closely you can come to getting to sleep while visualizing what the inside of that lock looks like at any moment in the attempts you have been making to pick it. It is often amazing the results that one can get with just 5 or 10 minutes' time in which to acquire a fresh viewpoint. The point is, you rest a little and go back; not back off and say "To heck with it!" and quit.

When a lock causes inexplainable trouble in picking, it is always a good idea to take it apart and see if there is some feature in its construction of which you were not aware, or which caused a problem even though it ordinarily does not. Often, taking apart one or two locks of a given type can avoid a whole series of disastrous picking attempts if the taking apart is done with an intelligent, inquiring mind. To be really superior in the fine art of lock picking, it is necessary to be able to visualize in your mind *exactly* what is going on inside the lock at any given instant in the picking process. This means an *absolute knowledge* of the principles with which the lock has been constructed, and that just doesn't all come from books, or word of mouth, or even Divine inspiration; most of it has to come from "dedicated perspiration."

Many people have asked me how I pick a lock. The answer I always give is, "I do everything but taste it, and if I thought that would help, I would try it, too!'' They think they are having their leg pulled or are being told in a roundabout way that it is none of their business, yet it is the exact truth. Everyone has five senses, including taste. Here are the uses of the remaining four, as they are helpful in picking locks of all types:

1. *Smell.* Your sense of smell will often tell you if a lock has been lubricated with some type of oil. In most cases, all oil should be removed from a lock before picking; the simple bolt throwing type of warded lock construction being the outstanding exception.

2. *Sight.* The first thing to do when one sees a lock which he is going to pick is to determine as much about it as he can before touching it. The shape of the keyway will often tell not only the type of lock a person is up against, but what size and shape pick will be required for the internal mechanism, assuming that one is familiar with the construction of that type of lock. At times, slight scratches or even spots near the keyhole, will provide a clue as to the direction of rotation. It is, then, this visual examination which first determines the method and tools to be used in the initial attack on the lock.

3. *Hearing.* In the course of picking many locks, one can hear sounds as certain parts are moved. Often, a click will denote a part moving into or out of place; or, a change in the character of a sound will indicate something of what is happening inside, provided one is sufficiently familiar with the type of lock to be able to interpret the sounds correctly.

4. *Touch.* The feel of a lock while it is in the process of being picked is the locksmith's most direct pipeline of information as to what is going on inside. Since one's sense of touch is the least educated of man's senses, it is this one which causes the most difficulties for the average person in trying to pick a lock. There is a difference in the quality of feel, as well as a difference in quantity (or intensity). For example, take an old lock pick or a short piece of spring steel. Hold it very lightly, so lightly, in fact, that at first you may not be sure that it is still on your finger tips. Now run your pick along the length of a mill bastard file with only a sharply rounded corner of the pick touching the file, moving very slowly and gently.

Then turn the file end-over-end and repeat; next pull the pick crosswise across the file. Notice the difference in the quality of the feel. Now, draw the pick across sheet iron, or sheet brass. Or, let the

pick slide off the edge of a thin piece of sheet metal. This is the feel which will be present in a pick on a much more diminutive scale when a tumbler clicks into place.

Not all tumblers click into place. With some, a sudden change in tension is required to move the pick. In other types of locks where a separate tension wrench is used, the feel of picking a tumbler may show up more noticeably in the tension wrench than in the pick which is actually moving the tumbler.

The specific feel involved in the picking of a warded lock is: first, feel a way around or through the wards; feel for a detent lever (the give of its spring will show its presence); feel for when it clears the detent post and for the beginning of the motion of the bolt as your turning force begins to take effect when the detent is released. A single lever lock is also picked in this manner. If the lock is of the spring held type, it is only necessary to feel for the turning notch in the bolt and to apply enough pressure to move it, much as one would in the previous example, but with greater turning force, and without having to bother feeling for the deadbolt detent lever.

For the springlatch type or the spring-apart padlock shackle locking mechanisms, the feel is very distinctive and the condition of the lock is very easy to visualize as the spring gives under the thrust or turn of the pick. The direction in which the pick must turn a lock is very important and can often be determined best by the feel of the small amount of "give" when one is attempting to turn it in the correct direction. The positioning of the bolt notch (determined by feel) is also of help in determining the direction of rotation of the key or pick. When you have one of these locks apart, notice the position of the bolt notch relative to the keyhole.

A sixth sense which could be referred to as *visually reconstructive imagination* is highly important in all lock picking. Familiarity with the principles of construction of the various types of locks permits one to visualize in the mind just what is going on inside it every moment during the picking process. This sense is highly developed among people who have long been blind. For the sighted locksmith to develop this sense, a good form of practice is to open a lock case, study the mechanism carefully, then close his eyes and try to reconstruct a picture of it in his mind. Opening his eyes and re-checking how many details escaped his mental picture will advise him when to progress to more difficult and complex locks.

Lever Tumbler Lock Construction

Lever tumblers are a natural outgrowth of the deadbolt detent lever and frequently carry a distinct resemblance to that part. Just as the deadbolt detent lever will not clear the detent post until it is lifted high enough, so also will the lever tumbler not clear its post (called a *fence*) unless it is lifted sufficiently to do so. Unlike the deadbolt detent lever, overlifting of the lever tumbler also causes the tumbler to fail to clear the fence and operate the bolt. Although the deadbolt detent lever customarily travels with the deadbolt, it is less common for the lever tumbler to travel with the bolt. Often it is the fence which travels with the bolt in a lever tumbler lock while the tumblers fill a passive function in barring the path of the fence until the tumblers are properly aligned by the key. The lever tumblers are usually pivoted on a pin which is attached to the lock case.

There are exceptions to this condition, however, and those situations wherein the lever tumblers travel with the bolt are referred to as a *traveling lever* construction. In this type of construction, the lever's pivot pin would, of course, be attached to the bolt, rather than to the lock case. In a traveling lever construction, the fence would be attached to the lock case in a fixed position.

TUMBLER GATING

Figures 95 and 96 show the essential elements of two common types of lever locks utilizing *closed gating* and *open gating*, respectively, and show the various configurations of the gating which may be used with each type. Note how the differently shaped gatings will each produce a distinctive type of tumbler action. The enclosed gating usually shows somewhat superior strength characteristics, while the open gating often exhibits somewhat superior characteristics with respect to the difficulty of picking and the accuracy requirements of the operating key. This is due to the increased distance of the bolt fence from the tumbler pivot, in relation to the distance of the key contact point from the tumbler pivot.

The increase in proportional distance of the fence from the pivot means that a similar motion of the key in lifting the tumbler will produce a correspondingly greater lift of the tumbler at the point where the fence contacts the gating of the tumbler. The implication is that, given gatings which clear the fence equally, much greater accuracy will be required of the operating key, as the distance of the fence from the pivot post increases while the distance of the key from the pivot post remains the same.

There are, however, practical limitations to this which include the accuracy with which it is possible to make the keys, and the effect of normal wear on such a super-accurate key. More importantly, the amount of clearance required at the pivot post to allow the tumbler to move as the key is turned would introduce sufficient eccentricities of operation to create a state of unreliability in such a lock.

In both the closed and open gating, the height of the tumbler during the progress of the fence through the gate must remain constant if the fence is not to bind against the bar of the gate. This necessitates

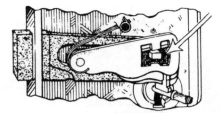
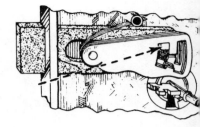

BOLT UNLOCKED. TO OPERATE, THE LEVER TUMBLERS MUST BE ALIGNED SO AS TO PROVIDE A "GATE" THROUGH WHICH THE "FENCE PIN" (ATTACHED TO THE BOLT) MAY PASS.

THE KEY HAS ALIGNED THE LEVER TUMBLERS SO THAT THE "FENCE PIN" IS PASSING THROUGH THE "GATE", AND THE BOLT IS THUS MOVED.

THE BOLT IS IN LOCKED POSITION AND DEADLOCKED AGAINST END PRESSURE BY THE "FENCE PIN", AS THE LEVER TUMBLERS FALL BACK INTO PLACE.

Figure 95. Closed gating of a lever tumbler. Courtesy of Eaton Yale & Towne, Inc.

that a *belly* be included on the under side of the tumbler at the point where it makes contact with the key. Unless the gating of the tumbler is loose, the curvature of the belly must vary with the depth of the cut at that point on the key. Remember, then, that a wide curvature of the belly would indicate a very shallow cut on the key, and that a very narrow curvature would indicate a very deep cut. A great deal of information can be gathered either by sight through the keyhole, or by feel with a pick having an L-shaped tip.

Knowing the general configuration of the key is of considerable help in picking these locks, and sometimes it may be possible to gather enough information to make a key for a lock utilizing only a few different depths of cut and not too many tumblers. To avoid sight reading, a good many locks provide a

curtain, that is, a fixed piece of sheet metal, either as part of the case of the lock, or as a separate piece inside of the tube of the lock, to prevent visual access to the tumblers. In this situation, the belly curvatures must be felt for with a pick. This curtain serves the additional purpose of restricting the size of pick and the amount of motion which may be imparted to a pick and thus transmitted to the tumblers.

Changes in the depths of tumblers are made by one of two methods. The most common method is by changing the height of the gating in the tumbler (that is, its position). The other method utilizes a gating height which is uniform while the position of the belly of the tumbler varies, and is more commonly employed when closed gating is used because of the restrictions imposed on the travel of the tumbler by the presence of the fence.

PICKING TECHNIQUE

As shown in Figure 97, it is common practice in lever tumbler locks to provide a small *window* in the lock case at the point where the fence enters the gating, making the mechanism visible after the lock has been dismounted from the unit in which it is installed. This window, if it is visible or can be made visible, is of great assistance both in making an original key for the lock and in picking it. Indeed, if the window can be made accessible by drilling a hole through to it at that point, it is usually quite easy to manipulate the tumblers into alignment. The lock is picked through the window, bypassing the keyhole. Often, a straight pick or a straight spring wire, and a *tension wrench** in the keyhole are all that is usually required.

The picking of a lever lock which is in the "in service," or mounted position, requires that tension

Figure 96. Open gating of a lever tumbler lock. Note the square hole behind the gating pocket. The hole permits the use of a "master tumbler" type of masterkeying in this kind of lock. Illustration courtesy of Eaton Yale & Towne, Inc.

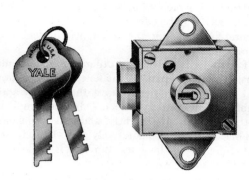

Figure 97. A lever tumbler lock. Note the window in the upper left part of this lock with the gating of the tumblers showing through it. Illustration courtesy of Eaton Yale & Towne, Inc.

be placed on the bolt all through the course of picking the lock. Those lever locks having a cam attached to the barrel of the lock will usually transmit sufficient pressure to the bolt for picking merely by turning the barrel of the lock from the front face of the barrel with an ordinary tension wrench.

Not all lever locks are equipped with a cam at the rear of the barrel to throw the bolt. Those locks having no cam rely on the bit of the key in direct contact to throw the bolt. Other locks have a cam which is joined to the barrel in a rather flimsy connection. Either of these latter two situations will require the use of a special tension wrench, shown in Figure 98, which penetrates the full depth of the lock in order to engage directly against the bolt or the cam. The use of this type of tension wrench is to be preferred in picking any type of lever lock, whether it has a cam or not, unless there is no question that the cam is strong enough to maintain its attachment to the barrel. A lever lock cam is provided with a slot into which the tip of the key fits, and it is into this slot that the tip of the tension wrench should apply its pressure.

With the tension wrench properly in place and exerting a turning pressure in the proper direction to throw the bolt, the fence will come against the bar of the lever. In lever locks of the most common type of construction, the fence is a comparatively small and rather delicate part of the lock; therefore, if an excess amount of tension is exerted, it will tend to

*Tension wrench: A device appropriately shaped and used to maintain a torsional or other force on a lock while its tumblers are being manipulated by a pick, with the objectives of both assisting in the picking of the lock and of operating it after picking has been completed. See Figure 98.

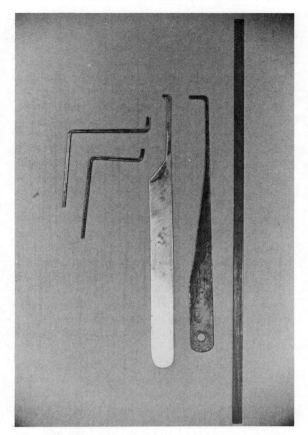

Figure 98. Tension wrenches and picks for lever tumbler locks. Left to right: two tension wrenches, two picks, one shaker pick.

spring the fence in tightly to the bar on the back tumbler and at the same time it will be sprung slightly outward, away from the bar of the front tumbler of the lock. For this reason, it is good practice to begin picking the lock at the back tumbler, raising the tumbler until one feels a slight click. This click will very seldom be audible, but is a matter of feel in the form of a sudden release of the amount of pressure required to lift the tumbler while dragging the bar of the tumbler across the meeting surface of the fence.

Variations of materials and design may make the click tend to become less sharply noticeable and act more as a progressive lessening of resistance to moving the tumbler. When either of these events occurs, further lifting of the tumbler is immediately discontinued; the pick is then slightly withdrawn to the next tumbler, where the process is repeated. Having progressively picked these tumblers from back to front, the lock should open; but if it does not, tension should be maintained while working one's way back into the lock one tumbler at a time, jiggling each just a trifle in case the alignment of one or more of the previously picked tumblers has been slightly disturbed in the process of picking subsequent ones.

As a general rule, it is necessary to exert considerable pressure on the bolt during the process of picking these locks because the strength of the lever springs is usually great enough to cause the levers to drop out of position, once picked, unless considerable tension is being maintained. As an optimum, the tension wrench, which is used in picking a lever lock, should be thick enough to substantially fill the full width of the keyhole and should project beyond the nose of the lock just far enough to clear and avoid hampering the work of the other hand with the pick. Also, that portion of the tension wrench which lies flat along the bottom of the keyhole should take up as little vertical space in the keyhole as possible, consistent with the strength required to exert sufficient turning tension. This is essential in order that room may be left for a good, strong pick to enter the lock.

The amount of motion of the pick required to pick a lever lock is somewhat less than one would be likely to imagine, because of the fact that common-practice in the keying of a lever lock calls for the maximum depth of the cut under any tumbler to be not more than half the width of the key, and in no case more than about two-thirds of its width. The pick used should be thick and sturdy, but need not have much hook on the end of it for that reason.

Strong picks are required in lifting the tumbler, because of the need to overcome both the strength of the lever spring and the friction of the bar along the fence (which *must* be greater than the force of the spring in order to accomplish the purpose of the tension). The resistance of the tumbler to lifting may sharply decrease as the fence enters the gate, but this is only momentary and will come back again if the tumbler is overlifted. Remember that overlifting of a tumbler means starting over with the picking of a lever lock.

Another indication that a tumbler gate is aligned with the fence may come in the tension wrench. Often the slight jump of the tension wrench as the fence enters the tumbler gate ever so little is more pronounced than is any feel coming from the pick. The actual picking of a lever lock may be a rather simple matter, for example, by taking advantage of a fence which is attached to the bolt on one end only. Normal picking procedure calls for feeling for the tumbler which is most tightly bound by the fence and is quite effective if the lock is not too accurately made for the difference in feel to be significant.

If there is no detectable difference in feel, more tension can be applied to the fence until it is noted that the back tumbler (nearest the bolt) is appreciably tighter than the others. This is accomplished as the bolt is sprung slightly, and makes picking of the back tumbler comparatively easy. Utilizing this technique, once the back tumbler is picked, it's down hill all the way.

It is very important that once the back tumbler is picked, tension should be backed off to about one-fourth that used while the back tumbler was being picked, because, instead of the tension being applied to the fence by the back tumbler in a position practically against the fence, it is appreciably further from its juncture with the bolt after the first tumbler has been picked, and it is the second tumbler which is applying the force to the fence. Tension should continue to be lessened as subsequent tumblers are picked, but not as rapidly as between the first and second tumbler, because the increase in leverage against the fence is not as rapid the further one progresses from the point of attachment of the fence (which serves as a fulcrum). This makes picking of these locks simply a process of lining up the tumblers one by one, in succession, from back to front, if the length of the levers is reasonably constant from pivot to the bars of the gating.

The feel of a lever tumbler as it is being picked may be in the form of a click, not necessarily audible, but more of a felt click. This click may manifest itself in the pick as the tumbler is lifted, but more likely it will be felt most distinguishably in the tension wrench, particularly if the wrench is of a rather solid type with little if any "give." A feel may also be transmitted through the tension wrench of the bolt moving just slightly toward the unlocked position as each tumbler is picked. The feel of these occurrences during the picking of a lock will be slight and requires the development of considerable sensitivity to this sort of thing in one's fingers. This doesn't come overnight or in any miraculous or sudden way. It is simply developed as one practices picking locks of various types.

Another good form of practice for developing sensitivity of touch and which may be done any time or place is to handle small, light objects, such as a pencil or coin. Attempt to handle them in such a way as to learn everything about them judging by touch alone. For example, imagine how a blind person learns to read Braille. This is the type of sensitivity which should and must be developed in the fingers to develop any real skill in picking locks.

Visually Reconstructive Imagination

The factor of sensitivity of touch developed to the necessary degree is only a means to the method of getting the lock unlocked. All the sensitivity in the world in one's fingers will not of itself induce skill in picking a lock. One must have sure knowledge of how a particular lock is built in order to interpret the messages of touch which come in through one's fingers. One must know what part is being manipulated at every moment and what objectives must be achieved with respect to that particular part into its proper relationship with another part. This is true whether the lock is of the level tumbler or of any other type, and is basic knowledge for lock picking.

If one happens to be gifted with the faculty known as visual imagination, that is, being able to visualize in the mind a mechanical device not visible to the sight, it will help one to interpret correctly the feel of a lock as it is being picked. With a well-trained visual imagination, it is not only possible but practical to follow in the mind the positions of the parts, and the cause-and-effect relationships through the entire process of picking a lock. Many people have this faculty but are unaware that they possess it because of lack of training. In any event, it is quite common.

Since nearly everyone dreams, almost every person has some form of visual imagination. The question then becomes how to train this extra faculty, not how to possess it. If this sounds like an imposing task, remember that other faculties may be trained, with no less a prospect of success. For example, one's hearing may be trained to the point where a certain number of vibrations per second are recognized as a specific musical note. Certainly, it is impossible to count the 5,000 or 10,000 vibrations per second which make up that particular note, but the ear recognizes it, nevertheless. The nose can distinguish between pot roast cooking and pot roast burning. With training, the tongue may be able to distinguish between various brands of coffee by taste alone. This, then, is the sort of thing we are talking about when we say to train your visual imagination; give it some organization, and a purpose.

Training of visual imagination is much like the training of any other faculty in that time, patience, and practice are required in degrees which vary with the individual. Since the matter has arisen in connection with the picking of lever locks, and since they provide the versatility necessary to illustrate the training of visual imagination, let us begin with a simple one-lever lock of any type. As the first step, remove the cap from the lock and study the interior parts without touching them, noticing every detail. When you think you have seen every tiny detail, close your eyes and describe it as exactly as possible, then open your eyes and see what you have missed. Once the features of the lock have become fixed in your mind, try doing the same thing, but instead of describing merely the shape of the parts, describe additionally their finish, any tool marks in evidence, the dirt present, and if there is any grease, where it is lumped and doing no good, or where it is scraped away.

Take the key and insert it into position in the lock just as if the cap were on and slowly turn it. Can you remember the infinite variations in the spatial relationship of the parts and the sequence of events in all positions of turning the key? Can you close your eyes and describe accurately what happens? Can you describe the feel of the key as it turned? Was there at any point an irregularity in the feel of the key? If so, at what point and why?

Do the same thing in picking the lock with the cap off, correlating sight and feel, memorizing both the sight and feel, and coordinating them to the point where either the sight or the feel at any given instant will evoke the memory of the other sense. Replace the cap of the lock and manipulate the lock with a pick (and tension wrench, if required), visualizing all the while what is occurring inside the lock as it relates to the feel obtained through the pick and tension wrench. This is visually reconstructive imagination.

Visually Constructive Imagination

Having thus established the basis for visual imagination, it is only necessary to have further practice with locks of varying degrees of complexity until one arrives at the point where it is impractical to develop the skill any further; but that point should not be considered as having been reached by a locksmith before he can take parts from one lock, and in his mind, fit them together with parts from another in an arrangement never intended by the manufacturer. A step further leads us to invention when one begins assembling the known, common parts with parts which have not yet been built. Nor is the value of visual imagination limited to locksmithing.

The ultimate refinement of visual imagination leads to the creation in the mind of parts and devices which may be based on scientific theory and which have never existed before, as in the case of Edison, or to the formulation of new theories themselves, as propounded by Einstein. All of this may seem pretty far afield in a discussion of locksmithing, but in the picking of the more complex types of locks, as well as in repairing them, it is extremely important to be able to visualize an optimum condition.

Lock Variations; Alternative Opening Techniques

Lever tumblers of the open gated type are sometimes equipped with a saw type edge with matching serrations in the contact edge of the fence. When this type of lock is encountered in picking, it can be recognized by the fact that the levers are completely immobilized when tension is applied to the bolt, or by a grinding noise and feel which takes place, even under very light tension, as these serrations grate across one another when the lever is lifted. If the saw edge and the matching serrations of the bolt fence are well matched and properly positioned on the face edge of the lever, such a lock is very difficult to pick by normal picking procedures. If, however, there is a slight mismatch in the serrations, or if these serrations are not precisely coordinated with the gating of the lever tumbler and, in some cases, with the stop on the tumbler, it becomes somewhat easier for them to be picked, although it is still an extremely difficult job.

One picking technique which will often work for lever locks which do not have serrated, or saw toothed tumblers, is the vibration method. Using this method for a lever lock is somewhat different from other types because of the comparatively stiff tension exerted by the tumbler springs. Choose a straight piece of spring stock of a thickness just sufficient to allow it to enter the keyhole without binding. The end which enters the keyhole is polished until very smooth and then lubricated with graphite or light oil.

The width of the spring stock should be either small enough to clear the curtain, or notched for the curtain (if there is one in the lock), but in no case must it be wide enough to lift any tumbler above the position of the deepest cut, even when moved about in the keyhole. Having prepared this shaker in a convenient length (8 or 9 in.), push it full depth into the keyhole and shake it vigorously up and down while keeping a turning strain on it at the same time (in the proper direction to unlock the lock).

One procedure for opening a lock of this type is to drill for the window location and either to manipulate the lock directly through the window, or use it as a visual aid in picking. As an alternative with some lever locks which have a fence that is attached to the bolt on one end only and is not too heavy to begin with, it is sometimes possible to apply sufficient force to the bolt to break off the fence at its juncture with the bolt. This will permit the bolt to be drawn, just as if the key had been turned.

In attempting this procedure, it is well to guard the lock against the possibility of breaking off the cam before the fence is snapped by inserting a cutaway key shaped something like the illustration (see Figure 99) and of dimensions suitable for the particular lock at hand. This cutaway should be inserted full depth into the lock in order to be sure to engage the cam position, and, in this position, should yet leave room for a stout screwdriver or special turning tool to be inserted deeply enough into the keyhole to permit application of the considerable pressure required to snap off the fence.

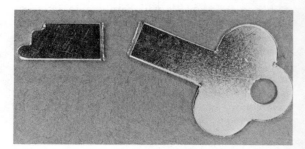

Figure 99. Fence breaking. A key may be cut away as shown for the purpose of breaking off the fence of a lever lock. Refer to the text for a description of its use.

Another method of opening these locks when used on such an installation as a safety deposit box door is through the use of a nose puller. While this does constitute forcible entry, just as does drilling for the window, only the lock is damaged, while the door is unscarred, if the job is carefully done. Nose-puller kits are available from a locksmith supply house, and are a good investment if you are interested in getting work from banks in your locale.

A word of caution is necessary at this point. Force of any sort should be used as a last resort, and then use only the minimum amount required to get the job done. The use of force, if unsuccessful, may create more problems than existed in the beginning, so one should learn to get by without force if at all possible. Learning to pick lever locks is not an easy job, and very few men in this country are exceptionally proficient at this phase of locksmithing. Therefore, their services are in considerable demand. There are not enough lever locks in many localities to warrant a local locksmith putting in much time in learning to pick them and keeping in practice; but in or within traveling distance of any of the larger cities, it would be definitely worthwhile for any locksmith to equip himself with both the tools and knowledge to cope with all types of lever locks.

MAKING ORIGINAL KEYS FOR LEVER LOCKS

Assuming that we have now successfully picked the particular lever lock for which keys are desired and removed it from the installation, the cutting or bitting of a lever lock key very much resembles the teeth of a comb, with the exception that the teeth of this comb and the valleys between them are of random lengths to correspond to the various dimensions of the tumblers. The key of a lever lock is unique in that both the high parts of the bitting and the spaces in between them are used to make the lever tumblers operational (with the exception of some lever locks being made in England).

Since the tumblers are of a uniform thickness, the spacing of the cuts on the key are likewise uniform, even though on some lever lock keys, the higher points may be thinned somewhat to allow the key to go between two deeper tumblers without binding them. This is necessary unless the tumblers have been thinned in a radius of arc about the belly of the tumbler in order to avoid binding of this type.

Thinning of the tumblers also is of benefit if the key eventually rolls the edge of the tumbler as it scrapes across it in turning countless times. For less expensive locks, relief on the key is provided by utilizing a cutter on the key machine which is slightly wider than the thickness of the tumbler in an arrangement which centers the *center* of the cutter with the location of the *center* of the tumbler. This will leave the higher (or shallower) cuts on the key somewhat thinner than the deep cuts and is necessary in order to permit the high cut of the key to pass between the two adjoining tumblers which demand deeper cuts in the key.

If one has access to the window of the lever lock, either by removing it from the door, or by drilling through the door, the key for a lever lock is not exceptionally difficult to make. In fact, if a set of *code keys* are available for that particular make and model of lock, it is possible to determine the depths of the cuts one at a time and duplicate that particular cut on the key blank, using whichever code key has the proper depth of cut in the correct position for that tumbler. The procedure is readily established by visual observation through the window of the lock.

This method should be used only with the lock in the *locked* position, as a single pocketed tumbler construction would bind the high cuts on the code key before the code key turned sufficiently to have any realistic value in determining the depth of cut of whichever tumbler one might be trying to cut on the key blank. This process is known as *decoding the lock* and tells one where to make the various cuts of the key and how deep to make each of them. Further information on code keys and their use in producing a finished key may be found in the section called Key Duplicating and Code Key Cutting.

If code keys are not available, and the gating is of the single pocket type or of the slot type, and the lock is in the unlocked position, the fence will permit the lever tumbler to rise little or no higher than the unlocking position of any one particular tumbler. Thus, if a key blank, that is, a key of the proper size and shape but without the tumbler notches cut into it, is inserted the full depth into the keyhole after having been smoked lightly, a small amount of turning pressure on the key will give the exact location of the warding cut first.

In a lever lock of the type which takes the flat steel key, this will be a cut where the nose wraps around the cylinder barrel. The second position to show a mark will be the curtain cut, and is also made to accommodate a fixed ward. When these cuts have been made, without permitting excessive play in the lengthwise positioning of the key in the lock, and the key is retained in the fully inserted position as it is turned, the next markings to appear on the key will be those made by the tumblers themselves. The key should be turned in the direction which would normally lock the lock until it makes sufficiently strong contact with one of the tumblers to leave a mark on it at the point where the first cut should be made.

Now, it is almost universally true that the first tumbler to mark the key will also be the deepest tumbler cut, therefore, the lever cut for this tumbler may be made fairly deep without hesitation. It is important that a warding file of a thickness nearly the same as (but no greater than) the thickness of the tumbler be used, and that the cut be made at an exact right angle to the edge of the key, in order that the lever cut may not cause any of the levers to bind against the key and thus make for erratic behavior of the finished key. Once the position of the first few tumblers has been noted and filed into the key, smoke may be discontinued as the tumblers will leave a better mark on the key bit without it after the position of the deepest tumblers has been ascertained and partially cut.

It should be noted at this point that as subsequent tumblers begin marking, the depth of cut required becomes shallower, and care must be taken that the depth of no lever cut is any greater than that required to permit the fence to pass through the gate; otherwise, the key will fail to lift that particular tumbler enough to clear the gate when the lock is in the locked position, although it may turn the lock from the unlocked to the locked position very well. If this situation should occur in the process of making a key, all is not a total loss; the key may be duplicated in a *key duplicating machine* (or *key machine*, as it is commonly called) if care is taken to leave any cuts which were made too deep just a little shallower on the new key than they were on the old one. The too deep cuts can then be worked down a little at a time until the key works properly.

The appearances of the marks on the key blank by the warding cuts may be either at the notched edge or along the side of the blank and, since warding parts are usually of steel or iron, the markings will be fairly plain unless a hard key blank is used. The markings for the lever tumbler cuts are usually made by brass tumblers, and as a result do not leave a mark *in* the key blank, but *on* it, in the form of a brassy-looking spot usually at the extreme corner of the key blank. It is often helpful in making these cuts if the edge of the key to be cut is first made smooth with a fine file so as to provide a sharp corner to receive the mark of the tumbler. In short, one looks for indentations for warding cuts (usually) and for brass deposits for lever cuts.

As in other phases of the craft, practice generates skill and makes the job of making original keys for lever locks considerably less tedious. It is quite common to find that the first time the original key throws the bolt into the locked position, it fails to unlock it again because there had been an upward pressure by the key on the bottom of the tumbler. When this upward pressure is relieved by the fence clearing the gate, the tumbler will jump up slightly, requiring that the cuts of the key (or some of them) be relieved slightly before the key will return the bolt to the unlocked position. If the key is being made while the lock is on the door, this means that the door must be open during the final stage if one

is to avoid having to re-pick the lock. This is by no means an uncommon pitfall, especially in the slot type of gating, since the tumbler can neither rise nor fall because of the fence being held in a fairly snug fit in the gating.

The key should always be tested with the door in the open position in any key making or lock repair work if one is to avoid a rather sticky problem of his own making. If this should happen in making a key for a lever lock, and none of the tumbler cuts have been made too deeply, it may often be corrected by duplicating the key in a key duplicating machine with a thickness or two of ordinary note paper behind the key blank from which the newer key is being cut. If the cuts produce a gating alignment which is uneven, individual lever cuts may be touched up with a warding file until even alignment of the gating is secured, with frequent visual checks through the window to avoid cutting any of the cuts too deeply into the blank.

If the lock should be of the type with a double pocket gating in the tumbler, the original key must be made by a different method, because in this event, it is possible to overlift the tumbler, whether it is in the locked or the unlocked position. Keys for the slot type or the single pocket type of gating may also be made by use of the same technique as is required for the double pocket gating. This method calls for cutting the key starting with the lock in the locked position. The ward cuts are made in the same fashion as previously described for single pocket gating. For the lever cuts, the smoke is retained on the key to mark the position of each of the lever cuts.

It is well, as an initial measure, to mark the key blank for the point of contact of each tumbler while the smoke is still on the key, since the double pocket arrangement (or the unlocked position) does not permit the key blank to come up against the tumbler hard enough to consistently secure a good, clear mark on the key, there being nothing solid to ground the levers against to produce a good mark to establish the position of each tumbler unless smoke is used.

If one or two levers (in a five or more tumbler lock) does not mark clearly even with smoke, it may be possible to locate the position of the missing cuts with reasonable accuracy by remembering that the tumblers are evenly spaced, and measuring for the position of the nonmarking tumblers. Having established the position and marked the location (with a scribe or knife point) on the key, one cuts the tumbler cut on the key which is nearest to the *bow* (or head) of the key enough to permit the corresponding tumbler to drop down into a position of alignment with the fence as the key is turned as far as it will go toward the unlocked position. Having made the first tumbler cut to depth, or nearly so, one proceeds to make a similar cut for the second tumbler.

As one works his way progressively back through the nest of tumblers, it will occasionally be found that after making a very deep tumbler cut, the cuts previously made are not quite deep enough. If this rise is not sufficient to obstruct visibility through the window, one may continue until the last tumbler is cut to depth, then return to the front tumbler and repeat the process of aligning gating and fence by progressively taking off a little metal from those tumbler cuts where it is necessary. One matter of some importance is that the bottom of the tumbler cuts be kept at right angles to the sides of the key bit constantly; otherwise the key will be erratic in operation and virtually impossible to duplicate. Final cuts on tumbler cuts should be very small in making the "touch-up" cuts.

There are certain lever locks, more commonly found in Europe, for which keys should never be made while the lock is in the unlocked position, nor should the lock be picked before making keys for it. These locks are the change-key locks wherein the depth of the tumbler cuts change to conform with the tumbler cuts of whatever key is used in locking it. If there are other keys which should continue to operate such a lock, or if for any other reason the combination should be preserved, one should remember that picking it, or locking it with an improper key, will destroy the existing combination to which the lock is set. Keys for this type of lock should be made with the lock in the locked position if one is to avoid destroying the key bitting which it requires.

Frequently these locks are part of a set of locks, all operating from the same key, so the importance of accidentally changing the combination in the process of making a key should not be minimized. Such locks have a sort of built-in memory device, mechanical in nature, which "remembers" the key which locked it and will yield only to that key; that is, it may be unlocked with a key quite normally, then locked again with a key having a different combination. It cannot again be opened by the first key, but only by the second. A key blank would lock such a lock, and it would open to another key blank of the same dimensions. If maintaining the original keying is not important, or if a change of com-

bination is desired, then it merely becomes a matter of picking or otherwise unlocking the lock, cutting a key using depths and spacing which are standard for it, and using the key to lock the lock.

These change-key locks may be recognized when the cover is removed by the presence of a sectional type of fence with a saw edge which engages a spur when the lock is in the locked position and disengages as the lock is unlocked by the proper key. Another method uses a geared wheel having a slot in it for the fence and acting as the tumbler gating. The lever tumbler engages this geared wheel in a particular tooth of the gearing, depending upon which depth of cut is pre-selected by the key in use for locking the lock. One of these wheels is used with each lever. The fence is not sectional in this case. The combination to which the lock is set is determined solely by the angle at which the gating is set with respect to the fence, the various angles being brought into alignment by the engagement of the teeth with the corresponding gear teeth in the end of the tumbler, with the positioning of that engagement having been determined at the time the lock was locked.

Yet other lever tumblers are variable and are made in sections with meshing serrations. The meshing is controlled by a separate and special change-key in the manner in which a change-key is used in changing the combination on a key change safe lock. This does not change the combination of the lock by merely throwing the bolt with a different key, so the danger of changing the combination inadvertently is not present with this type.

The variations which may be observed in lever locks is almost endless insofar as locks of British manufacture are concerned; however, these variations are somewhat less common in locks of American manufacture, although they do occur to some degree. One of these observable variations is the "detector lock." They are so-called because they possess a feature which "detects" any attempt at picking them. They contain a special "detector lever" which engages all the other tumblers and holds them immobile until the lever is released by use of the proper key. This takes a variety of forms in the internal construction of the lock, but essentially it contains one lever which, when overlifted, is either itself caught up and held in that overlifted position, or a mechanical device is engaged which freezes the operation of all the levers as previously mentioned.

Detector locks are extremely difficult to pick, but it can be done by locating the detector lever and lifting it into the picked position last. The detector lever is determined either by feel, or by deliberately causing it to engage, then releasing it by putting a strong turning tension on the bolt to force it to extend a little further from the lock case, thus releasing the detector.

Another factor which may be of significance in the construction of an original key for a lever lock is manifested if the lock is master keyed. It is essential that the original key being made possesses none of the characteristics of the master key for that particular lock; otherwise a master key for the entire series of locks, of which this particular one is a part, may be inadvertently constructed instead of the individual key desired; or, a sub-master key which operates a number of locks within the series may result, depending on the nature of master key system involved.

Mechanical Aspects of Masterkeying Lever Locks

Masterkeying of lever locks is accomplished by a wide variety of methods, virtually all of which have one thing in common: it is necessary to make the master key system at the factory when the lock is in the process of manufacture. It is true that small master key systems involving a few locks can usually be done in the shop by widening the gates of some tumblers and filing out a second gate in some others, but this is limited and usually none too satisfactory, with the resultant diminishment of protection offered by the lock.

If special tumblers or other parts are obtainable from the factory, this situation is alterable, but a considerable variety of tumblers is required at best in order to accomplish a specific job. Even then, some types of lever locks are just not intended to be masterkeyed in any way, and where such masterkeying is required, a different lock case is utilized which has a provision to accept a master key system, as for example in the case of some locks having two sets of tumblers and two key holes.

Other locks utilize a masterkeying which is based on the use of similar tumblers in the lock, but with a different warding arrangement from one lock to another, and the master key so cut as to bypass all wards. This represents the skeleton key approach, while yet allowing lever tumbler protection against keys which are entirely outside of the master key series (or set). Ordinarily, this type of masterkeying is confined to lever locks of the bit key or barrel key (pipe key, Brit.) types, although it can be done with regular flat steel keys as well.

Individual keys for locks of this construction should have the warding cuts made as close fitting as possible in order to prevent unplanned interchange of operation with another lock in the set. These wards may be in the form of internal wards or of keyway wards, or both. Within limitations, masterkeying of this type may be done in the shop; however, the quantitative controls involved in producing it will make a shop-produced master key system normally much more limited than would be a comparable system from the factory.

Warding differs may be provided in the bit or barrel key type of lock by drilling a hole through the lock case to one side of the key hole and inserting a bit of brass or steel rod of appropriate size and allowing it to protrude a given depth into the lock in the path of rotation of the key bit. The rod may be secured in place by welding, riveting, using a press fit, or by drilling and tapping the case and using screws which are inserted and cut off at the appropriate depth, and secured by punching the threads, or by a drop of solder. Small screws of steel are preferable for this purpose.

Another form of masterkeying which is used with lever locks involves the use of double gated tumblers. This is possible only in tumblers with a slot type of gating, and in some instances where cuts are rather close to being the same depth, one gating will run into the next, producing a wide gated tumbler. The setting too closely together of the gating in double gated tumblers is the reason why the pocketed type of tumbler is impractical to use, as the pockets tend to run together even before the gating does. (See Figures 176 and 177.)

In order to eliminate the objectionable features of the slotted type of tumbler construction, and to better control the deadbolt function, one or more tumblers of the pocketed type may be used in a masterkeyed lock of this type if all of the keys of that set, including the master key, have that one particular cut in common. The smaller the master key system contemplated, the more of the tumblers may be left with the pocket intact, and all key cuts for those tumblers are the same.

Wide gated tumbler master key systems for lever locks are more or less an adaptation of the double gating, but the wide gating requires that more or less closely adjoining depths of cut for the same tumbler be used on both master key and change key, since the same gate must pass the fence for both depths, and yet provide some protection against wrong keys and picking. For example, a master key may have a No. 3 depth of cut for a particular tumbler, and a No. 2 depth of cut for one change key and a No. 4 depth for another, different change key. Now, retaining that same No. 3 cut for the master key depth, if we use a change key having a No. 2 depth of cut and another change key having a No. 1 depth, the lock having a gating wide enough to pass the No. 1 and the No. 3 will also pass the No. 2, resulting in an unplanned interchange of operation by change keys.

In order to determine which key cuts are *master depths,* it is most desirable to have one of the master keys available in order to avoid hitting that depth of cut on any tumbler in a change key under construction. If one has a sufficiently large number of individual keys for other locks in the series, it is possible to determine by the process of elimination which are master and which are change key cuts, but ordinarily it is unnecessarily time consuming.

Occasionally lever locks of the springbolt type are masterkeyed by the use of two sets of levers, each of which operates a latch lever to draw the latchbolt of the lock. All of these methods are rather cumbersome for a locksmith to masterkey, or to make any changes in a master system.

A somewhat less cumbersome system of masterkeying lever locks which is currently rather common in the United States in the less expensive varieties of lever locks, involves the use of a *master tumbler.* Locks having the master tumbler type of masterkeying utilize one or more perfectly common tumblers in a series differentiation function; that is, these tumblers are used to keep one master key from interchanging with another. Then, a master tumbler is used which most likely has no gating in it, but stops short of the deepest depth of the gating of the other tumblers. Instead of having gating, this tumbler has a post or pin protruding from its side and the pin extends through a hole which is cut in the change key tumblers in such a way that the change key tumblers are free to act independently of the master tumbler through the normal limits of the depths of cut of a key.

In addition to the hole for the master tumbler pin, these change key tumblers have a gating, just as would any normal lever tumbler. However, as the master tumbler is lifted by the master key into its operating position, its pin picks up all of the change key tumblers and carries them with it up to the position where the fence will pass through the gating of all of the change key tumblers. This makes for an extremely versatile master key system, one in which change keys may vary widely from one key to the next, or may be cut very close to one another. A large number of key changes can be operated by a single master key. The principal difficulty with this type of masterkeying is that in picking the lock, once the series tumbler(s) have been picked, the entire lock may be picked by merely lifting the master tumbler to the correct depth.

Change keys for such masterkeyed locks are made by cutting away under the master tumbler (so that it does not lift the change key tumblers) and cutting the change key tumblers to correct depth. Master keys are made by cutting away (or skeletonizing) under the change key tumblers and cutting only the master tumbler and series tumbler(s) to correct depth. Tumblers for masterkeying this type of lock may be available from the factory, if changes in keying are required.

The Disc Tumbler Lock Mechanism

Essential Elements and Usage

THE disc tumbler locking mechanism is simple in basic design, but the production of the mechanism had to wait until the twentieth century to become an actuality. Only with the advent of die casting, early in this century, did it become practical to form the rather complex shape required of the disc tumbler plug. The stamping process by which the disc tumblers themselves are formed is likewise a part of the twentieth century's production technology. Despite this comparatively short span of its existence, the disc tumbler lock has become an important part of the lock industry; indeed, nearly everyone today who has keys has at least one key to a disc tumbler lock in his possession. This is partially accounted for by the inexpensiveness of the disc tumbler lock, and partially because of its extreme versatility and adaptability, which is rivaled only by the pin tumbler mechanism.

Although the protection furnished by the disc tumbler lock is usually not as great as that afforded by the fine warded locks which were produced in the past, nor as great as that afforded by moderate cost lever or pin tumbler locks, the cost factor and the almost universal adaptability of the disc tumbler lock for use in various mechanisms, and to perform various operations, are such that it has found widespread use. With this acceptability to the public has come various refinements in the design and construction of disc tumbler mechanisms to the point where some of them may be considered as offering very good protection.

The basic disc tumbler mechanism consists of four essential parts which fulfill the locking function: the *tumblers* (usually referred to as *disc tumblers,* but also as *wafer* or *plate* tumblers*); the *spring,* which provides for seating the tumbler against the key and for automatic return of the tumbler when the key is withdrawn; the *plug,* which carries the disc tumblers and contains the keyway; the *hull* or *cylinder,* which houses the plug, and in which the plug rotates. Other elements exist, of course, to retain the plug in place and to coordinate its action with the mechanism which it must operate, but these are the four parts by which the locking operation is accomplished.

The Disc Tumbler

Although the disc tumbler has undergone a great many modifications in both design and application, we shall consider only the basic disc tumbler design at this time in order to avoid any confusion arising from the subspecies derived from or related to the disc tumbler design. After the principle has been established, we shall discuss some of the more important derivations of the disc tumbler principle.

The basic disc tumbler is customarily stamped from sheet metal, usually brass, of comparatively thin stock. If one can imagine a circular disc of sheet metal as the beginning point for these tum-

---

*This multiplicity of names for the disc tumbler is simply a result of their wide acceptance by the public. For our purposes, and in the interest of avoiding confusion, we shall continue to refer to them as disc tumblers.

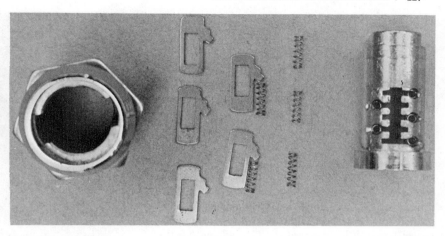

Figure 100. The four essential elements of a disc tumbler locking mechanism. Left to right: cylinder, disc tumblers, springs, and plug. An additional set of tumbler springs is installed in the plug to show their location on alternate sides of the keyway.

blers with two straight cuts made across it, parallel and equidistant from the exact center of the disc, one comes up with a piece of sheet metal which is roughly rectangular in shape, but with rounded ends on the narrow sides of the rectangle. A rectangular hole is punched, centered between the flat sides of this stock, through which the key is inserted to operate the tumblers.

One additional feature is required to make the tumblers operable: a spring tension to provide for an automatic return of the tumbler to the locked position. This is accomplished by leaving an arm extending out from one side of the tumbler under which the tiny compression-type coil spring thrusts to provide the needed tension. In common practice, a disc tumbler lock may contain anywhere from three to six of these tumblers, with five being the customary number. Each of the tumbler positions may be provided with a tumbler requiring a depth cut into the key, which is one of as many as five such standard depths that may be cut into the key bit, yielding a theoretical maximum of 5^5 (five to the fifth power) possible different keys on one keyway of a lock using five tumblers with five different available depths of cut.

Practical keying limitations such as avoidance of involuntary key interchanges within a given *key series** reduce this number drastically from the theoretical maximum of 1,875 possible combinations down to around the 250 combinations used in most five tumbler lock series utilizing disc tumblers. This

---

*Key series: A planned grouping of key or lock combinations arranged to accomplish specific purposes. An *involuntary key interchange* is the unplanned operation of a lock by a key which was not supposed to operate it, but which nonetheless works because of excessive manufacturing tolerances, poor master-keying, or, in this instance, poor key series planning.

is brought about partially by the fact that manufacturing tolerances are customarily such that at least one tumbler in any given lock must be at least two depths of cut (if four or five depth tumblers are used) from the next closest key to that combination, if any potential involuntary interchange is to be avoided. Another factor is that certain combinations are inherently undesirable. For example, a key bitting where the cuts are 11111 (all the shallowest possible) would yield a lock operating properly with merely a key blank.

A key with all tumbler positions cut to the same depth is considered undesirable, partially because of the difficulty in keeping such a key in the lock until it is once again securely locked; likewise, a combination such as 23345 would permit a worn key to be withdrawn without ensuring that the lock is actually in the fully locked position, unless a considerably high point is left between the bottoms of the adjoining cuts.

The difference in the disc tumblers used to produce variations in the depth of cut on the key does not lie in the outside dimensions of the tumbler itself. These tumblers all have the same outside dimensions. Nor does it lie in the size of the rectangular hole through which the key passes. Instead, the sole difference lies in the positioning of that rectangular hole relative to the two ends of the tumblers. As seen in Figure 101, the distance between the rectangular hole and the end of the tumbler which is opposite the spring arm is considerably smaller for tumblers used to produce shallow cuts on the key than is the corresponding distance on tumblers which are used to produce the deeper cuts in the key. It is solely by means of the position of this rectangular hole in the tumbler that disc tumblers are differed.

The Disc Tumbler Lock Plug

The disc tumbler lock plug provides the housing for the tumbler and the spring, and contains the keyway. Since the tumblers and springs are housed in the lock plug, the disc tumbler lock plug is a completely self-contained unit insofar as keying requirements are concerned. For this reason, the disc tumbler unit is extremely compact and permits greater design flexibility than almost any other system of locking.

In order to contain these eccentric shaped tumblers and the springs which actuate them, the configuration of the plug in those areas in which the tumblers are housed must be quite intricately shaped insofar as building them by any machining process would be concerned. The advent of the technique of die casting has, however, made this intricate shaping not only possible, but practical. In contrast with the similarly intricate warded locks of the past, the combination of the disc tumbler lock requires no intrinsic time consuming changes in the plug itself, but merely a change (substitution) of the disc tumblers.

The tumblers are standardized to such an extent as to be readily replaceable (with few exceptions). Holes for the tumblers in the plug are flat, and rectangular in cross section completely through the plug, corresponding to the rectangular cross section of the tumbler. To the side of this through hole is another rectangular area housing the spring arm of the tumbler. This section of the tumbler hole does not go completely through the plug, and terminates in its outermost portion in a small, round hole (which also does not go completely through the plug) providing a housing for the spring. The spring thrust is against the blind end of this hole, and the opposite end of the spring thrusts against the tip of

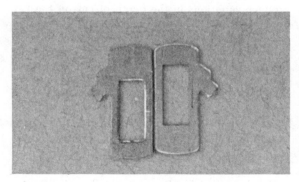

Figure 101. Disc tumbler differs. The difference in disc tumblers requiring a deep or a shallow cut lies in the positioning of the rectangular hole through which the key passes, not in its size.

Figure 102. Secondary shouldering of disc tumbler lock plugs and cylinders. Rotation of the lock plug may be restricted by the use of secondary shoulders on the shoulders of the plug and housing (cylinder). Note the use of the plug retaining tumbler in the foreground on the lock plug.

the tumbler arm. This provides for the return of the tumbler to the locked position when the key is withdrawn.

Other Features

At the extreme front of the disc tumbler plug, ahead of the row of tumblers, is a shoulder which fits into a matching shoulder in the housing, thereby keeping the plug from being pushed on through the lock, and preventing the insertion of shims or other small tools through the crack between the plug and its housing. As shown in Figure 102, within the structure of the shoulder there is frequently a secondary shoulder on the plug. This combines with another secondary shoulder or a stump within the shoulder of the housing. These secondary shoulders combine to restrict the amount of rotation of the plug to those limits required for whatever associated locking device is used.

The rotation of the plug may be restricted to a one-half or one-quarter turn, or may be restricted only by the associated mechanism, or may be completely unrestricted (in which event no secondary shoulder is used at all). At the opposite, or rear end of the plug there is a protrusion which moves the parts which fulfill the actual locking function, or to which such parts are attached. These protrusions, or drive cams, may be of any of a wide variety of shapes and sizes, depending upon the nature of the locking device to be driven. In the event that the disc tumbler locking mechanism is in the form of a cylinder, it is entirely possible that the drive cam may be more on the order of a cylinder cam and be applied to the plug as an *applied cam*.

An applied cam may be attached to the plug by the fitting of a shaped hole in the applied cam over a correspondingly shaped protrusion or stump at the rear of the plug, and held in place by either a screw and washer arrangement, with the screw fitting into the center of this stump (also by custom, the center of the lock plug); or, the stump on the plug may be riveted over the top of the cam after it has been applied to the stump. In some instances where the applied cam is riveted, two stumps are used, one on either side of the keyway.

Quite frequently in the performance of the locking operation, an applied cam of this type is merely turned between matching oblong holes in two pieces of sheet metal in order to lock the pieces and prevent their sliding across one another, as is often done in a metal desk drawer. A slot is cut in the sheet metal lip of the drawer, and a corresponding slot in the lip under the top of the desk, with the applied cam turning in and out of these two matching slots in the sheet metal parts of the desk.

There is an alternative to the applied cam, used where the plug is intended to drive a secondary locking mechanism, as for example in a desk lock having a vertical rising bolt rather than a rotating bolt (cam). The cam at the end of the plug may assume the form of a stump which resembles a round pin driven into the plug (in some rather rare instances, this actually is done), but which is usually formed by being extruded from the plug in the process of casting. Other usages may call for the stump to be dual in order to actuate two compensating parts (which move in opposite directions). In other applications, such as a springbolt or padlock, this cam may be required to perform a wedging function; in this case, the cam is likely to assume a shape resembling half a pie, with the corners of the shaped cam supplying the thrust as the cam is rotated. Somewhat less than half of the plug is usually used in this type of application, but

the shape remains roughly the same.

Other usages to which the disc tumbler lock may be adapted include the rim lock cylinder and the automotive lock cylinder (which is often of a type similar to the rim door cylinder); that is, the rear of the plug is adapted to receive and retain a cylinder connecting bar of a cross sectional shape which is suitable to drive the associated locking mechanism. Sometimes this assumes the form of using a special retaining cap applied with screws to the plug, the cylinder connecting bar being held in a slot or otherwise shaped configuration within the confines of the rear of the plug.

At other times, the rear of the plug is shaped to form, and the cylinder connecting bar retained by means of a pin pressed through the cylinder connecting bar (sometimes with an associated cap device). Usually this latter type is restricted to automotive usage and the shaft is either square or substantially so. These are, of course, only some of the more common applications of the disc tumbler lock plug, yet they include by far the greatest number of such locks in use today.

We have seen how the plug is kept from being pushed in through the housing by a shoulder at the front, but have touched only briefly on the means by which it is kept from being pulled out of the plug housing as the key is withdrawn. The simplest of these methods is by the use of a cam or connecting bar retainer, which is slightly larger than the major diameter of the body of the plug, and is attached to the rear of the plug. Being larger than the plug itself, this applied part cannot be pulled through the plug housing.

Another method, shown in Figure 102, utilizes a *plug retainer,* or *plug retaining tumbler,* in a position on the plug which is spaced one tumbler interval behind the last tumbler in the plug. The plug retaining tumbler is not operated by the key; it is driven by either one or two springs. It operates at right angles to the operation of other tumblers at times, and at other times thrusts straight up, away from the uncut side of the key. (The latter is the most common arrangement.) Unlike the key tumblers, it never protrudes on the other side of the plug. When this retaining tumbler is retracted, it covers nearly half the circumference of the plug, and comes just flush with its surface, allowing withdrawal when the retainer is in the retracted position.

When the springs push the retainer out of the plug and into a slot in the inside of the housing, the plug is locked into place. At times, it is necessary to drill a hole in the housing opposite the retainer in order to depress it so that the plug may be withdrawn. But some locks provide a hole in the housing in such a position that the retaining tumbler is accessible through the hole only when the lock is in the unlocked position.

Other locks using this method of plug retention may have a hole in the face of the plug through which a wire may be inserted when the plug is partially turned. If the wire has a point which is tapered from one side (not from the middle), the wire will wedge between the housing and the retaining tumbler, forcing the retainer into the plug. In such an arrangement, the wire usually travels down the tumbler chamber in the housing. The plug must be turned approximately 90 degrees before the wire can be inserted through the housing to the depth of the retainer.

Snap rings are occasionally used to retain the plug in the housing. When this is done, the expanding type of snap ring is fitted into a recessed ring near the rear of the plug, and when the plug is inserted full depth into the housing, the ring opens up into a matching recess, winding up partly in the plug and partly in the housing. If the snap ring is accessible from the rear of the housing (as it usually is), then it can generally be compressed back into the plug far enough to remove the plug by using two very small screwdrivers at the same time and maintaining an end pressure on the plug. If the snap ring is not accessible, it is unlikely that the plug can be removed without destroying the lock.

Somewhat less common methods of plug retention include the use of a retainer of sheet metal inserted through a slot in the housing and penetrating through it into a slot running around the rear of the plug. The retainer is held in the housing by crimping the housing tightly behind it, or by a sleeve around the plug housing. A round pin may be used in a similar application, but this method is less common. A pin may be used through the housing in a far off-center position, far enough that the center part of the pin runs in the slot in the plug. This usage is extremely rare.

The Disc Tumbler Lock Housing

The housing of a disc tumbler lock is extremely simple. As has already been mentioned, the front of the housing customarily contains a shoulder to match the shoulder on the plug, with a secondary shoulder being present at times when it is necessary to limit the rotation of the plug by some other means than the limit of travel imposed by the

mechanism which the plug operates.

Along the sides of the hole in the housing which receives the plug are two grooves directly opposite one another, which receive the rectangular portions of the tumbler ends. One of these grooves houses all of the tumbler ends when the lock is locked and no key is in it. As a key is inserted, some of the tumblers will be moved out of the groove in the housing and into the opposite groove as key cuts which are too shallow for that particular tumbler pass over it. When the tumblers are all in line with the surface of the plug and no part of any tumbler is in either of the housing's tumbler chambers, the plug can be turned freely. These grooves must be sufficiently deep to permit the full travel of the tumblers under any circumstances of keying of the lock and without regard as to whether the key being inserted into the lock is the one to which the tumblers have been combinated.

The tumbler chamber must be above the tumblers in order that the tumblers may rise above the surface of the plug and into the chamber when no key is in the lock, or when the key tried is cut too deeply for one or more tumblers. The tumbler chamber opposite to that one, or beneath the tumblers, permits an uncut key to be inserted through that tumbler which is cut to the deepest depth possible. This is particularly important as it would otherwise be impossible to insert a key which had a very shallow cut near the tip and a very deep one near the bow.

Variations

Within the limits imposed by the foregoing requirements, the plug housing may assume almost literally any size, shape, or form. It may become a lock cylinder, a padlock, an automotive ignition switch, a desk lock, or any of a host of different applications.

Picking the Disc Tumbler Lock

The basic disc tumbler lock just described is one of the easier locks to pick, although considerably more difficult to accomplish than is the probing technique used in picking warded locks. It is recommended to the beginner as the basic lock on which to learn the fundamentals of picking tumbler locks, since most of the techniques used in picking other types of locks will also work on disc tumbler locks. For developing skills and techniques with those tools commonly used in nearly all classes and methods of lock picking, there simply is nothing better than practice with the disc tumbler lock.

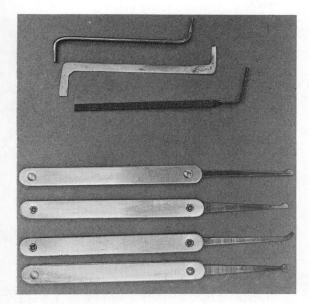

Figure 103. Disc and pin tumbler tension wrenches and picks. Although other types of picks are used in picking either disc or pin tumbler locks, those shown are the ones most commonly used for picking both of these types.

Although a number of methods are considered standard technique in the picking of disc tumbler locks, the basic method calls for the use of a tension wrench and a pick. The tension wrench does not reach the full depth of the keyway, as in picking lever locks; instead it is confined to that portion of the keyway which is ahead of the first tumbler in the lock. Usually, the best results are obtained by the use of a single tang tension wrench inserted in the portion of the keyway which contains the paracentric warding. This frees the more open part of the keyway for inserting and maneuvering the pick, resulting in considerably more freedom in manipulating the tumblers.

The tension wrench should be sufficiently large at the point where it enters the keyway to prevent its slipping off the wards it engages as a turning force is applied; at the same time it must not be so bulky as to hamper the insertion of the pick into the keyhole. In addition to the selection of a tension wrench for the previously mentioned factors, the beginner is advised to select one which is quite flexible in the shaft portion, as this will be of considerable assistance in learning to avoid the pitfalls of applying too great a turning tension. Too much turning tension in picking a disc or pin tumbler lock is a common fault and is about as wise as using a hammer to swat a fly on a baby's head!

A large number of disc tumbler locks have an internal shoulder at the point where the shoulders of the key enter the plug, instead of merely coming to a stop on the surface of the plug as they do with locks not having internal shouldering. If the internal shoulder extends deeply enough into the keyway, a turning tension may be applied in the shoulder recess by using a spring loaded straddle type of tension wrench (shown in Figure 103).

A tension wrench of this type is commercially available and works very well even in many locks without internal shouldering, as the space which it requires in the keyway is quite small. It is especially valuable in working with those locks which do not contain enough keyway wards for other types of tension wrenches to engage properly without using one so large as to unduly restrict the use of the pick.

Several types of picks are used in the picking of disc tumbler locks, and their merits are widely debated. But the more commonly used types are the half-diamond, the half-round, and the rake pick.*

The half-diamond pick, so called because of the triangular shape of its point (half of a diamond shape), is usually considered the best choice for picking the disc tumbler mechanism, due to the fact that it may be used to rake the tumblers into place with a back and forth scrubbing motion. Pay very little attention to the feel of the tumblers on the pick, but maintain a uniform pressure against the tumblers as the scrubbing motion is carried out. The half-diamond pick may also be used effectively to pick the disc tumblers individually without the need for changing picks should the scrubbing technique prove ineffectual, although the scrubbing or raking technique is usually quite effective in picking a disc tumbler lock.

The scrubbing technique for disc tumblers calls for a rather moderate amount of pressure, turning in the direction required to unlock the lock. The direction in which the plug is turned by the tension wrench can usually be determined by exerting a turning pressure, first in one direction and then the other. If, as usual, there is a shoulder stop on the plug, it will be found that in attempting to turn the lock plug in the direction in which the shoulder stops it, the stop will be very abrupt and solid in feel. When turned in the opposite direction, the plug will be stopped only by the tumblers engaging the side of the tumbler chamber. This direction of

---

*The rake pick is so called not because of the technique with which it is used but because of the resemblance of the manipulating tang to a common garden rake. The rake pick is intended for use in picking tumblers individually, not with a raking motion, as are the half-diamond and half-round picks.

rotation comes to a stop which is usually a little further from center, and the feel of this stop is not so abrupt or firm feeling (that is, the stop seems to be made gradually, often in increasing stages of tension up to the point of a firm stop). The gradual stopping of the plug indicates that this is the direction in which the lock should be picked.

When the direction of turning has been established, a moderate tension is applied, and the pick is rather lightly held. The pick should be free to move up and down in the keyhole as the resistance of the tumblers to being moved by the pick changes. That is, the pick should be held lightly enough to allow it to rise over tumblers which have already been moved into the picked position, and yet be firm enough to depress those which are not yet in that position.

The holding of the pick the right way is the most critical part of this type of picking operation, although the turning tension is also very important, as in other picking operations. A basic principle of this type of picking is that a uniform tension be maintained on the pick, irrespective of its position in the keyway (vertically) as it is inserted and withdrawn in a scrubbing motion. Customarily, less tension with the inward motion of the pick is used in order to make it easier for the point of the pick to clear the higher tumblers.

The feel called for in this type of picking, insofar as the pick is concerned, is one of maintaining a uniform tension under conditions of motion. It does not call for the feeling of the individual tumblers being picked into place as do other picking methods, even though it will probably be noticed that as the lock is picked, the tumblers do acquire a more solid feel. This is a sign that picking is proceeding successfully. In the successful use of this technique, it will be found that the pick will pivot slightly in the fingers at that point where it is grasped and which is nearest the face of the plug.

The principal feel is manifested in the tension wrench and in a slight turning of the plug, or easing of the tension as the tumblers come into alignment. The tension may be varied until scrubbing of the tumblers yields some indication that this is occurring, although this indication may be very slight and requires considerable practice to develop to a point of reliability. Either the half-diamond or the half-round type of pick may be suitable for this type of picking.

The author usually prefers the half-diamond type, because the half-round pick does not taper quite as fast from the high point of the pick as does the half-

diamond. This means that a shallow cut tumbler, with a deep cut tumbler immediately adjoining it, may not be readily depressed far enough to place into the picked position when using the half-round pick, because in depressing the shallow tumbler far enough for picking, the high tumbler may be over depressed into a position well beyond the picked position. While this same situation can and does occur using the half-diamond pick, it does not happen as often. The rake pick may be used with a similar technique in these locks, but, because of its shape, it is necessary that it be lifted completely out of contact with the tumblers as it is going back into the lock, making the operation comparatively slow.

Although the scrubbing technique is usually the fastest means of picking the disc tumbler lock, and for this reason is usually tried first, there are disc tumbler locks which are quite reluctant to yield to this method. When these are encountered, revert to the individual tumbler method of picking, using a turning tension which is anywhere from moderate to very light. Then start using the pick to depress the tumblers, one by one, until that tumbler is felt to click into place in the picked position. This click may be sufficiently loud to be audible, or may become softer by stages until the click is noticeable only as a sharply increased resistance to the further motion of the tumbler. However it may be manifested, the tumbler is depressed by the specific amount necessary to obtain such a reaction, rather than by maintaining a specific pressure on the pick as in scrubbing.

Having progressed through the entire sequence of tumblers, you may find that the lock still will not turn; in this case, it may be necessary to backtrack, repicking those tumblers which may have slipped slightly in the process of picking subsequent tumblers. The ordinary disc tumbler lock will usually yield after a couple of passes through the sequence of tumblers in this manner. If any tumblers are overlifted past the picked position, the tension must be relieved until that tumbler drops back. Repicking is started from that point.

Some disc tumbler locks may not yield readily to either of these techniques. The tumblers may all be held solidly in position, yet the plug will not turn. This is indicative of one of two situations: 1. You may be attempting to turn the plug in the wrong direction. 2. The lock may contain notched tumblers. If the lock is being turned in the wrong direction, it is remedied simply by reversing the turning tension. Identification of the trouble is the greatest problem here. In order to make the proper determination, it is necessary to understand the notched tumbler construction.

In the notched tumbler construction, the disc tumblers have a V-shaped notch in each of the four corners of the tumbler, and the tumbler chambers have undercut sides which match the V's in the corners. The idea is that the V of the tumbler will become engaged with the matching undercut in the tumbler chamber. These locks can be identified by using a very light turning tension and pressing a single tumbler down with the pick. As the tumbler is depressed, the plug will turn slightly in the direction *opposite* to that on which the turning tension is applied. This is due to the tapered surface of the V notch sliding over the tapered side of the tumbler chamber.

Picking the tumblers one at a time, using a very light turning tension, will usually do the trick with these locks. Likewise, a scrubbing technique is often successful if the turning tension is kept whisper light; just enough to rock the plug is quite enough. Two or three attempts may be necessary to get the lock entirely picked, releasing all tension when it has become apparent that one attempt may have failed; this is necessary in order to release tumblers which have been depressed too far and are held up in the V notch on the opposite end of the tumbler. Vibration picking may be used if other methods are not effective.

If all other methods fail, one may use a key blank which will suitably enter the keyhole of the lock and from which a key could be made. After starting the tip of the key blank into the keyhole, and exerting a moderately strong turning tension, the blank is driven in by tapping it *lightly* on the head with a screwdriver handle or a very light hammer. Care must be taken to avoid excessive damage to the lock in this process, but usually the driving of the key blank into the lock in a series of very light taps will suffice to take the sharp corners off the tumbler notches to the point where picking is quite readily possible. This method should be held in reserve as a last resort measure only. At best, the tumblers are slightly damaged.

If one is sure of turning the plug in the right direction, and is still unable to feel the tumblers properly, or the tumblers fail to hold the picked position after having been placed there, the presence of lubricant should be suspected. Any such lubricant should be flushed away to help restore the picking "feel" of the lock. Many experts make this the first order of business in starting to pick a lock as a matter of routine, although it is often not required. The author, however, does this only when necessary to the effective picking of this type of lock, as there are times when he believes that a small amount of lubricant aids the smooth turning of the plug, which is also a factor in picking.

Vibration picking, or impact picking, is usually extremely successful with the more common varieties of disc tumbler locks. This method of picking simply consists of maintaining a *very* light turning tension on the lock with a tension wrench, and inserting a straight tool much like a half-diamond pick with the point cut off. Usually this tool is perfectly straight, with no curves, points, wrinkles, or gimmicks. It is just a perfectly straight piece of metal, which is caused to vibrate in the keyhole.

One means by which this vibration is induced is called the safety pin pick (see Figure 105). Assuming that the lock is in the position in which the key would enter the keyway with the cut edge of the key up, the safety pin pick then would be inserted with the shortened end (the end which does not enter the lock) upward, and the end which enters the lock would be at the bottom. Holding the safety pin pick in place with the tang in the lock, and just barely touching the tumblers, the operator uses his thumb, which he rests on the top edge of the pick, to depress the top loop of the pick. The thumb is then permitted to slide off the compressed part of the safety pin, allowing it to snap back, and, in the process, strike a light blow against the lower edge of the tang which extends into the lock.

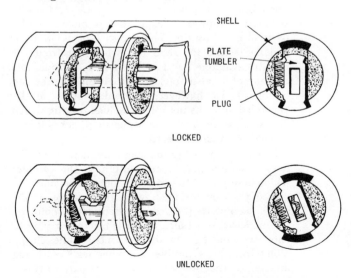

Figure 104. Notched tumbler construction of disc tumbler locks. Note the matching undercut in both upper and lower tumbler chambers. Illustration courtesy of Eaton Yale & Towne.

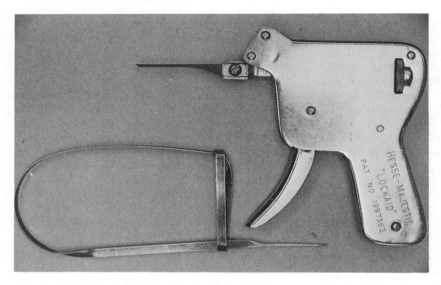

Figure 105. Safety pin pick and Lockaid tool. Both of these picking devices utilize the vibration method to align the lock tumblers in the picked position.

This produces an impact against the tumblers, causing them to jump up, and be caught and held by the irregularities induced by the shear line (the line formed by the surface of the plug) and the higher friction produced on the tumblers at that point; and caught also on the corners of the plug and the housing's tumbler chamber. Repeating this snapping of the safety pin pick, while exerting a very light turning tension on the plug, will usually result in speedy picking of the lock.

The other vibration methods of picking vary from this mostly in the means of producing the vibration. The vibration may be induced by a lock picking gun such as the Majestic Lockaid tool, or more recently, by an electric vibrator. The electric type of vibration picking is fine for shop use where electricity is readily available, and will save a considerable amount of time in picking either the disc or pin tumbler locks; however, it should not be totally depended upon, as not all of the lock picking which one is called upon to do is in the shop, nor is it in a place where there is electricity.

In any of the vibration methods of picking, the theory is to impart a momentum to the tumbler which is sufficient to carry it to, but not beyond, the picked position (or flush with the surface of the plug), keeping its progress arrested at this point by the increased resistance which it meets in trying to enter the housing's tumbler chamber while sliding along the plug's tumbler chamber wall, or being arrested by the interpositioning in its path of travel by the tumbler chamber's corner. This requirement demands that the tension be carefully regulated, and also that there be some regulation of the

strength of the impact used in the vibration methods of picking. In the case of the picking gun, either electric or manual, control is provided by an adjustment on the gun itself. In the safety pin type of vibrator, it is provided by simply not depressing the free loop of the safety pin quite so far before releasing it.

It is important to stress here that it is this type of lock which the beginner should select in his initial stages of picking practice. Indeed, it is good for practice picking at any time, but never more so than in the initial stages of learning the art. It teaches the proper use of the tension wrench, during what is probably *the* most critical phase of lock picking. The proper handling of the tension wrench is even more important than the proper handling of the pick; if the tension wrench is mishandled, the operator might as well take his pick and go home. No amount of manipulation of the pick will compensate for too much tension being applied to the lock. Too little is usually not as bad as too much, but the natural tendency, especially for the beginner, is to apply too much tension, rather than too little.

Making Original Keys for Disc Tumbler Locks

We have emphasized that different depths of cut are required on the key in order to align a disc tumbler properly so that it is entirely within the plug, with none of it protruding into the tumbler chamber, thus assuring that the plug will turn. Although some disc tumbler locks use three, four, or six tumblers in their construction, the customary number of tumblers used is five. Any of these five tumblers may require that the key be cut to any one of three, four, or five depths. Once again, the customary number of possible depths for any one tumbler position is five. The interval between the various depths of cut may be anywhere from .0125 in. (125/10,000 in.) to .030 in. (30/1,000 in.), and are established in the tumbler by the position in the tumbler of the hole through which the key is inserted.

The shoulder of the tumbler will stop against the inside surface of the housing if no stop is provided for it on the plug itself (such a stop is usually but not always provided), since the shoulder portion of the tumbler is too wide to enter the slot which is the tumbler chamber. The bottom portion of the tumbler slot (away from the shoulder) will come to rest, because of the uniformity of this shouldering, at a different position for each different depth of cut used, as may be seen in Figure 106. This makes it

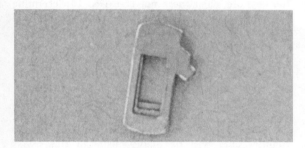

Figure 106. Disc tumbler configuration. The position of the bottom of tumbler slots relative to the shoulder portion varies according to the depth of cut required by the tumbler. This is the basis for sight reading of disc tumbler locks.

possible to look into the keyhole and determine the depth of the cut for each tumbler by observing its position relative to the others, and to the convolutions of the warding of the keyhole.

It may be necessary to depress the higher tumblers toward the front of the lock in order to observe the position of shallower positioned tumblers behind them. This is done by the use of a "straight pick" (having no shaped point) or a piece of wire to depress the tumblers in succession, until the position of all have been determined, in accordance with the standardized depths which are proper for that particular lock. This can be accomplished without having had the lock apart, or having done much more than just looking into the keyhole.

It is not necessary, by this method, that the lock be picked to make a key for it, but it is necessary to remember that all of the positions of the tumblers in

the keyhole will be somewhat higher, and the difference between one cut and the next will be somewhat smaller than if the lock were picked. This technique does, however, allow the front tumblers to be moved to observe the position of tumblers further back in the lock.

Good practice for this method is to pick a lock while looking down the keyhole with a good light. Point the light from below the tumblers and look in from above them. The position of most of the tumblers can be seen from this angle; those unseen are tumblers that are very shallow in cut, and close behind, they have a deeper cut. Make a key based on these observations, remembering that in the picked position the tumblers will be exactly in the position in which they should be held by the key. Whether this key works or not, observe the position of the tumblers in the locked position, comparing their position with that which they occupied in the picked position.

Make a second key based on this observation. Quite frequently it will be found that, even though the difference observable between depths is lesser, and the tumblers are in a higher position in the keyway, the ease with which they may be seen will be even more helpful than picking and having them aligned in operating position. The tumblers may be depressed one at a time by the straight pick from front to back and read in order, but it is usually easier to start with them all depressed; then to release the back one, reading its depth; then to release the next one, etc., until all are read. When using the latter method, it is important to remember that the cuts as they are read will be from the tip of the key to the bow, rather than the customary sequence of from bow to tip.

If it is desired to sight-read these locks by picking the lock first, and not all of the tumblers can be seen, a half-diamond pick may be slid straight in over the high tumbler. Keeping the pick perfectly level in the keyhole, note the position of the pick as it crosses the high tumbler, then note how much it drops down as the point of the pick is felt to slide across the top of the unseen tumbler. This will help to establish their relative heights, assuming that the pick is kept exactly level at all times.

This "sight-reading" method requires considerable practice to master thoroughly, but once mastered, is the quickest and easiest way of making keys directly from disc tumbler locks. It is sufficiently important that proficiency in this method of key making should be required of any master locksmith or journeyman. It is part of acquiring a mastery of the trade, as it is an alternative to the often tedious method of disassembling the lock merely to make a key.

Some disc tumbler locks have an attached cam which is attached to the plug by riveting, in which case the flange made by riveting must be carefully filed away, without disturbing the original material of the plug. The remaining original material of the plug must be used to re-rivet the cam back on after the work is completed. If a key is made by disassembling the lock, it is only necessary to make cuts positioned exactly beneath each tumbler until they are cut to such a depth that all tumblers are flush with the surface of the plug.

A great many locks of the disc tumbler type are provided with means by which the plug may be quickly and conveniently removed. This is usually accomplished by the use of a retainer tumbler behind the last key tumbler. It is spring loaded, fitting into a groove in the plug, with the spring action extending the retainer into a mating groove in the housing. Sometimes this groove is provided with a small hole through which a straight piece of wire may be pushed to depress the retainer tumbler when the lock is in the unlocked position. The plug may then be pulled out of the lock quite readily by hooking a half-diamond or half-round pick over some irregularity in the keyhole, and using it to withdraw the plug from the housing.

Other locks are provided with a similar hole in the flange of the plug into which a wire with a tapered point may be inserted to travel through the housing's tumbler chamber to wedge the retaining tumbler back into the plug. In this type of construction, the retaining tumbler is positioned 90° out of line with the operating tumblers, and the lock must be turned about this amount before the wire will be able to reach the retaining tumbler. Whether the retaining tumbler is in line with the keyhole or at right angles to it, the retaining tumbler almost invariably is exactly one tumbler interval behind the last operating tumbler. That is, in a lock with five operating tumblers, the retaining tumbler occupies the place where one would expect to find a sixth operating tumbler, although not necessarily in line with the operating tumblers.

Like everything else, when one is tempted to make flat statements, variables usually come home to roost, and this is no exception. Certain automotive glove box locks, and a few others, use a different arrangement utilizing the front operating tumbler chamber for the retainer tumbler, or a retainer tumbler considerably further back on the

plug than the sixth tumbler position.

It is a simple matter to make keys for a disc tumbler lock when the plug has been removed if one has proper guide keys or code key cutting facilities and is thus equipped to cut the keys to standard depths and spacings. By inserting a key blank in the plug, it is possible to visually gauge the depth of cut required for each tumbler position, because the further the tumbler protrudes above the surface of the plug, the more deeply the key must be cut.

Unless one is sure of having judged the depth correctly, it is well to make the cut on the key the shallowest of the possible depths which the position of the tumbler indicates. This saves key blanks and time, too, if one has to go back to recut those cuts which have already been made correctly, putting them onto a fresh blank because the old one is spoiled with too deep a cut. If one has a set of guide keys* which are cut on the proper key blank to enter the keyhole, then these may be used to determine the depth of the tumblers by inserting them into the lock one at a time until each of the tumblers has been aligned with the surface of the plug.

If one is not equipped with the means for cutting a particular key by using these standardized depths and spacings, it becomes necessary to cut them by hand filing. The first step in this procedure, after the plug has been removed, is usually the smoothing off of the top of the key blank (the edge where tumbler cuts will be made), using a flat Swiss pattern file in a No. 2 or No. 3 cut. This removes the hard surface glaze which most key blanks have. The key blank is then inserted into the lock until its shoulders come against the plug. The highest tumbler may then be tapped *very lightly*, using a special hammer with a soft brass head weighing an ounce or two (the end of a light piece of brass or aluminum rod may be used if such a hammer is unavailable), and tapping just hard enough to mark the position of the tumbler on the key.

The key is next filed to the proper depth for this tumbler, using a round Swiss pattern file, No. 3 or No. 4 cut, and cutting just a little, then trying the key in the lock until the tumbler is brought flush with the plug. The trying of the key is more frequent as the cut for the tumbler approaches completion. It is important that the tumbler finishes up *exactly* in the bottom of the cut with the key inserted full

*Guide keys: A set of keys cut to standard depths for a specific lock. Each key of the set is cut to the same depth in all tumbler positions (except for keys cut on both sides). For example, disc tumbler guide keys would be cut 11111, 22222, 33333, 44444, 55555 if the lock uses five different tumbler depths.

depth into the lock and that the sides of the cut are properly sloped to permit other tumblers to slide over the finished cut in both directions as the key is inserted and withdrawn.

If it is desired to mark all of the cut positions on the key blank at the same time instead of one at a time as described here, it can be done by scribing. One reaches through the gap between the tumblers which commonly exists in disc tumbler locks and marks by using a very fine scribe. The mark is made at the point where the edge of the tumbler nearest the head of the key touches the key blank. It should be remembered that this point is *not* the center of the tumbler position and subsequent cutting of the key must take this into consideration. Both sides of the tumbler are not marked because of the ease with which so many markings become confused. Likewise, the tip side of the tumblers is not marked because the last tumbler often rests on the tip of the key.

Important points to remember in the hand cutting of disc tumbler keys are:

1. Each tumbler must be well centered in the bottom of its cut on the key.
2. The sides of the cut for each tumbler must be well sloped at such an angle as to permit easy insertion and withdrawal of the key and yet not be sloped enough to intrude upon the position on the key in which the next adjoining tumbler rests.
3. No tumbler may protrude above the surface of the plug when the key is finished.
4. Cuts must be square across the blade of the key and smoothly finished.
5. The finished key must be capable of duplication in a standard key machine without requiring touching up by hand.

Disc tumbler locks may be masterkeyed by two different methods, one of which utilizes the regular tumbler in an application similar to the wide gating of lever tumblers. The other utilizes a special keyway which will accept two different complementary key blanks having blade portions for meeting the tumblers on opposite sides of the keyhole, and making use of "stepped" tumblers to correspond with the variance between the cuts of the two different keys which will operate the lock. This will be covered in detailed instructions for both methods in the section on masterkeying.

Variations

A number of variations exist in the disc tumbler mechanism, including two which merit consideration on an individual basis. These variations include the Bell lock and the Briggs and Stratton side bar lock, well known through its use since 1935 on General Motors automobiles.

The tumblers of this lock do not protrude at any time from the plug of the lock, but merely move back and forth within the plug. Hence they do not enter the slot in the housing. Rather, a bolt is set into the side of the plug, with its position controlled by the tumblers. This bolt or *side bar* is spring loaded in such a way as to press gently against the side of the tumblers. The edge of the side bar which is in contact with the tumblers is V-shaped in cross section, with the V portion entering a complementary notch in the side of the tumblers when the tumblers are all properly aligned by the key. Due to the spring loading of the side bar, it moves into the notches of all six tumblers simultaneously, bringing the opposite edge of the side bar down flush with the surface of the plug, thus allowing the lock to turn.

The spring tension which is exerted on the side bar is light. This means that any turning tension applied to the plug before these tumblers have all been aligned with the side bar will result in the latter being caught and held in the housing slot, even though the tumblers may all be aligned subsequent to the application of turning tension. Since one cannot apply tension which can be transferred to the tumblers, no tumbler feel of any significance on any one tumbler is possible, nor can it be held in any position in which it is placed. How, then, does one pick such a lock?

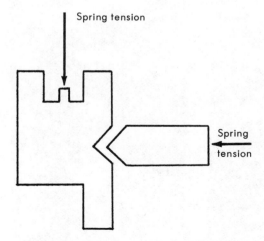

Figure 107. The side bar principle. Note the side bar at right with the tumbler in position for the bar to enter after other tumblers have been aligned.

While it is true that there is no feel to the tumblers when they are moved one at a time, there is a very delicate feel when all of them are properly aligned at once. In the case of the side bar lock, this alignment does not need to be critically perfect in order for the side bar to drop sufficiently into the plug to clear the housing.

It is possible to pick the lock with a "snake pick" (see Figure 108) by imitating the effect of the complete bit of a key in the lock. The trick of this method is to know when the bit of the key has been properly simulated and the side bar is down, since no turning tension may be applied. This is manifested by a barely perceptible increase in the amount of tension required to further depress the tumblers, and by a lessened back pressure against the pick by the spring tension of the tumblers when the side bar is in place in the notches of the tumblers. When this occurs, the plug may be turned. One should not be discouraged if the plug does not turn the first time it is tried, as it may well be that

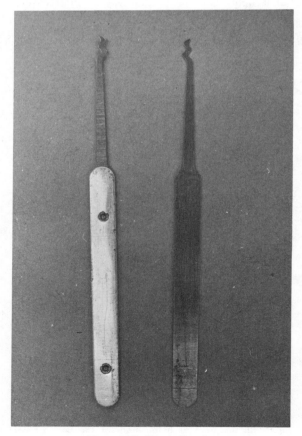

Figure 108. "Snake" picks.

the position of the pick is only nearly correct. When this point has been achieved, only minor adjustment in the position of the pick is usually required to allow the plug to turn, as the side bar is dropping, nearly, but just not quite enough to clear the housing.

Various sets of pick keys, or so-called master keys, for such locks are available. In fact, many police departments agree they are all *too* available.

These keys are not actually master keys at all, but keys which are deliberately miscut to half-depths to take advantage of manufacturing tolerances which allow a half-depth cut to operate either of the adjoining depths. They are usually close enough to operate the lock, particularly if the key is jiggled a bit.

Yet another picking method involves the use of a set of three keys, precut to selected half-depths, and milled on the flat side of the key bit to form a notch in that side of the key through which a half-diamond pick may be inserted to manipulate the other tumblers. With much practice, this method is highly effective, at least on the older models using four tumbler depths.

Masterkeying of these locks may be done on a very limited basis by the use of two notches on the side of one or more tumblers. As with most types of masterkeying, this somewhat diminishes the amount of protection offered by the lock, so any such masterkeying should be restricted to an absolute minimum.

The Bell lock is perhaps not a disc tumbler lock at all, in the strict sense, but more of a bar lock in that it has a row of bars spaced along both sides of the keyway. As seen in Figure 109, each of these bars has a round stump projecting from the side of the bar into the keyway, the keyhole being perfectly rectangular in shape. As the key is inserted into the lock, this pin projecting into the keyhole is caught up in a rather snaky-looking groove in the sides of the key blank. The stump of the tumbler which protrudes into the keyhole follows this groove, moving the bar up and down in its channel in the plug.

In the process of inserting the key, the bar goes in and out of the plug, until the key inserts to its full depth, at which time the key has correctly aligned the bar to be fully contained within the plug in much the manner of the disc tumbler alignment. The lock customarily utilizes 6 tumblers in each row of bars, for a total of 12 tumblers, with only the small stumps projecting a short distance into the keyhole from each side to operate the tumblers.

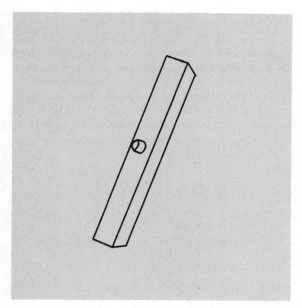

Figure 109. Bell lock type of tumbler. Depth requirements for the key are established by the position of the stump protruding from the side of the tumbler. Since the aligning stump works in a slot on the side of the key, such a tumbler is held firmly in position by the key and no tumbler spring is required.

This type of lock is rather difficult to pick, and because of the nature of the cuts in the key, it requires a special machine even to make a duplicate key. However, a code number is usually stamped somewhere on the lock. With the proper equipment, one can make duplicate keys for them, but this is a highly specialized field. As with most other aspects of the trade, patience and practice will prevail.

Double Bitted Lock Mechanisms

A double bitted lock mechanism may be described as one utilizing a key which is cut on both edges, with both of the cut edges operating simultaneously to align tumblers so that the key may operate the lock. One form of the double bitted mechanism has all of its tumblers in separate tumbler holes in the plug with the tumblers arranged so as to work off alternate opposite edges of the key. Such a lock mechanism is described and illustrated in the discussion on keying Schlage wafer tumbler Locks.

Other double bitted lock mechanisms have a single tumbler hole in the plug which is of sufficient size to contain all of the tumblers used; and although enough tumblers are used to fill the available space, they are free to slide across one another. The tumblers of this type of lock corre-

spond in general to the type of the common disc tumbler, but the criteria and mode of operation are somewhat different, thus requiring a different configuration and dimensioning.

The key is inserted through the hole in the center of the tumblers, but the key blank will not go through the hole until it is cut to the proper depth on both edges. The cut key always fills the hole in the tumbler. This holds the tumbler immovable any time that the key is in position in the lock, thus making it unnecessary to have the tumblers separately sprung. Indeed, many such locks use no springing in the tumblers, although most use a single spring or pair of springs to provide "at rest" positioning of the tumblers when the key is not in the lock.

Because of the way in which the cut key fills the hole in the tumblers, it is necessary that the key maintain a uniform distance between the cut edges, measured at right angles to the key bit for as much of its length as may be required to penetrate the tumblers. Thus, if one side of such a key bit is cut "x" high at a given position, the opposite side must be cut correspondingly low.

Picking of the double bitted lock mechanism is not greatly influenced by this type of construction, and the technique is similar to the one described for more conventional disc tumbler locks. The technique usually most effective is "scrubbing," but it is applied to opposite sides of the keyhole on alternating strokes of the pick. A round or full diamond point pick is often more convenient in applying the technique to these locks, since such a pick may often be used on both sides of the keyhole without needing to be turned over to change sides.

The need to pick both sides of the keyhole requires that a tension wrench be chosen which will permit the necessary availability of both sides of the keyhole to the pick. This usually means the tension wrench used is one which straddles the full width of the keyhole with the pick being operated through the cut-away center of the tension wrench. Pressure exerted upon the pick, not the distance the pick moves, is the important thing in picking this type of lock; once this is remembered, the picking of these locks will usually proceed smoothly.

Key duplicating machines are available which will duplicate these keys quite well, by gripping both the pattern key and key blank in one of the warding grooves, rather than seating them against an irregularly cut edge. The most successful method of making original keys for such locks is also a duplicating process, and consists of duplicating the proper cuts from a reference set (usually from 200 to 500 keys per set) according to the manufacturer's specifications for the code number, type, and keyway involved. Such reference sets are usually called *code sets* or master sets (a misnomer), and are generally available through locksmith supply houses. Other methods may be used, such as cutting from guide keys, but with variable success. Here again, a notable exception to this is provided in the Schlage wafer tumbler lock (see index).

REVERSIBLE KEYS

Reversible keys, even though usually cut on both edges of the key, often do not require that the lock be a double bitted mechanism. Frequently such keys are for simple single bitted locks of either the disc or pin tumbler type, and have a keyway of such a pattern that the key rests upon a shoulder in the keyway rather than on the bottom of the keyway hole in the plug, as in more conventional locks.

The Pin Tumbler Locking Mechanism

Since the pin tumbler locking mechanism combines a unique versatility with a reasonably good degree of protection at a rather moderate cost, the use of the pin tumbler mechanism in the United States has become extremely widespread. Therefore, a complete and thorough mastery of this type of mechanism is essential to anyone in the locksmithing field. Although the pin tumbler locking mechanism is basically simple in concept, the variations of it and the keying potentialities of the pin tumbler lock mechanism are numerous, to say the least.

Essential Elements

With this in mind, let us proceed with the examination of the basic pin tumbler mechanism. Essential elements of the pin tumbler lock mechanism are the *plug*, or *cylinder plug;* the *hull, cylinder hull, cylinder,* or *shell;* the *key pins,* or *lower pins;* the *driver pins,* or *upper pins;* and the *tumbler springs,* or simply *springs*. The plug contains the keyway, usually of the paracentric type, and a number of tumbler holes (usually from four to seven) which must be very accurately drilled in exact alignment and diameter and depth to function properly with corresponding holes in the hull.

The key pins go into these holes in the plug point first, with the flat or slightly rounded end at the top. With the proper key in position in the lock, these flat ends come just flush with the surface of the plug, permitting it to be rotated. The driver pins fit into the corresponding holes in the hull, and when the key is withdrawn, the drivers are pushed downward, partially into the plug by the action of

gravity coupled with the thrust of the springs which are directly on top of the driver pins. As can be seen in the illustration (Figure 110), the overall length of the holes in plug and hull combined is limited, but uniform from one tumbler to another.

In order to ensure optimum performance of both lock and key, the overall length of the key pin and the driver should be uniform. This combined length must be great enough that a pick or other tool may not be used to lift both the driver and key pin up into the hull and thus permit the plug to be rotated without a key, or by picking the lock. Conversely, these tumblers must not be so long in combined length that the drive spring will not compress sufficiently to allow the key pin to be lifted completely out of the keyway portion of the plug. This would be very necessary in the event that a key carried a shallow depth of cut near the tip of the key. Such a situation would result in damage to the spring, even if the key could be forced past such an over-length *tumbler stack*. Therefore, if a possible lock-out is to be avoided, the overall length of the tumbler stacks should be uniform from one tumbler to the next, and of a definite overall length for each specific type of pin tumbler mechanism.

Customarily, although not universally, the cylinder plug is provided with a flange at the front end of the plug, with a corresponding recess provided in the hull to introduce an offset in the contour of the plug. This serves a triple purpose: 1. It provides a positive stop for regulating the alignment of the pin chambers in the hull and in the plug. 2. It prevents the plug from being pushed on into, or through, the

SIDE VIEW   END VIEW

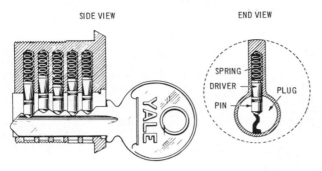

WRONG KEY
HERE THE PINS AND
DRIVERS ARE IN IRREG-
ULAR POSITIONS, FORM-
ING OBSTRUCTIONS WHICH
PREVENT ROTATION OF
THE PLUG.

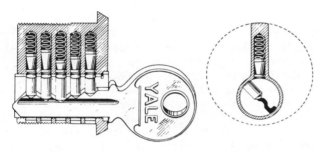

RIGHT KEY
NOTE HOW THE PROPER KEY LINES UP THE PINS AND DRIVERS AT THEIR
INTERSECTIONS SO THAT THE PLUG MAY BE TURNED TO OPERATE THE LOCK.

Figure 110. Essential elements and operating theory of the pin tumbler lock cylinder, as described in the text. Illustration courtesy of Eaton Yale & Towne, Inc.

hull as the key comes into operating position. 3. The offset shoulders provided by the flange and recess prevent the insertion of thin strips of metal (called *shims)* as an aid in picking the locks from the outside, a technique known as "shimming the lock." This will be considered in somewhat greater detail in the discussion on picking the pin tumbler lock.

The plug is retained in the hull by a variety of methods, depending upon the specific application in which the locking mechanism is being used. For a mortise cylinder, the retention of the plug is accomplished by the presence of the *cylinder cam,* which usually is attached firmly to the plug. For some other types of usage, the plug may be secured by the use of snap rings, retainer clips, or by a long pin similar to a driver pin which protrudes from the hull down into a groove in the plug.

The long pin method does not utilize a spring, but fills the upper chamber completely, being retained (usually) by a cap pressed on from the outside of the hull. Such an assembly, in addition to providing plug retention, may be used to limit the rotation of the plug in the cylinder to less than a full revolution of the key by the simple expedient of milling out the groove in the plug only part way around the plug. This method of plug retention demands the same high degree of accuracy of alignment as is necessary for the tumbler chambers in plug and hull.

Essentially then, four basic methods are used to align the tumbler holes in plug and hull:
1. Shouldering at the front of the plug.
2. Retention by the addition of an accessory part, such as a cam, at the rear of the plug.
3. Retention by a snap ring.
4. Alignment by use of a retainer pin and groove.
(Methods 2 and 3 are contingent on the use of a shouldered plug.)

Certain variations occur in the key pins between locks of various manufacture, and even of different models of locks of the same manufacture. The configurations of these variations, particularly as they concern the point, or key, end of the key pin will affect the shape of the cut required in the key to operate the lock properly. Taking the point end of the key pin in cross section, as shown in Figure 111, the configuration may show the point to be completely rounded in a hemispherical form. Keys to fit locks using this type of key pin may utilize a cut, the bottom of which is either rounded to a radius slightly greater than that of the point of the key pin itself, allowing the key pin to bottom properly in the cut, or it may use a flat bottom cut with an exceedingly wide flat.

Other locks may use a key pin whose point end is in the form of a truncated cone; that is, the sides of the pin are tapered, but instead of coming to a sharp point, the point is flattened off. Such a key pin requires a flat bottomed cut in the key, with the flat sufficiently wide to accommodate the flat of the truncated cone. A third type utilizes a combination of these two types; it is essentially a truncated cone, but the sides of the cone, instead of being straight, are curved in a convex form. A tumbler of this type may be used with either a curved bottom or flat bottom cut on the key; if, however, the flat bottom cut is used, either the flat must be wider, or the sides of the cut must form a greater angle in order to accommodate the bulging sides of the point of the key pin.

Still another variation of the key pin is one much

like the driver pin, being flat on both ends with a small ball bearing between it and the key. The ball bearing key pins will work with either the round or flat bottom cut in the key, just as will the hemispherical pointed key pin, and have the same requirements as far as the cuts in the key are concerned. A lock using the ball bearing key pins will usually have this type only in the front end where the wear is heaviest from the insertion and withdrawal of the key. Their use is usually avoided in the innermost one or two tumblers to provide protection against lifting all tumblers until the balls come flush with the surface of the plug, thereby allowing the lock to be picked.

Figure 111. Key pin configurations. Left to right: truncated, cone, modified truncated cone, hemispherical, truncated cone with curved sides, and ball bearing.

The upper end of the key pin, that is, the end which comes flush with the surface of the plug, may be domed slightly, using a radius which corresponds closely to that of the plug; or it may be perfectly flat, in which case it is usually necessary for the surface of the plug to be flatted for the width of the pins unless the driver is curved sufficiently to compensate for the curvature of the plug where it rises above the flat of the pins. The top of the key pin may represent a compromise between these two types, being flat on the top with the corners beveled on both key pin and driver. This allows the key pin's flat top to rise somewhat higher in the pin chamber than would be possible than if the beveled corners were not used.

Both ends of the driver pin are usually alike, being formed in one of the styles which is used on the top end of the key pin. The shape of the ends of the driver pin which is used with a given key pin will depend upon the configuration of the top of the key pin. For example, the ends of the drivers may be perfectly flat, provided that they are used with key pins of the domed type, but a flat ended driver may *not* be used with a flat or beveled end key pin unless the top of the plug has been made flat for the width of the flat on the end of the key pin.

The flattening of the plug is a highly undesirable thing in those locks where masterkeying is involved, for reasons which will become apparent in the section on masterkeying of these locks. Drivers with bevel-edged flat ends may be used with key pins having either the beveled flat end or the domed end, just as may those having the domed end. In addition, the domed end drivers may be used with the flat end key pins, *provided* that the edges of the driver clear the pin chamber along the longitudinal axis of the plug and no portion of the key pin rises above the surface of the plug.

Pin Tumbler Lock Cylinders

Rim lock cylinders of the pin tumbler type are almost universally equipped with a flat cylinder connecting bar which is attached to the rear of the plug in a jointed attachment, permitting it to flex somewhat. If the cylinder plug is installed somewhat out of line with the cam which the cylinder connecting bar must operate, the slight degree of flexibility provided will permit the key to operate without binding, assuming that the extent of the flexibility is not exceeded nor limited by the thickness of the cam or the tightness of fit of the cylinder connecting bar in the lock cam.

Mortise lock cylinders of the pin tumbler type are equipped with a cam, the operating portion of which is usually, but not universally, of a roughly rectangular shape. Even the rectangular shaped cam may vary in dimension. The cam, alternately, may be fan shaped or shaped roughly in the form of a cloverleaf, depending on the nature of the mechanical devices which it must operate.

When the pin tumbler mechanism is used with desk locks, file cabinet locks, or padlocks, the cam may be cut directly from the material of the plug itself, or it may take the form of one or two stumps extruded from, or set into, the rear of the plug. The rear of the plug may be extended beyond the back of the hull. It may have a slot in the end to engage a bar which is rectangular in cross section and which operates the remaining mechanism of the lock in a manner quite similar to that used with disc tumbler plugs.

PICKING THE PIN TUMBLER LOCK CYLINDER

Because the pin tumbler lock enjoys widespread popularity, and because it offers greater resistance to picking than do either the disc tumbler or the inexpensive warded locks of current production, probably more different means and methods for picking have been developed for them than for any other type of lock.

Figure 112. Mortise cylinder cams. These represent only a small fraction of the various configurations one may expect to find in use.

The first step in picking the pin tumbler lock is to determine in which direction the plug must rotate in order to unlock the mechanism. While this may sound like a rather trivial matter, the beginning locksmith (and, alas, many of the old timers) will waste many hours in fruitless endeavor trying to pick a lock in the wrong direction.

While there are exceptions to almost any set of rules which could be made for the direction in which to turn the plug, there are certain rules which will help you avoid a great deal of wasted effort and will cover by far the greatest majority of locks of any one type. For example, cylindrical locks, padlocks, and file cabinet locks are almost universal in turning clockwise, or in both directions, to unlock. Most mortise deadbolts require that the plug be turned in such a direction that the top of the keyhole (the cut edge of the key) is turned toward the edge of the door to unlock; however, there are exceptions to this, notably mortise locks of Corbin and Russwin manufacture. The same rule holds true, in general, for the mortise deadlatch and springlatch functions, although the rule is slightly less reliable for these. A mortise panic exit device usually may be unlocked by turning the cut edge of the key toward the edge of the door.

Accessory types of pin tumbler mechanisms, such as cabinet or desk locks, or locking handles, may turn in either direction (not both) to unlock; however, in many cases, when the lock is in either the locked or unlocked position, part of the plug comes up against a firm stop in order to align the tumbler holes for convenient withdrawal of the key.

When this is the situation, there is a slight difference in the feel of the lock between the direction in which it is free to turn and the direction in which it will not turn. The direction in which the plug will turn, and should be turned, will have a feeling of slightly more "give" than will be noticeable in the direction in which the plug encounters a

firm stop. This happens after the tension wrench is inserted, but before any tumblers are lifted or any picking attempt is made. The direction in which the plug *should* be turned comes to what the author terms a "soft stop." In other words, the stop is not hard and sudden, as it is in attempting to turn the plug in the wrong direction. This is, of course, by no means a universal panacea for these problems, nor is it always correct, but the percentages of success for this procedure warrant trying it first.

Before going any further with an attempt to pick a pin tumbler lock, it should be determined if the pin tumbler mechanism is clean and free (insofar as possible) of all dirt and lubricants, because the picking of these types of locks is dependent to a considerable degree on friction.

THEORY OF PIN TUMBLER LOCK PICKING

As nearly everyone is aware, perfection in the production of mechanical devices is an illusory thing. Perfect accuracy particularly in the production of locks may be approached rather closely, but never attained. There is always a certain amount of tolerance, such as plus or minus .0002 in. (2/10,000 in.), but this variation, however small, is still there.

The variation within tolerances means that in attempting to turn the plug in the lock without the proper key, one tumbler will be caught up and become tight before subsequent tumblers are. Therefore, when a turning tension is exerted on the plug, and the tight tumbler is lifted with a pick, there will be either a clicking feel or a sudden momentary relief in the tension which such a tumbler generates on the pick as the pick lifts the tumbler. The sudden release of tension will occur at the time that the key pin is brought up even with the shear line (surface of the plug). At this point, lifting should be discontinued.

Another factor operating to assist in picking is that a certain amount of space must be left in the tumbler chambers between the walls of the chamber and the tumbler itself. This is necessary in order to keep the tumbler from sticking in the tumbler chamber. This clearance, as it is called, will permit the tumbler to tip slightly within the chamber, with the result that as the upper pin is lifted beyond the shear line, the plug will turn ever so slightly, and the top surface of the key pin will strike against the hull at the corner of the driver pin's chamber.

Care should be taken not to over-lift a tumbler being picked, because this brings the key pin into the driver pin chamber. Bringing it down again can be a rather tricky job at times unless one releases tension entirely and begins again. The procedure to avoid over-lifting is to use just enough force on the pick to lift it against the friction and spring tension involved, being careful not to overcome the additional resistance encountered when the key pin contacts the hull. Working the key pin back down again after it has been caught in the driver pin's chamber may be done by slowly easing off on the tension, meanwhile running the pick lightly back and forth over the key pins to "vibrate" them into slipping downward. During this time the driver is caught at the shear line by the slight offset between the respective pin chambers.

Although the same types of picks are used for disc tumbler locks, the selection of the tension wrench and pick to be used will depend considerably on the size and configuration of the keyhole. The tension wrench must enter the keyhole and apply a turning tension without slipping and without obstruction to the free action of the pick in the keyhole. The tension wrench should be one which has a certain amount of spring in it for the best results under ordinary circumstances. A moderate tension may be used when the tumblers are picked one at a time in the fashion outlined above, beginning with the tumbler which is bound the most tightly.

A method of picking which is usually easier and faster involves the use of a light tension applied with a tension wrench that has considerable spring in it, and a half-diamond pick. (A half-round pick may also be used, but the half-diamond is usually more efficient.) With light tension being applied, one "scrubs" the tumblers with the point side of the pick, moving inward and outward in the keyhole.

Instead of lifting the pick to a certain position in the keyhole, one applies a certain amount of *tension* to the pick, allowing it to move up and down in the keyhole in response to the varying tensions of the different tumblers as it comes across them. In this fashion, tumblers which have been lifted into the

picked position, and which have the driver holding up the corner of the key pin chamber, will not be over-lifted, because it would be necessary to move the key pin against not only the corner of the driver chamber, but also against the simultaneous application of spring tension to the driver by the tumbler spring. If the pick is permitted to bounce up and down sufficiently with the varying positions of the picked tumblers, and if tension is correctly applied to the plug, one should encounter few problems in picking the common pin tumbler lock by this method.

Vibration picking may be used, utilizing the "safety pin" type of pick, the hand-powered picking gun, such as the "Lockaid" tool, made by Majestic, or (for bench work) an electric pick of the vibrating type. In any of the vibration picking methods, a light turning tension is applied. If immediate results are not produced by one turning tension, or one intensity of vibration, either of these methods may be varied until results are attained.

The theory of the vibration method is that just enough tension is applied to the lock plug to give sufficient momentum to the lock pins to move them against the friction generated by the turning tension. When the top of the key pin meets the increased resistance caused by contact with the corner of the upper pin chamber, it will not penetrate the chamber, but will fall back down into the plug (as the upper pin is caught on the corner of the plug's key pin chamber). The trick to this method of vibration picking is to achieve the proper balance between the turning tension applied and the momentum supplied to the pins by the picking device.

It is possible that, in the course of picking a pin tumbler lock, the plug will turn slightly as if it were going to unlock, and then firmly hang up with the tumblers all firm and solid, as if picked. Yet, the plug will not turn more than 2° or 3°. When this happens, it means that one has encountered spool or mushroom type drivers. These are simply drivers which have had the center portion cut down to a smaller diameter than the ends, in a fashion similar to that shown in Figure 113, leaving a shoulder on the end nearest the key pin. The shoulder will hook under the corner of the upper pin chamber when the tumbler is lifted with a turning tension on the plug, not clearing the plug's pin chamber and obstructing the shear line. A construction of this nature makes the picking of the pin tumbler mechanism by any of the methods previously described much more difficult and tedious.

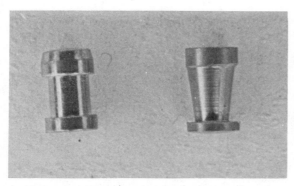

Figure 113. Spool and mushroom driver. Usually one or two innermost drivers are **not** of this type because of the alignment problem they would create and the resultant difficulty of inserting a key as the mushroom corner of the driver engages the cylinder hull. See Figure 110 top and bottom, right.

The job can be simplified considerably when this type of construction is encountered by going into the keyhole with the flat side of the pick uppermost (or with a straight pick) and lifting all tumblers to the top of the keyway and holding them in that position by the subsequent application of a strong turning tension. The turning tension must be done with a rigid, or nearly rigid, type of tension wrench in order to obtain adequate pressure to hold all of the tumblers simultaneously in the fully lifted position. While maintaining extremely heavy turning pressure, reverse the half-diamond type pick and gently scrub the tumblers with an in-and-out motion. While doing this tumbler scrubbing, the tension on the plug is very slowly slackened off.

It will be found that scrubbing will vibrate the lower tumblers sufficiently to cause them to drop to the shear line. If the tension is released slowly enough, the upper pin will not enter the plug, thereby picking the lock in exactly the reverse manner to that which would be used with an ordinary pin tumbler lock without mushroom drivers.

If the pin tumbler mechanism is in a padlock, it is often possible to vibration-pick it by using a tension wrench. Apply a turning tension to the plug. At the same time, strike the lock sharply in the area immediately above the upper pin chamber with a hammer handle or a stick of similar size and weight. Care must be taken in doing this, however, to avoid damaging the lock. This furnishes a momentum to the lock while the tumblers remain relatively immobile due to inertia. It could be thought of as a sort of backhand method of vibration picking, and the result and principle are the same insofar as the relationships of tumblers, plug, and hull are concerned.

Picking a padlock cylinder by this method involves a bit of juggling in learning how to hold the padlock correctly in the palm of the hand while maintaining a proper amount of pressure on the tension wrench with the fingertips, and at the same time not striking your fingers with the hammer handle. This method, though effective, is definitely labeled "Approach with caution!"

If the lock to be picked is a pin tumbler cylinder which has been removed, one may use a similar method by first removing the cam, placing the thumb against the end of the plug, and pressing outward on it rather firmly. It is important that one finger be kept ahead of the front end of the plug to keep it from sliding outward. Otherwise the drivers will fall into the next plug pin chamber as the plug is pushed forward when the tumblers have all been picked.

The plug will press forward in picking instead of turning because this is the direction of the tension. This is not just an academic possibility. It will happen unless conscious provision is made to avoid it. If desired, a piece of adhesive tape may be substituted at the front of the plug, but it may prove slightly less reliable than a finger. When, by rapping or the wielding of a pick, the lock has been picked and the plug moved slightly forward, it may be turned and pushed back against the shoulder, thus maintaining the picked position and avoiding a chance of one or more of the drivers falling into the wrong plug chamber. The hazard of banging fingers in rapping a cylinder is almost as great as in rapping a padlock, so care should also be taken here.

In picking detached lock cylinders where the keyway is restricted, or the operation of a tension wrench is ineffectual, tension may be maintained on

Figure 114. "Rapping" open a mortise lock cylinder. Note the positioning of thumb and fingers.

the end of the plug while a pick is used in the keyhole. At times, this straight line tension makes picking considerably easier than it is with a turning tension, therefore the method may be considered even though no keyway problems are encountered.

Another method often used to pick detached cylinders (removed from the lock they serve) which do not yield readily to picking otherwise, is the process known as "shimming." This involves the use of a strip of metal *shim stock*, possibly 1/4 in. wide, and very thin so that it will slide into the clearance space between the plug and the rear of the cylinder hull. Usually this amounts to shim stock from .001 in. to .003 in. thick. The shim is pressed into the crack between the plug and the cylinder hull directly in line with the tumblers until it comes up against the rearmost tumbler.

While maintaining a light inward pressure on the shim stock, the back tumbler is lifted with the pick until the shim stock will slide between the upper and lower pins. Then it is slid forward until the next tumbler in line is contacted, and the process is repeated until the first tumbler is reached; then, when the first tumbler is lifted, the plug is free to turn. Alternatively, if one has lifted all but two or three of the tumblers, and has difficulty with the shim stock collapsing, the remaining tumblers can usually be picked quite easily in the conventional manner, using a pick and tension wrench, or a half-diamond pick, with end pressure supplying the tension for picking.

MAKING ORIGINAL KEYS FOR PIN TUMBLER LOCKS

Often, a *code number* is stamped on the lock cylinder or elsewhere on the lock from which a key may be cut, using the method described in a subsequent heading entitled "Code Key Cutting." Where possible, this method should be followed, as it is nearly always more convenient and efficient than other methods, and produces a better configuration in the cut of the key than do other methods. For most locks, the accuracy of the finished key cut by code is superior to that of those produced by the hand fitting methods. This section will, however, concern itself with hand fitting methods under the assumption that no intelligible coding exists for a particular lock, since a great many pin tumbler locks do not incorporate any code number whatsoever.

Disassembly

After picking the lock and leaving it in a partially turned position, one selects a *plug follower*. The

Figure 115. "Shimming" a pin tumbler lock cylinder.

plug follower consists of a short length of bar or tube the same diameter as the plug. It may be notched on one end to accommodate any protrusions which are integral with the rear of the plug. Having chosen the plug follower, remove any retainers, either screws, snap rings, or pins, which may serve to hold the plug into the cylinder.

It should be remembered that there must be no notch or gap, in either the plug or the follower, into which the driver tumblers may drop in the course of pushing the plug out of the front of the cylinder when using the follower to press with. Also, it is imperative that the plug be upright as it emerges from the hull so that the key pins are not spilled. Usually if the plug is turned about one-eighth of a turn (that is, about 45°), and if the follower fits snugly against the plug at that point, one can press the plug very carefully out of the cylinder without catching on any of the drivers. If a retainer pin is used in a notch in the plug to hold it into the hull, turn that plug sufficiently to clear the notch in which the retainer pin operates, or the drivers will

Figure 116. Plug followers.

drop into the notch and create a difficult situation in getting the plug out of the cylinder.

If the notch goes completely around the plug, and the plug is free to turn through the full 360° before the pin is removed, or if a snap ring is used on the rear of the plug, and the notch for the snap ring is sufficiently wide for the tumbler to drop in, it becomes necessary, before pushing the plug out at all, to turn it 180° away from the locked position. This puts the open bottom of the keyhole directly under the drivers; in fact, the upper pins (drivers) may well drop into the keyhole and prevent it from turning any more. This is not objectionable. The follower is then placed against the end of the plug in such a way that no gap exists at the point where it meets the keyway. The entire unit is then turned in such a way that the key pins will not fall out of the plug as it leaves the hull. The follower is pushed into the cylinder hull and the plug is pushed out.

As mentioned, the upper pins may have dropped down into the keyway to such an extent that they will not clear the follower as it is pushed through the cylinder hull. If this is the case, the drivers may be moved out of the way by inserting a straight pick and moving them flush with the shear line while pushing the follower into the plug. The drivers are readily accessible through the bottom of the keyway to a straight pick or a piece of straight spring wire. Almost anything which will enter the keyhole will suffice to lift them back into their chamber in the hull and permit the passage of the plug follower. This business of removing the plug from the hull calls for doing a great many things with both hands, seemingly at once; however, practice quickly leads to proficiency in this operation.

Once the plug follower is completely through the hull, and the plug has been removed, the plug is immediately placed in a *plug holder,* as in Figure 117. To lay it down would risk its overturning and spilling the key pins, losing the existing combination to the lock. A key blank is inserted into the keyway very slowly until contact is made with the first pin in the lock. The pin should be raised slightly above the shear line by the action of the key blank; when it is, it should be grasped with a pair of tweezers and lifted out of the plug. It is then deposited in an appropriate section of a *pin tray* so that it may be replaced in the lock in the same chamber from which it was removed.

The reason for lifting the pins out of the plug with tweezers when they have been started far enough to grasp is that the lock may have been masterkeyed, in which case there would be more than one tumbler

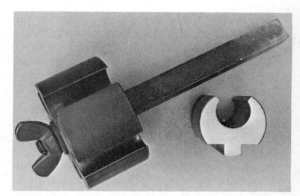

Figure 117. Plug holders or setting-up thimbles. The rotary type may be used for various sizes of lock plugs.

present in each hole. If this is the situation in any of the tumbler holes, the pins must be placed in the pin tray in the same order that they were removed from the lock, thus preserving the combination.

When extra pins are found in any of the tumbler holes as the tumblers are removed from the plug, there is a possibility that additional tumblers, or *split pins, master key chips,* or *master tumblers* (they are known by all three terms) may be found in the upper pin chamber in the hull. This will necessitate the removal of the tumblers which are otherwise ordinarily retained throughout the procedure in the upper pin chamber.

These tumblers have to be removed one at a time from the upper pin chamber. It can be a rather tricky job to pull the follower out of the plug in

Figure 118. A pin tray. A less elaborate one may be made by nailing bottle caps to a board.

order to uncover only a part of the tumbler. Hold the tumbler down with a pair of tweezers until you can completely remove the follower from the tumbler hole; then allow the spring tension to slowly push the tumbler out of the hole while holding the tumbler in place with the tweezers. If a second tumbler is detected under the first, the follower must be pushed back over that tumbler as soon as the first has cleared the hole. A key must not be made for any lock containing these master key chips unless the master key is available so that the key which is made is *not* inadvertently cut to the same depth as the master key on any tumbler.

Assuming that the key pins are all removed from the plug and no master pins are present, the first step is to be sure that the tumblers are all clean and in good condition, showing little or no key wear. Extreme key wear on the tumbler shows up as a tumbler whose point is worn into a chisel-shaped point, as shown in Figure 119. Wear is more noticeable toward the front part of the lock where more of the key passes under the tumblers. Such a

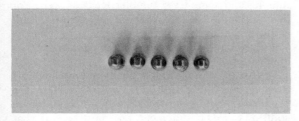

Figure 119. Worn key pins. Note the gradual evolution of the chisel point.

tumbler, once its original length has been established, should be immediately discarded. The cut for the new tumbler should be made by the use of standard depth charts,* a code machine, or code keys for that particular lock. If the worn tumbler falls between two depths, the new key should be cut to the deeper of the two. This is done on the assumption that the tumbler is worn to whatever degree it lacks of meeting the required length for that depth.

If none of these aids is available, two or more of the less worn tumblers may be measured for length with a micrometer to establish the interval between cuts of the key; that is, the variation between one depth and the next adjoining depth. Once this is known and the proper depth of cut for one tumbler

* Standard depth chart: Chart showing the standard depths and spacings to which keys are cut for various locks; usually found in catalogs which include key cutting machines.

has been determined, the depths of all cuts may be plotted by simple addition or subtraction from the established depth. For example, one may find two tumblers which vary in length by .030 in. Since most pin tumbler locks have an interval of from .015 in. to .020 in., and .030 is an even multiple of .015 in., it is a good bet that the interval is .015 in. If measurement of a third different length tumbler is also a multiple of .015 in., the interval may be considered confirmed. New tumblers to replace the worn ones may be selected to conform with the confirmed tumbler intervals.

Having established that the tumblers are in satisfactory condition, and having cleaned the plug, you should inspect the tumbler chambers in the plug for wear. Chamber wear in the plug is revealed by the chamber assuming an oval shape. If the oval shape shows an eccentricity amounting to more than 10% of the diameter of the pin, wear should be considered extreme and the entire cylinder replaced, if possible. If the customer insists that the same cylinder be used, he should be told that the results may not be satisfactory because of this condition; therefore, no guarantee can be made. If the tumblers and plug are in good condition, these factors naturally have no significance, and one may proceed with cutting the key.

The proper key blank is selected and inserted into the keyway to the full depth, with the shoulder stopping against the face of the plug. The first tumbler is inserted into its proper chamber, point downward, and tapped lightly with a small hammer, just hard enough to produce a slightly visible mark on the top of the key. It is often helpful in obtaining the initial marking if one takes a fine file (a flat Swiss pattern file in No. 2 to No. 4 cut will do) and files off the polished finish on the top edge of the key. This helps to make the marking produced by the tumbler more easily visible without having to "swat" the tumbler so hard that it is actually distorted in the process.

Having established the exact position in which the cut will be made on the key, one may proceed, using a round Swiss pattern file in about a No. 3 cut, stopping when the tumbler is in the proper operating position in its chamber. For domed end tumblers, or for flat end tumblers in a plug which has been flattened in the area of the tumbler chambers, the proper operating position is with the top of the tumblers flush with the surface of the plug. For flat end tumblers with a beveled edge housed in a plug which has not been flattened in the area of the pin chambers, the beveled edge should be flush with the surface of the plug. This will leave the flat end slightly below the surface of the plug, but is compensated for by the corresponding bevel on the end of the upper pin (driver). Progressive cutting of each tumbler along the length of the key bit in this manner will yield the completed original key.

A word of caution is in order at this point. In tumbler positions which require a rather deep cut in the key, it may be necessary to take a triangular shaped file, or a diamond shaped file, and taper the sides of the cut in order to permit the key to be readily inserted and withdrawn without the tumbler holding it up in the plug, otherwise the slow insertion and withdrawal of the key might cause such a tumbler to jump out of the plug.

This, then, is the old-time method by which original keys are made for a pin tumbler lock; it is not, however, the best way to make the keys, because the round bottom cut produced by a round file often is not a configuration which duplicates well. In addition to this factor, there are a number of types of key pins having flattened points with which a round bottom cut in the key does not yield the best results in terms of durability of the finished key. A more modern and efficient method would be to make the original keys in a code key cutting machine, just as if one had a code number to use, and the proper depth numbers to cut them spelled out in the code book; the difference being that one would have to take the key down one depth at a time after getting close to the proper depth. Code keys are also available to establish the standardized depths and configurations of cuts, although they do not generally work as well or as fast as does the code machine.

Careless workmen sometimes try to fit a key by filing down the tops of the tumblers when they come almost flush, this being easier than cutting the key to fit them correctly. This may produce a key which will work properly for a time, but it hastens the wearing out of the lock, ruins the configuration of the top of the key pin, and if the lock contains master key tumblers, it is just about disastrous.

Master key tumblers in a lock which has had the plug filed will tend to wedge between the plug and the cylinder hull as the plug is turned if the master tumblers are rather short, as many of them are. If other locks use the same key, the original key made by such a method will not fit them. It will fit only the lock for which it was produced. Anyone who is later called on to make a key for the lock without having a key to go by, or who is trying to make the key work properly by cutting a new one to standard depths, will have difficulties because the standard depths will no longer operate.

The Impression Systems

Briefly stated, the impression systems are systems of making original keys for a lock where no keys exist by means of causing the key pins to bind at the shear line, and moving the key in such a way as to leave a mark, or impression, on the key blank at the point where key pin touches the key blank. When that tumbler is cut to the proper depth, it will no longer mark the key blank.

In order to utilize either of the impression systems correctly, it is essential to understand a bit of the theory behind the methods used. Since the impression systems were developed for and are primarily used in making keys for pin tumbler locks, the basic theoretical discussion will be oriented in that direction, although it should be remembered that the impression systems may also find a degree of usefulness in the making of keys for some disc tumbler and some lever tumbler locks as well. For this reason, a certain amount of theory is inextricably interwoven with the methods by which it is utilized and will be referred to in context with its application.

The first step in making a key by impressioning is the selection of the right key blank. Of course, the key blank should enter the keyway properly, but in addition to this requirement, one must be sure that the key blank is long enough to adequately engage all of the tumblers for which cuts are to be made. For example, if one has a six tumbler lock and uses a five tumbler blank, the blank would be too short to engage the back tumbler. Additionally, the key blank selected for impressioning must be neither too hard nor too soft, if maximum results are to be obtained. Too soft a key blank will twist and break too readily with the pressures involved in making a key. Conversely, too hard a key blank will not mark readily and makes the location which must be filed unduly difficult to establish.

Any lubrication in the pin tumbler mechanism should then be flushed away with a solvent* which will evaporate quickly and completely. For safety's sake, a non-flammable solvent should be selected, but even this should be used with discretion as the fumes can be toxic in a confined area. It is a good idea to subject locks about to be impressioned to his flushing whether lubricants are present or not,

---

* Methyl chloroform (preferred) or carbon tetrachloride; liquid freon may be used to hasten the drying.

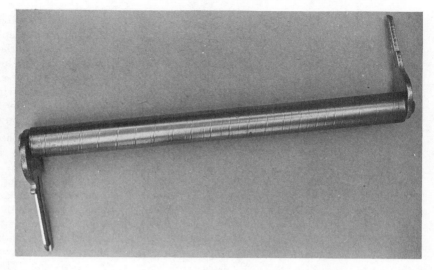

Figure 120. Wiggle system tool with key blanks attached. Screws in opposite ends of the tool are of different sizes to suit different sizes of holes in the bow of the key.

not only to remove lubricants, whether or not observable, but also to clear the keyway of any loose dirt or other foreign matter. It is of the utmost importance that all of the lubricant which can possibly be removed be flushed out of the lock before attempting to make a key by the impression systems, since they depend upon friction of the tumblers. A lock which is clean and free of all lubricants will impression much more readily than one which is lubricated. The lock will dry out quite acceptably while the key blank is being prepared for use.

Most key blanks, as they come from the factory, have a hard glaze on the surface of the edge which contacts the tumblers. This glaze should be removed. A flat hand file of the Swiss pattern type, with a No. 3 or 4 cut, is suitable for the purpose. Two or three light strokes of the file across the top of the key bit is usually sufficient to remove the glaze without cutting into the metal deeply enough to interfere with the shallowest depth of cut on the key bit.

The key blank, thus prepared, is attached to the wiggle system tool as shown in the illustration and the key blank inserted into the lock until it stops at the shoulder of the blank. A strong turning pressure is then applied and maintained while the head of the key is rapped lightly, on both the top and bottom edge, two or three times with a small ball pein hammer. Pressure is then released and reapplied, turning in the opposite direction this time, and the rapping procedure repeated. The key is then removed from the lock and a tumbler cut is made at each point where the key bit has been marked by a tumbler.

It may be that initially every tumbler will mark; if so, all spaces should be filed just a little bit. It may be that only one tumbler will mark; in this case only that one tumbler cut should be filed. It is imperative that no cut be made unless a mark is visible. The proper file to use in making these tumbler cuts is a round Swiss pattern file, preferably 8 in. long, with a No. 3 cut. Although various people will prefer a No. 4 cut or a No. 2 cut for some keys, a No. 3 cut will be found satisfactory in nearly all cases.

The marks by which one files a key by the impression systems may not be what one would expect to find, since the mark is not truly an impression or imprint. Instead, it is a change in the texture of the surface, resulting in changed light reflecting characteristics. For this reason , it appears as a somewhat more shiny spot, more mirror-like than the surrounding area, and only at the point of actual contact with the key pin's tip. Hence it is very small in area. The No. 3 cut Swiss pattern file leaves a great many tiny, almost microscopic striations in the surface of the metal.

When extreme pressure is applied to one spot on the surface by a key pin, those tiny striations are smoothed out, or burnished, into a mirror-like finish. For this reason , it might be more technically correct, and far less misleading, to say that we are making a key by the burnishing method, rather than by the impression method. To assist in familiarizing oneself with the type of appearance presented by the true impression mark on a key bit, it may be well to create a similar situation on a larger scale and observe the effects of the pressure of a smooth object against such a microstriated surface.

In order to obtain an idea of the characteristics of

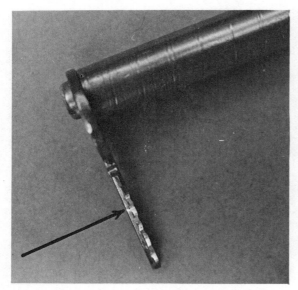

Figure 121. Wiggle system mark on a key. Note the position of the mark in the bottom of the cut and the small, round shape of the mark. It is also noteworthy that marks on the key made by either of the impression systems may appear either more dull or more bright than the surface of the unmarked part of the cut, depending upon the angle at which the light strikes the mark. An oblique light is frequently helpful in locating impressioning marks. At other times a direct light is better, therefore the key should be examined in several lighting positions if the mark is not readily visible.

the marks made by a key pin on a key blank, choose a small piece of brass some 2 in. square and sand it thoroughly, using a piece of 180-grit emery cloth, and keeping the strokes all in one direction so that they do not crisscross. When finished, the surface should be quite clean and have a dull metallic sheen. Take a medium sized ballpein hammer with a smooth ball end, and place the ball end on the prepared surface. While maintaining a heavy downward pressure on the hammer, draw it a short distance across the prepared surface at right angles to the direction in which the sanding strokes are made. Note that the dull metallic sheen has been burnished to an almost mirror finish where the ball end of the hammer has passed over it. This mirror-like finish mark is typical of that made by the key pin in the course of impressioning a lock by the pull system, except, of course, that the mark produced by a key pin will be much smaller.

Next, raise the hammer about 3 in. or 4 in. above the prepared surface and allow the ball end to drop onto the prepared surface, not hitting it a sharp blow, but merely allowing the end to drop. It will

be seen that a similar mark is made: it will have the same mirror-like appearance, and yet not produce an indentation in the metal itself. This is the type of mark which is produced by the wiggle system which we have been discussing. Notice that the mark produced by the wiggle system pinpoints the location of the tumbler, as well as indicates for which tumbler to file a cut.

Having established the identity and location of the marks, one proceeds to complete the cutting of the key in much the same fashion until the key turns in the lock, making cuts for such tumblers as are marking, and making *no* cut where no mark appears; reinserting the key in the lock, turning it, and rapping it. Frequently, just prior to the key turning in the lock, the markings of the tumblers on the key blank will be much heavier and more pronounced than heretofore. This is because the last tumblers needed for alignment at the shear line are already beginning to slip under the body of the hull and the turning of the key itself is producing a downward thrust on the key pin. At this stage, one files with particular care, usually cutting away only enough metal to remove the mark made by the key pin.

It is not necessarily true that the key is completed because it turns. It may be that the key will turn too tightly and drag all the way around as it is rotated. In this event, the notches on the key will show markings where the tumbler is too tight, even though the key turns. Such a tumbler should have its notch filed down until the key turns quite freely in the lock.

It is by no means infrequent in making keys by the impressions system to find that one or more notches have been filed just a little deeper than they actually should be. This situation is manifested by a "jump" in the turning of the cylinder at the time when the key first begins to turn. If this jumping should occur, or a clicking sound is noticed when the plug first starts to turn, it is a good indication that one or more of the tumbler cuts are already filed too deeply. When this situation occurs, the simplest remedy is to duplicate the key as outlined under the procedure for duplicating worn keys. This duplicate may then be touched up a bit in those positions where it is necessary as indicated by tumbler markings on the key.

Impressioning keys by the pull system is somewhat different from the procedure used with the wiggle system. The basic principle is, of course, the same in that the key pin is locked in place by friction, jammed between the upper and lower pin chambers as the plug tries to turn. The key pin,

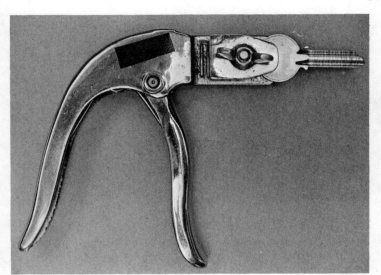

Figure 122. The impression system tool with prepared key blank in place. This tool utilizes the pull system. Remember that the key being cut should be pulled out of the lock about 1/32 of an inch. Pulling the key out further will enhance the tendency of the cuts to creep toward the tip of the key; less pull is likely to result in unsatisfactory marking.

however, is moved against that friction by pulling on the key, rather than by an up and down wiggling motion of the key.

In order to make the key pin produce a mark on the key, it is necessary that the initial cuts, marking the positions of each tumbler, be a bit deeper for the pull system than is necessary for the wiggle system. This is required in order that a means be provided for lifting the key pin sufficiently to leave a readable mark on the key. Since the marking is left on the inclined surfaces which are toward the tip of the key, as shown in Figure 123, rather than being in the base of the notch as is the case in using the wiggle system, there must be enough of a shoulder to receive the mark.

In making a key by the pull system, one marks the position of each tumbler similar to the way in which the locations are marked for the wiggle system, albeit somewhat deeper. The key is then inserted full depth into the lock and a rather strong turning pressure is applied. Then the key is pulled outward from the lock about 1/32 in. The direction of turn is reversed and the process repeated. The key is then withdrawn and examined for impression marks. This turning and pulling of the key may more conveniently be done by the use of an Impression System Tool attached to the head of the key at the eye, as shown in Figure 122, but also may be accomplished by using pliers for turning and a screwdriver under the pliers to pry the key outward.

As has been previously noted, the impression marks will appear as burnished, oblong spots on the inclined surface of the cut which is nearer the tip of the key. One does not file the position in which the

mark is made by the pull system, but the cut should continue straight into the key, regardless of the position of the mark; otherwise, the position of the tumbler cuts would tend to "creep" toward the tip of the key. The marks will be removed to make space for fresh marks, even in the process of filing straight down, *if uniform filing practices are used.*

If the tumbler positions are allowed to creep toward the tip of the key, it may become impossible to make the key by impressioning at all, or the finished key may be ill-fitting and loose in the lock. The tendency of the cuts to creep toward the tip of the key is one of the things with which the beginner usually has trouble when he makes keys by the pull system. In fact, the wiggle system usually provides more accurate positioning of the cuts, for beginners or others.

The impression system tool consists of a clamp for gripping the head of the key and holding it on a special peg, and a plunger which operates along the side of the head of the key to press on the plug. The key is pushed out of the plug as the plier-like handles of the tool are squeezed. The usual mistake in using this tool is pulling the key too far out of the plug. Results are misleading if the key is pulled much more than 1/32 in. Aside from remembering to file the notches straight down, rather than where the mark appears, making a key by the pull system is exactly the same as making it by the wiggle system. File a little bit at a time where the marks appear, and not where no mark appears, until the key turns in the lock. Some corrections or final touchups may be necessary, even after the key turns, just as with the wiggle system.

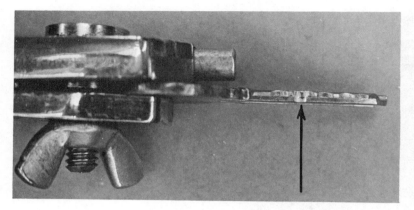

Figure 123. Pull system mark on a key. Note the position of the mark on the inclined edge of the cut which is nearer the tip of the key. The oblong shape of the impression mark is characteristic of the marks produced by this system. Cutting is done in the bottom of the cut, not at the point at which the mark appears.

Key pins having a hemispherical tip will work well with keys properly made by this method, provided that the cuts on the key are kept exactly in alignment with the pins in the lock. If, however, the pin should stop on the sloping side of one of these cuts, the key made by this method will not prove particularly durable, since a small amount of wear will cause the tumbler to drop an appreciable distance and the key will behave badly. For the same reason, such a key will not duplicate well.

Key pins having the truncated cone type of tip are somewhat less compatible with keys made by one of the impression systems, since the impression systems leave a round bottom in the cut which does a somewhat less than perfect job in matching up with the flat point of the truncated cone. Only a relatively tiny portion of the key pin is actually in contact with the key, resulting in an accelerated wear rate. This factor is significant in direct proportion to the frequency with which the key is used.

If the hull, plug, or tumblers of the lock being impressioned are of a comparatively soft alloy such as the softer brasses or die cast, the extreme forces concentrated around the pins may well cause these elements to acquire wear which would not otherwise occur except over a period of months or years. For this reason, if, for no other, the author prefers to make keys by other methods whenever it is practical to do so.

In most modern practice, *mushroom*, or *spool-type tumblers* will be found only in the upper pin chamber. Since it is the lower, or key pin which is gripped between the two chambers in impressioning a lock, this poses no obstacle whatsoever to the use of the impression systems. Some locks of older production, however, do incorporate these spool-type tumblers as a key pin. In those cases, the impression systems may still be used, although there will come a point at which the plug will turn

slightly, and then hang up again. This hanging up may be ignored and impressioning continued just as if the spool tumblers were not present, because all that is necessary is that they be gripped between the upper and lower pin chambers. This is precisely what a spool tumbler does when it is used as a key pin.

If this partial turning occurs and further cutting of the key will not permit the lock to turn, it is safe to assume that cutting on one or more tumbler cuts has already gone too deeply into the key bit and the mushroom driver is what is engaging and hanging up in the two pin chambers. Should this occur, there is no alternative but to begin anew. If impressioning has been carefully done to begin with and no cuts made unless there is a mark, it is unlikely that this will pose any problem.

Two factors frequently are overlooked in the impressioning of a lock, even by a good many men who have become quite adept with one or both of the impression systems. One is failure to take reasonable precautions to ensure that the key made by the impression system is not a master key or a sub-master key. This, naturally, is not likely to occur in residential locks; but in apartment houses, hotels, industrial plants and their subsidiaries and satellites, and in some business locks, masterkeying may be present, and this factor should be taken into cognizance in making a key by the impression system. Second, in any given community, there is likely to be a preponderance of one, two, or three keyways used in various types and functions of locks.

In the majority of pin tumbler locks, the shallowest cut (the very one which poses the greatest hazard to overcutting in impressioning) can usually be determined in advance by going into the keyhole with a very sharp-pointed probe and feeling to see whether or not the junction between the key pin and driver is evident, as it is in tumbler positions 4 and 5 in Figure 124. In most pin tumbler locks, the shallowest cut on the key will permit this juncture between key pin and driver to be visible in the keyhole. In other locks, one, two, three, and occasionally even four tumbler lengths may be observed in this fashion. Lifting the tumblers for visual examination through the keyhole will often disclose this information, and frequently saves considerable time and frustration in the impressioning of a key.

It is, however, necessary to be familiar with the appearances of the various tumbler lengths in the context of a familiar keyway in order to have any assurance which depth of cut is represented by a given visible tumbler length. One knows that it is a shallow cut if the juncture is visible, but practice with a given lock is required if one is to be able to look at a given visible tumbler length and say, "This is one of three visible cuts. The juncture is too far from the top of the keyhole to be a No. 3 cut, but it doesn't look to be far enough away from it to be a No. 1, nor does the key pin look short enough to be a No. 1; therefore, it must be a No. 2 cut."

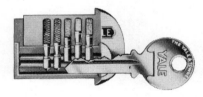

Figure 124. Shear line accessibility. The junction between key pin and driver occurs within the keyway portion of the cylinder plug in tumbler positions 4 and 5 in this illustration, and may be determined by probing as described in the text. Illustration courtesy of Eaton Yale & Towne.

Regular types of disc tumbler locks, but not including the Briggs and Stratton side bar lock, may be impressioned in a similar manner to the methods used for pin tumbler locks. It should be firmly borne in mind that the disc tumbler lock is a com-

paratively fragile mechanism, both with respect to the plug and the tumblers, and either may be quite easily damaged by impressioning, since the internal forces generated in the impressioning process are of a very considerable magnitude relative to the small size and frailty of the parts to which these forces are applied. The torsional strain of impressioning may cause disc tumblers to be distorted in such a way that the rectangular hole through which the key is inserted assumes more the shape of a parallelogram, resulting in damage which is sometimes irreparable. Impressioning of disc tumbler locks is, for this reason, not a recommended procedure, even though it may be successful if carefully done.

The risk of damage to a disc tumbler lock is considerably greater than the risk to a pin tumbler lock, but this risk may be reduced somewhat by filing the cuts with the file held at a rather steep angle across the bit of the key, rather than horizontally in the customary fashion. A key produced with these angled cuts must be duplicated, the duplicate being used and the original thrown away if a serviceable key is to be produced by this method.

In those instances where it is absolutely essential to impression a disc tumbler lock, the wiggle system would be the recommended method, since the forces generated by this method are all along the width and length of the tumbler, rather than across the thickness of the tumbler, as would be the case if the pull system were the one in use. The wiggle system may also be used with some types of lever locks, those which have a sufficiently strong force to apply the necessary tension on the tumblers.

While the impression systems have their limitations, and are certainly no panacea for all key problems, their usefulness is substantial and the cultivation of the art is highly recommended. There are situations where no other method will serve a key-making purpose as efficiently, as effectively, and as quickly as will the impression systems, *provided* that their limitations are recognized and the proper precautions associated with those limitations are taken.

IV
Utility Locks

LESSON 15

Padlocks

Locks covering a wide range of special purposes are usually referred to as utility locks. Some utility locks are designed and intended for only one specific usage, while others cover a wide range of usages and are equally functional in a multitude of applications. Included in this broad range of utility locks are padlocks, desk locks, filing cabinet locks, window sash locks, interlocks (controlling a sequence of electrical or mechanical operations), and locks with numerous other specialized applications.

Whatever the use to which utility locks are put, it is abundantly clear that they are becoming more and more important as the complexities of modern living increase, and the demand for them grows. Vending machines and other coin-operated facilities have been responsible for a major increase in the number of utility locks in use.

Because of the vast numbers of these locks, their importance to the locksmith is beginning to assume major proportions. Although they may differ considerably in outward appearance, the basic principles behind their operation are few, as virtually all of their functions and applications are derived from either the basic locking mechanism (such as a lock cylinder) or from equally basic principles of camming and levering, which are exemplified in a few specific types of utility locks. Some of the locks which illustrate matters of principle are those which we shall examine in this chapter.

Notwithstanding the proliferation of diverse types of utility locks and dramatic increases in their usage, the padlock remains the most common form of the utility lock. The types of keying used in padlocks is not significant here, except as the requirements of the warded, the disc tumbler, the pin tumbler, and the lever tumbler mechanism dictate internal construction changes.

These keying systems have been discussed previously and their principles remain the same, regardless of the type of lock unit in which they are packaged or used. What is of concern to us at the moment is the construction principle by which a particular keying system is adapted to a particular purpose. In the instance of padlocks, it is controlling the locking dogs which secure or release the shackle of the padlock.

CYLINDER TYPES

The locking mechanism associated with padlocks may be said to fall into one of three classes: integral, inserted, or removable. In the integral class of locking mechanism, the warded type of construction may be considered almost typical, although it is not truly a cylinder. The keying mechanism is an intrinsic part of the case of the padlock and incapable of being removed without either losing its identity or destroying the lock.

If any part of the locking mechanism is part of the lock case, it may be said to be of the *integral keyed type*. Most disc tumbler padlocks would fall into this classification, by virtue of having both the upper and lower tumbler chambers cut from the material of the lock case. Pin tumbler locks with the upper tumbler chambers drilled into the material of the case would likewise be considered of integral construction, since only a portion of the keyed mechanism could be removable. This would be true even though the lock plug for either the disc or pin tumbler locking mechanism may be retained in the padlock by means of a screw, filed smooth with the case to conceal it, by a rivet, concealed or not, or by a retainer pin or disc, operating in a ring or groove in plug or case.

Padlocks using a snap ring to retain the plug would be included in this classification as well. Some lever type padlocks also fall into the classification of integral locks, since the levers are mounted inside the case on a pivot pin which is attached to or is part of the case, and the bolt of the lock is often similarly constructed. Not all of the locking mechanism is removable because of these factors; therefore, the lock is of integral construction.

Work on integral padlocks, either repairing or combination changing, is often out of the question.

Figure 125. Integral keying in a pin tumbler padlock. Note the position of capped upper tumbler chambers in the edge of the lock. The screw protruding from the lock at the left is the plug retaining screw and is filed or sanded off flush with the lock case when keying is completed.

Indeed, these locks are frequently so inexpensive that they are seldom worth any more time than it takes to give them a squirt of oil or dry lubricant when they are not working properly. Padlocks with a disc or pin tumbler lock plug in which the plug is retained by means of a screw, a rivet, or a retaining wafer which is accessible, may be serviced by drilling out the retaining pin or screw if sufficient care is taken to avoid damage to the plug in the process.

A number of pin tumbler padlocks with integral keying may be serviced for combination changing or masterkeying through the holes which have been drilled in the padlock case to form the upper pin chambers. These holes are drilled through that edge of the padlock which is indicated by the top part of the keyway. The holes thus formed are capped after the tumblers have been installed by driving in a tight metal plug to wedge firmly in the top of the hole. The excess protrusion of these plugs is filed off or ground off flush with the lock case. This makes them quite invisible to casual inspection if the material of the plugs matches that of the lock case or if the lock case has been painted over after the filing is done.

Such plugs may usually be located by use of a strong oblique light shining across and nearly parallel to the surface in which they are located. If they are not visible by this means, a bit of polishing will usually make them more easily seen. (A fine wire wheel or cloth buffing wheel works best for this.) These plugs can be detected by the hairline joint where the plug and the case of the lock join, or by a slight dissimilarity in the color of the metal. Care must be taken in the drilling to avoid damage to the upper pin chambers.

It is recommended that a drill substantially smaller than the size of the plugs be used. Many of these plugs are quite simple to remove if the drill is allowed to penetrate well into but not through them. A sudden heavy pressure applied to the running drill will often cause its point to gouge into the plug to such an extent that the plug will start rotating. When this occurs, it is only necessary to keep the drill running while withdrawing it slowly, and the plug will come along with it.

This type of construction will often be found even though the lock plug may be removable by means of a concealed screw or by drilling out a rivet. In closing the holes after keying has been completed, one should be just as careful to obliterate all traces of having entered the lock as the manufacturer was in assembling it. Materials for the new plugs should be chosen to match as nearly as possible the material of the padlock case, and care should be used in refinishing to make them as nearly invisible as possible. Smoothening with a rather fine grade of sandpaper is usually best for this purpose, although a No. 2 Swiss pattern flat file will often do nearly as well.

Lever tumbler locks having the keyhole on the flat side of the lock are almost universally of the integral keyed type, with the lock case made in two halves. The tumblers are mounted inside the case, and the case closed (usually by riveting). Such locks are not only tedious to take apart, but difficult to rekey after they are apart, since tumblers for them are not ordinarily available.

The *inserted type* of lever padlock construction usually has the levers and locking dogs contained in a sub-case, with the base of the sub-case forming the bottom of the padlock (often detectable without use of an oblique light). In removing the sub-case from the padlock, it is first necessary to remove the rivets which retain the sub-case in the case proper of the padlock. The next step is to either turn the key in the lock, or to pick it and hold the tumbler mechanism in the picked (or opening) position while pulling the sub-case from the padlock. The locking dogs must be disengaged from the shackle before the sub-case may be removed.

Once the sub-case is removed, the levers are exposed and keys can be made, or the levers may be removed and replaced in different order for a combination change. These locks can contain a surprising number of lever tumblers, usually in matched pairs, so that a key cut on both sides may be inserted either side up and still work. Therefore, care and patience are required if no key is available.

Warded padlocks with keying of the inserted type may be found among some extremely old padlocks utilizing the box-of-wards construction, which are not currently included in production padlocks.

Pin tumbler padlocks of the inserted keying type may have the complete cylinder in the lock case and fastened by means of a retaining rivet or pin, or a screw which has been filed flush. They have these factors in common: the entire keying is done in a complete cylinder outside of the padlock, and the cylinder is inserted into the padlock as a complete unit. Some cutting, or other disassembly of the padlock, must be done in order to remove one of the inserted type of cylinders. For example, some padlocks have a cylinder that is retained in the lock case by a pin which protrudes from the lock cylinder and is held in the protruded position either by

spring tension, or by the action of the cylinder plug. Such a cylinder plug has a recess into which the retainer pin may be pressed when the plug is turned slightly after drilling for the pin.

To qualify as an inserted type of cylinder, it should be removable without irreparable damage to the padlock. Thus, locks having a complete, identifiable cylinder, like the laminated pin tumbler Master padlock, would become locks of the

Figure 126. A removable core padlock. This padlock is used with a removable core of the type shown in Figure 188. The shoulder of the control sleeve shown in that illustration locks behind a complementary shoulder of the padlock, above, except when the control key is in position and the sleeve is turned to retract the shoulder to a position within the body of the core. Illustration courtesy of Falcon Lock Co.

integral type of construction as cylinder removal almost invariably results in destruction of the lock. The differentiation which we are trying to establish here demands that the complete lock cylinder be removable, and that all of the keying be possible and practical, in order to be classed as the inserted type.

Let us now suppose that a lock having the recess in the plug into which the retaining pin may be pressed has a hole already in the lock case above the position of the retaining pin, so that no drilling or cutting is required; instead, it is only necessary to turn the plug, and press the pin with a wire or punch inserted through the pre-existing hole. Such a lock may be said to have *removable keying.*

No disassembly must be done in order to remove this type of cylinder. Cylinders that have a retaining pin accessible through the open end of the shackle, which may be pressed to remove the cylinder from the lock case, would qualify for the removable cylinder classification. Those using a retaining screw in a similar application would qualify, although requiring some disassembly, because the disassembly required would not be in any way destructive.

Locks having *control key*, or the so-called *removable core*, construction as shown in Figure 126 are most certainly in this classification, since the turning of a special key in the lock is all that is required to disengage the removable core from the lock case. The turning of the control key moves a part of the lock core out of the path of a special obstruction or groove in the case, at which time a pull on the control key will remove the cylinder from the case.

This removable core construction is by no means confined to padlocks, as the same removable core, in several of the versions of its manufacture, may be taken from a padlock and placed in a mortise or rim lock cylinder, a cylindrical lock set, or various other types of locks. Removable core construction will be discussed in greater detail in the section on masterkeying, since removable cores are used principally in masterkeying locks.

LOCKING DOGS

Another essential feature of padlocks is the locking dogs, which are used to secure the shackle in the locked position. The locking dogs most commonly found in padlocks are of the bolt or latch type, either single or dual. The single locking dog construction locks only one side of the shackle, usually the side which opens and leaves the case entirely, whereas the dual locking dog construction has two dogs which engage notches in both sides of the shackle. These bolts, whether of the square bolt type, or more like the tapered latch bolt of a mortise lock , engage in notches provided in the shackle and are actuated by the direct action of the rotating lock plug, whether the lock be of pin tumbler or disc tumbler construction. This mechanism is much less commonly used with warded types of padlocks, but a version of it may be found occasionally in lever padlocks.

The locking dogs may be found in two common versions, one of which has the bolts spring loaded, with enough play left in them for the bolt to be pushed back into the opening position by the pressure of the closing shackle, even though the key has been returned to the locked position and withdrawn from the lock. This is the springbolt construction. The obvious weakness of this type of construction is, inferably, that most of the bolts may be pushed back by the insertion of a sufficiently thin shim through the crack between shackle and case, and the shim either pressed or turned to disengage the locking dog from its notch in the

shackle. That is particularly true if the locking dog is of the tapered type. It is more difficult, but not always impossible, if the locking dog is square.

Two locking dogs increase the protection afforded by this type of padlock, but it may still be opened by using the shim in one side only (usually the opening side), and by maintaining a firm pulling tension on the shackle and rapping the same side the shim is already in. The remaining dog may be jarred back out of its notch in the shackle, allowing the lock to open.

Locks having a single dog may be opened by maintaining a tension on the shackle while rapping the side opposite that in which the dog engages the shackle. Frequently, a lock having dual springbolts may be opened by maintaining the shackle tension while rapping the two sides alternately. If this method is used, no shim is required on either side of the shackle.

Another mechanism used to lock the shackle into the locked position is the ball-and-wedge type of locking mechanism. In its simplest form, the ball-and-wedge arrangement utilizes an extension on the lock plug passing between two ball bearings which are held extended into the shackle hole, with room being provided for them by matching notches in each side of the shackle which accommodate them when they are so extended. Turning the key in the lock rotates the plug to a position where its extension has been either flattened or hollowed out. This allows the balls to drop out of the shackle notch and into a corresponding notch in the plug. When the key is rotated back again, the wedging action of the plug forces the balls back out into the shackle. A deadlocking arrangement results when a padlock is thus equipped, therefore there is no means of withdrawing the key when the lock is unlocked. The key must remain in the lock until the shackle is closed in most deadlocking padlocks.

The ball-and-wedge locking mechanism shown in Figure 127 is, however, an exception to the rule. In this instance, a spring loaded coupling arrangement is provided to allow the plug to be rotated back to a position in which the key may be withdrawn without the need for locking the shackle. In this type of arrangement, when the shackle is moved into the open position, the ball on its fixed end engages the notch of the sectional part of the plug extension and holds it in position as the reverse rotation of the key back to the withdrawal position builts up a spring tension on the floating wedge portion of the plug extension.

Pressing the shackle back into the lock case

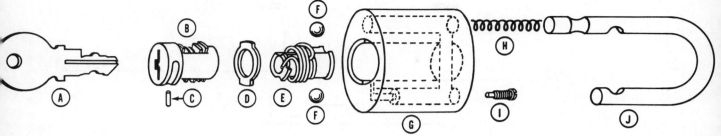

A. Key — Solid Brass, and double-bitted for extra security
B. Cylinder—Ten Brass Blade Tumbler cylinder, contained by "v" spring
C. Cylinder Retaining Pin
D. Retainer Washer
E. Retainer Assembly with Coiled Spring
F. Hardened Steel Balls
G. Case—Solid Brass, Hardened Steel, or Die Cast, depending on model of lock
H. Shackle Spring
I. Retainer Screw
J. Shackle or "Staple"

Figure 127. Ball and wedge locking in a padlock. This version of the ball and wedge mechanism permits the key to be removed with the shackle in the open position. Courtesy of the American Lock Co.

allows the retaining ball to enter its notch in the fixed end of the shackle, relieving pressure on the wedge, which allows the wedge to rotate under the spring tension provided by the withdrawal of the key. This forces *both* balls out into the shackle notches, and the padlock is then deadlocked just as much as if the plug extension were entirely part of, and joined to, the plug, instead of merely being joined in a spring loaded arrangement.

The rotary type of locking dog has been used rather widely with disc tumbler padlocks. In this type of construction, a lever wide enough to engage the sides of the shackle is attached to the rear of the lock plug in either a fixed attachment, or in a spring loaded arrangement similar to that described previously in the ball-and-wedge arrangement. In the latter situation, one end of the lever rests against the fixed end of the shackle when the lock is open, with the spring tension moving the lever into the notches of the shackle when the shackle is returned to the position where the notches will come into alignment with the locking lever.

This type of locking lever does not, of course, withdraw from the shackle notches in a straight line as do most other types; instead, it is withdrawn with a rotary motion, as the lever is rotated with the turning of the lock plug by the key. Such rotary motion demands a somewhat thicker lock case than would otherwise be necessary, since the rotation of the locking lever requires room beside the shackle to contain the lever when it is withdrawn from the shackle notches. If the locking lever is firmly attached to the lock plug in an immobile attachment, then the key may not be withdrawn until the

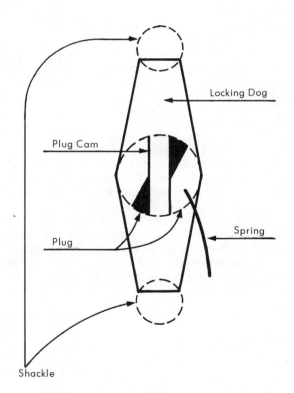

Figure 128. The rotary locking dog of a padlock application. The spring shown on the dog and the loose fit where it joins the lock plug denote the springlatch type of operation. For a deadbolt function, the spring would be omitted and the joining of the locking dog with the lock plug would be made much more tightly.

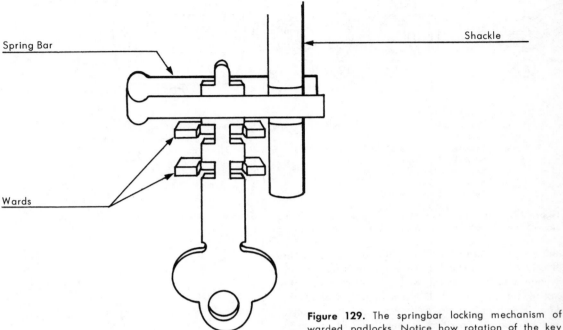

Figure 129. The springbar locking mechanism of warded padlocks. Notice how rotation of the key will edge the arms of the springbar apart.

shackle is reclosed and locked.

The remaining common locking method for padlock shackles is the springbar. The springbar dogging mechanism is restricted in use to the warded type of key construction. The springbar itself consists of a piece of flat spring stock bent into a shape resembling a lady's bobby pin.

The key is customarily of the flat or corrugated steel type, and is inserted between two prongs of the springbar. Rotation of the key expands the springbar, causing the open ends to move apart, out of two corresponding notches which are usually in the fixed end of the shackle. This unlocks the lock. When the lock is in the unlocked position and the key is withdrawn, the springbars rest against the full diameter of the fixed end of the shackle, their tendency to resume their normal shape and position keeping them in a state of tension against the shackle until such time as the shackle is returned to its locked position in the lock. This brings the shackle notches back into alignment with the two ends of the springbar, allowing their spring tension to move them back into the notches, relocking the shackle.

Desk, Drawer, and Cabinet Locks

ALTHOUGH these locks may be found in a wide variety of types, three types are most commonly used, and are, therefore, considered basic. Because our space is not unlimited, and because mechanical construction within some of these types may vary to a considerable degree, our discussion will be limited to three basic types.

ROTARY CAM TYPE

Simplest of the basic types is the rotary cam type. It is commonly found in association with the disc tumbler locking mechanism in a lock shaped much like an 0, with two opposite sides of the 0 flattened to prevent the lock from rotating in the drawer or door in which it is mounted. The cam usually consists of a lever which is attached to the lock plug, either by riveting or by means of a screw. As the lock plug is turned, and the attached lever, or cam, turns with it, either into or out of the notch which the lever enters to lock the door or drawer to the body of the unit in which it is housed.

Some of these locks are provided with an extra set of tumbler chambers so that the key may be rotated 90° and removed. Others have only one pair of tumbler chambers, and the key may only be removed in the 180° increment of rotation. Locks made with the 180° key rotation for key removal may often be equipped with a stop (usually a pin) on the lever, working in a notch in the rear of lock case to limit the rotation of the lock to 90° rotation. When this is done, the effect is that of limiting the withdrawal of the key to those times when the lock is in the locked position. Yet others are provided with only the single set of tumbler chambers, and the stop is arranged to permit 180° rotation of the plug. In such an arrangement, the key is removable in both positions.

A variation of the rotary cam construction has a hook at the end of the rotary cam, which works in an action much like a screen door hook in order to lock a sliding door, or to prevent jimmying of an ordinary drawer or door. Others have a cam reminiscent of the old fashioned rotary type of window sash lock, called a *draw cam* because of its action in drawing two parts tightly together. In any of these locks, however, the cam is directly attached to the rear of the lock plug with such a shape at the point of attachment that both it and the lock plug are forced to turn or remain immobile together. In other words, the hole in the cam, and the stump at the end of the plug on which it is mounted, are of

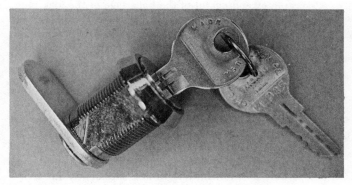

Figure 130. The rotary cam lock. This type of lock has a strong resemblance to the mortise lock cylinder.

corresponding shapes, and other than completely round.

Yet another variation utilizes a pin tumbler lock cylinder, and this necessitates a slightly different type of cam attachment. The cam is retained by the case, and driven by, but not actually attached to, the plug, if the key is to be withdrawn in the unlocked position. A drive mechanism much like that of certain automobile cylinders allows the plug to turn either 180° or 270° before the cam is picked up to be turned. This is usually accomplished by the use of an intermediate driver disc between the plug and cam. After it has turned a few degrees, the plug picks up the disc and starts rotating it; then the disc turns so many more degrees before it picks up the cam. But there are several arrangments of this, all with the basic principle of allowing the plug to rotate so many degrees before the cam is picked up and started to rotate.

Yet another pin tumbler version has the cam firmly attached to the plug, and either must be rotated the full 360° or turned back to the original position before the key is removable. This requires that the door or drawer in which the lock is mounted be locked before the key is removable.

SLIDING BOLT TYPE

The sliding bolt type of lock is similar in appearance to the lever lock in many of the versions in which it is produced. Indeed, such a lever lock is commonly used in this function. Pin tumbler locks are occasionally used, but disc tumbler locks are by far the most common in this type of application.

When a pin tumbler lock is used in the sliding bolt, the bolt is notched very much like a mortise deadbolt is notched, and is moved by a cylinder cam which is either attached to or part of the lock plug. Most commonly, the cam is in the form of a pin extending out of the back face of the lock plug, with the bolt retained in either the locked or unlocked position by means of a spring which engages one of two notches in the bolt at either position. The position of the bolt determines which of them will be engaged by the spring, one notch holding the bolt in the locked position, the other notch holding it in the unlocked position.

At other times the cam of the lock is so arranged that it, too, will engage the bolt and lock it in the locked position. The spring serves merely to maintain the intermediate position of the bolt until such time as the cam has turned sufficiently far to re-engage the bolt notch (since the cam actually serves no holding function at one point in its rotation, as

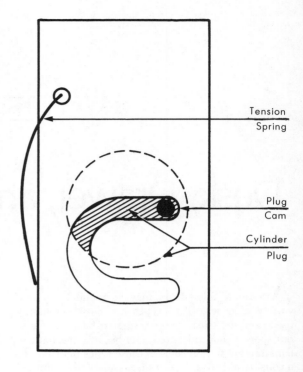

Figure 131. The U notch type of bolt construction permits use of a pin tumbler lock cylinder plug with its 360° of rotation.

may be seen in Figure 131).

The bolt notch in this type of lock resembles a U turned on its side. If the bolt is moving vertically up or down in its motion, it could slip and the plug cam would re-enter the same side from which it had just emerged, unless a spring or other friction generating device is provided to prevent it. Note that the bolt arrangement permits the key to complete a 360° turn in the lock.

An alternate to this type of construction is the use of a deadlocking lever exactly like that employed in mortise deadbolt construction. The sliding bolt construction commonly used with the disc tumbler locking mechanism may be somewhat more simple than that described for the pin tumbler lock, because of the fact that the disc tumbler lock need be turned only 180° in order to remove the key. If a double set of tumbler chambers are used, the lock rotation need be only one-quarter of a turn, or 90°.

The most common method of operating a bolt actuated by a disc tumbler lock plug is by means of a pin which protrudes from, and is usually cast integral with, the rear face of the lock plug. The operating pin protrudes into a square notch in the sliding bolt and remains in the notch in the bolt at all times. The pin moves the bolt up or down as the lock plug is rotated, through the camming action of the pin's eccentricity with respect to the center of the lock plug.

The direction in which a lock of this type is rotated will be governed by which side of the bolt is notched. For example, if the notch for the pin cam is located at the left side of the bolt, as in Figure 132, the plug will rotate clockwise in order to extend the bolt; conversely, it will rotate counter-clockwise in order to retract the bolt. If the notch for the pin cam is located at the right side of the bolt, the directions of rotation would be reversed.

Lock plugs in a lock of this sort are almost universally retained in their housing by means of the retainer disc type of construction. In a five tumbler lock plug, the retainer disc would be located in the position in which one would expect to find a sixth tumbler. The pressure point at which the retaining disc may be depressed fully in order to allow the plug to be removed is usually at the top of the keyway. A few locks, however, have a retaining disc with a pressure point which is 90° right of the keyway. If the lock is a six disc tumbler lock, then the retaining disc would be found in a position in which one might otherwise find a seventh tumbler. Such construction is not particularly common.

In order to remove the plug of any of these locks, it is usually necessary to drill only a small hole (about 1/16 in. or smaller) at the pressure position and to use a straight pick, paper clip end, or a bit of wire, to depress the retaining disc, while keeping a light pulling pressure on the lock plug. A pick hooked in the keyway may be used to exert this pulling pressure if a key is not available.

If the lock is installed in such a way that the position of the pressure point is covered, the hole may be drilled directly through the housing with the lock in either the locked or unlocked position. If, however, the pressure point is not covered by the cabinet in which the lock is installed, the pressure point access hole should be drilled in such a position that the retaining disc is only accessible through the hole when the lock is in the unlocked or partially unlocked position. This avoids circumventing the protection afforded by the plug through removing it from the housing while it is yet in the locked position.

It is always a good idea to drill the hole with the retaining disc turned away from the direction of the drill in order to minimize the possibility of damage.

Also, even with this precaution, it is still essential to use extreme care at the time when the drill is penetrating the housing to avoid damage to the retaining disc, its notch in the plug, or the springs which make it operable. After drilling, the lock plug is rotated until the retaining disc is aligned in such a way that pressure straight into the hole will move

Figure 132. A disc tumbler lock using a notched bolt. If the notch in this bolt were in the opposite side of the bolt, the direction of rotation of the lock plug in locking and unlocking the lock would be reversed. Note the retainer disc behind the last operating tumbler position and the hole in the housing through which it may be depressed.

the retaining disc in a direction parallel to the motion of the releasing pressure of the tool which is being used to depress it.

There are also tumbler locks of this type which utilize a similar means of actuating the bolt; however, these locks have keys which are not removable in the unlocked position.

PLUNGER TYPE LOCKS

Another type of lock commonly found on cabinets with sliding doors is the plunger type. It is installed

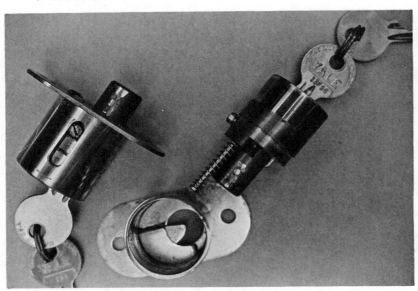

Figure 133. A push type of sliding door lock.

in the outermost door, and the lock unit itself is pushed in so as to extend a bolt out of the back of the lock to engage the other sliding door, thus locking the doors together. These locks are usually of the pin tumbler type, although disc tumbler locks are made in the same manner for the same usage. The locking mechanism used in this type of lock is a true lock cylinder. A retaining screw is used to hold the lock cylinder aligned in the housing. It is found at one side of the lock cylinder, installed in a notch in the housing. This allows the entire lock cylinder to slide in and out of the housing for a distance which is controlled by the diameter of the screw and the length of the notch in which the retaining screw operates.

The locking bolt is, in effect, nothing more than an extension of the lock plug. A spring is provided behind the lock cylinder so that it will pop out into the unlocked position automatically when the key is turned, retracting the bolt into the lock case, and withdrawing it from the other door. Locking is accomplished by the use of a notch in the extension of the lock plug into which a spring loaded lever fits. The spring loaded lever drops into the notch in the plug as shown in Figure 133.

When the plug is turned, its full diameter comes into contact with the lever. Instead of the lever bottoming in the notch, it is forced out to the full diameter of the plug, thus freeing the plug to move outward with the spring pressure, which is applied on the rear of the lock cylinder. This type of lock is not usually locked in the unlocked position; if the key is withdrawn, the lock may be re-locked by pushing on the lock cylinder. When the notch has traveled to the position of the lever, the lever will drop into the notch automatically to lock the lock.

LESSON 17

Filing Cabinet Locks

Filing cabinet locks are almost universally of the highly standardized, oval, cross section type, utilizing a spring loaded bolt which extends out of the oval cylinder body to engage a fixed obstruction in the filing cabinet, holding the lock in the pushed-in position. There is a notch or slot in the rear of the lock, through which there is a retaining screw for attaching one of a variety of locking mechanisms. These lock the cabinet when the lock is in the pushed-into-the-cabinet position. Length and width of this slot may vary, as may the overall length of the lock and the type of the bolt which is used with the lock. The bolt protrudes from the filing cabinet lock, usually from the edge indicated by the top of the keyway, although this bolt protrudes from the opposite edge of the lock in a few alternative situations.

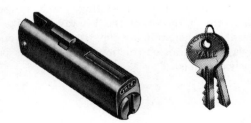

Figure 134. A filing cabinet lock. Variables in filing cabinet locks include: overall length; depth and width of the attaching slot at the rear; the shape of the bolt, its position in the length of the lock, and rarely, a position on the bottom of the lock. Illustration courtesy of Eaton Yale & Towne, Inc.

The bolt is always spring loaded in this type of lock in order that it will extend from the body of the lock of its own accord when the obstruction behind which it locks has been cleared by pushing in (and it can spring out behind this obstruction, which is part of the filing cabinet case). The bolt may be either round or rectangular in cross section. Part of the bolt which remains in the lock is usually other than round, and has a hole in the bottom to accept a compression type of coil spring to provide the tension required to extend the bolt.

There is a notch at the point where the bolt is opposite the cam on the end of the lock plug, and into which the cam extends to operate the bolt. This notch is sufficiently wide for the lock plug to be turned into either the retract position or the extend position without forcing the bolt to extend, and yet is so positioned that it will retract the bolt fully when the key is turned.

Some variations of the bolt may utilize a rectangular cross section bolt on the outside of the lock. The part which penetrates as far as the plug cam is round, with the rectangular portion of the bolt always at least partially within the body of the lock in order to maintain the alignment of the notch with the plug cam. Figure 135 shows the construction of one of these bolts. The mounting slots at the rear of the filing cabinet lock, by which the lock is attached to the mechanism which locks the drawers of the filing cabinet, varies, as previously mentioned, in the depth of the slot and the thickness of the slot; the positioning of the hole for the retaining screw may also be subject to some variation.

169

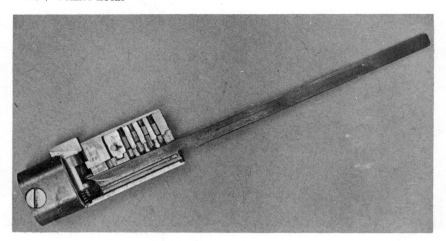

Figure 135. Open keyway construction of filing cabinet lock. Note the provision in the lock plug for use of the long pin method of keyway guarding, although it has not been used here. The straight pick shown in the keyway may be used to retract the bolt without picking the tumblers because of the lack of a keyway blocking pin. This method of bypassing the tumblers may be used with padlocks having a similar construction.

These variations do not occur from one lock manufacture to another so much as from one filing cabinet manufacturer to another; therefore, locks of the same manufacture and keyway may vary in bolt style and mounting slot dimension, even though the keyway may be the same.

Keyways for the pin tumbler lock mechanisms commonly found in filing cabinet locks may be of either *open* or *blind* type of keyway construction. Open keyway construction has a keyway which goes completely through the length of the lock plug with no solid obstruction, as shown in Figure 135. This means that one may work into the keyhole of the lock with a straight pick, lift all of the tumblers out of the way, and directly engage the notch of the bolt with the point of the pick. Exerting a downward pressure with the point of the pick will then pull the bolt, and unlock the lock without picking the tumblers, just as may be done with a number of padlocks utilizing a similar type of construction.

In order to overcome this problem, most manufacturers have resorted to a blind keyway type of construction. Since filing cabinet locks are die cast, a number of manufacturers are using a keyway which does not go completely through the length of the plug. This type of blind keyway completely seals off access from the keyway to the bolt. Others

may use an extra pin; for example, a four pin tumbler lock would actually contain five pins; the fifth, or innermost pin, would be an extremely long one, with no pin chamber above it into which the long tumbler could be lifted, thus effectively barring the access through the keyway to the juncture of cam and bolt. The lock shown in Figure 135 has a provision for such a pin, although none is in place.

Some others use a pressed-in pin crosswise in the keyhole, so positioned as to obstruct the insertion of a pick through the keyway to the locking bolt. The cam which operates the locking bolt is nearly always part of and molded right into the rear face of the lock plug. It is commonly of a half-moon shape, located at the top of the keyway. Sometimes the round pin type of cam is used to operate the locking bolt; this, too, is usually at the top of the keyway.

The only instance of a cam being located at the bottom of the keyway of a lock plug occurs when the locking bolt operates out of the bottom edge of the lock, instead of from the top edge. Either of the shapes of cam commonly used on these lock plugs will usually hold the operating notch in bolt sufficiently far down in the keyway to make it accessible to a straight pick through an open keyway, unless an obstruction is provided as previously described.

V

Environmental Servicing

LESSON 18

The Greater Enemy – Wear or Corrosion?

The material which follows represents an attempt by the author to place in proper perspective some of the problems which confront locksmiths with increasing regularity in the urbanized and highly industrialized society in which we live today. It is limited in scope and the problems encountered are spoken of in generalities. To detail the almost limitless subdivisions necessary to ensure complete accuracy would require that this work fill many volumes and be in language so technical as to make it almost valueless to a working locksmith. Such detail is unnecessary.

The many specific treatments and procedures suggested in this chapter have worked well for the author. It is the author's hope that his experience will enable the reader to reason out solutions to his specific problems, based on the principles shown in the text. To adopt a specific solution to a specific problem and follow it slavishly would represent stagnation. Experimentation, based on a firm foundation of logic and approached with all due regard for safety, will contribute to the progress of the trade and enhance its public image as well.

Some thought about the specific treatments and procedures suggested here will show the reasons behind those suggestions, and those reasons are the important things to remember. Once they are remembered and understood, the treatments become obvious. The author asks the reader to stop occasionally and ask himself "Why?"

There are three things which affect the serviceability of a lock: wear, corrosion, and abuse. We, as locksmiths, can do little to offset the factor of abuse.

That is up to the manufacturer, in the way his product is designed to withstand that type of treatment. We can make sure that it is properly installed and is operating properly at that time, but beyond that, we can really do very little about the way it is used.

WEAR

The factors of wear and of corrosion are, however, a vastly different matter. To deal properly with these factors, we must have at least a basic understanding of what they are, and of their interrelationship; therefore, let us first analyze wear. There are three types: (1) direct abrasion, or cutting; (2) galling, or tearing; and (3) pressure, or hammering.

Direct Abrasion Wear

Abrasion is, as previously indicated, a cutting type of action. As such, it is governed and limited by the same criteria as for any type of cutting action: (1) The hardness of the cut surface in relation to the hardness and sharpness of the cutting edge. (2) The amount of pressure applied between the cutting edge and the cut surface. (3) The area of the cut (the amount of material being removed with one stroke). (4) The machinability, or cutting characteristics of the material being cut.

The first factor, relative hardness of the cut surface, may be illustrated by taking a piece of square aluminum bar and whittling at the edge with a sharp pocket knife, and then whittling with the same knife at the edge of a file tang.

The second factor, applied pressure, may be shown by the same aluminum bar test as above. Little pressure yields a small shaving, while increased pressure yields a larger one.

The third factor, area of cut, may be shown by cutting the flat surface, rather than the edge, of the bar with your pocketknife. It is easy to learn here that taking a wider cut is much more difficult than making a smaller one on the edge.

The fourth factor, machinability, may be illustrated by filing in one place for a number of strokes on the aluminum bar. The first few strokes will come away with the cut surface clean and reasonably smooth, but as the file loads up with the cuttings, the surface becomes more and more ragged, showing gouges, tears, and pushed-up metal. Now try the same test with a clean file on a piece of mild steel. Note that the file does not load up nearly so readily and the cut remains comparatively smooth and clean. The steel does not tend so greatly to have a "cold flow" away from the cutting edge, as is the case with aluminum.

Abrasion by gritty substances such as sand, sandpaper, or emery is much the same as shown in the above examples, except that many minute cutting edges are involved, rather than a single large one.

Galling Wear

Galling is caused by the flow of metal from one moving piece to another, with sufficient bonding action between the two pieces for one to become lightly welded to the other. This may be the result of either overheating or a cold flow from one piece to the other. The illustration in the paragraph on machinability shows such a cold flow gall. Most locksmiths have seen a shaft on machinery which has overheated in the bearing and galled. The appearance of the metal is, in either case, nearly identical. They both resemble a spot weld which has been pulled apart.

A gall, however it may have been produced, is the result of the molecular properties of the two metals at the point of contact. Technically, when a group of molecules acquires an affinity for an external object which is greater than their affinity for adjoining molecules, those exposed molecules will attach themselves to the foreign object rather than maintain their original position. If the power of the movement is greater than the power of the parent metal to halt the movement, a galled, or torn surface is the result.

Pressure, or Hammering Wear

This type of wear is similar to a forging action, except that the term *forging* implies a redistribution of metal from one form into another, more desirable shape. Pressure wear merely redistributes the metal into a less desirable shape. Pressure wear does not usually remove as much metal as it moves aside. Hammering is pressure which is applied with great intensity for comparatively brief periods.

This form of wear may be illustrated by placing a square aluminum bar in a smooth, steel-jawed vise in such a way that the jaws bear directly on the sharp corners of the bar. Tighten the vise as much as possible, then remove the bar and examine it. You will find that the corners are no longer sharp and square, but are flattened. A closer examination will reveal that the metal which once made up the square corners has migrated to the sides, and there is a small, but distinct ridge where this metal has risen above the surface plane of the flat sides of the bar.

It may also be observed that there are traces of the aluminum lying about the area of the vise in tiny flakes, or clinging loosely to the bar. This is primarily the result of two factors: the tendency of pressure to cause rearrangement of the molecular structure of the aluminum and the attempt of the molecular clusters of aluminum to move to areas of lesser pressure. Therefore, some of these molecular clusters escape in a random pattern, to an area where other clusters are not going, and where the support of surrounding molecular clusters is lacking. The effect of this is either free flakes of raw material, or poorly supported ones.

A similar effect may be observed by center punching a flat aluminum or brass surface. By laying a straightedge over the area, it will be observed that there is a raised ring completely around the punch mark. This type of wear may be commonly seen in a mortise lock on the brass hubs or other points where there is a camming action. Note the pronounced ridge where the metal has moved away from the pressure point. Note also the presence in the lock case of small flakes of raw brass which have flaked off the hubs in the process of this flow of metal. Wear of this nature occurs when materials having what is called a "low coefficient of elasticity" are used at points of intense pressure. This simply means that when the metal is distorted from its original shape, it has but a slight tendency to return to that original shape. To illustrate this tendency, take an 8-in. piece of solid copper wire and a similar piece of spring steel and bend them around until the ends touch. When released, the spring steel will return nearly to its original form, whereas the copper will not. Copper has a low coefficient of elasticity, while that of spring steel is high.

CORROSION

There are a number of forms of corrosion which affect locks, although there has been little or no recognition of this fact in the trade. In some cases, corrosion is not even recognized as such and is referred to as something else, if at all, yet it is one of the most damaging factors which limit the useful life of a lock. This section will attempt to place this factor in proper perspective.

There are four principal types (and many subtypes) of corrosion which affect locks. This section will discuss the four principal types only:
1. Direct chemical attack (including atmospheric)
2. Fretting corrosion
3. Stress corrosion
4. Electrolytic corrosion

Direct Chemical Attack (including Atmospheric) Corrosion

Everything is composed of chemical elements, and to a varying degree, all of them will react with the others. This is the basis of all corrosion: a chemical reaction is involved. Common rusting is such a chemical reaction. Iron combines with oxygen in the air to produce iron oxide (rust). The tarnish on copper, brass, aluminum, zinc, or even silverware is such a reaction.

In modern urban living there are traces of many other corrosive agents in the air in addition to oxygen. For example, coal, oil, gasoline, paper, and all burnable rubbish contain carbon, which, when burned, forms carbon dioxide in the atmosphere. Carbon dioxide may then combine with water vapor in the air to form carbonic acid which, in the presence of small amounts of moisture, is more corrosive than either atmospheric oxygen, or atmospheric oxygen with water.

Also present in the atmosphere is sulfur dioxide from the burning of coal and gasoline (both of which contain small amounts of sulfur). This combines with atmospheric vapor and oxygen to form sulfuric acid, a very powerful corrosive. A car battery contains sulfuric acid. Note what it does to the battery cables and metal immediately adjacent to the battery.

Other things, such as household bleaches and cleaning preparations, which may come in contact

or in close proximity with locks also contribute their share of corrosive agents to the seemingly innocuous atmosphere which we breathe. What makes these factors particularly damaging to locks on outside doors is that the lock is almost always somewhat cooler than the air on one or the other side of the door. This allows a small amount of the atmospheric humidity to condense inside the lock; then, as the moisture is re-evaporated into the air, the corrosive agent is either left behind or has already reacted chemically to cause a minute bit of corrosion. In some cases, as exemplified by carbonic acid (from carbon dioxide), portions which have not reacted escape into the atmosphere as carbon dioxide, only to go through the whole cycle of carbon dioxide to carbonic acid, and back to carbon dioxide again.

Corrosion by direct chemical attack is usually recognizable by the uniformity of its effect on the surface which it attacks. It is important to remember that water is the catalyst for this type of chemical attack because it brings the elements closely enough together to interact. Chemical corrosion merits considerable attention not only because of locks in chemical plants and other places which routinely handle corrosives, but because it attacks locks in such prosaic places as business houses and private dwellings as well. It is also frequently associated with stress and electrolytic corrosion, although either of these may occur without association with any other similar type of corrosion.

Fretting Corrosion

Fretting corrosion, like all corrosion, is a form of chemical attack, but with the addition of special conditions, which direct the brunt of the attack to a particular spot and have a particular effect.

Fretting corrosion, as the name implies, shows its effect at the point where one part rubs over another. A film of corrosion (or tarnish) forms on a piece of metal, and when another piece of metal. slides across it, the tarnish is rubbed off and the raw metal is exposed to become tarnished once again. This process, repeated many times, is fretting corrosion. See Figure 136. Note that the corrosive attack is intergranular and loosens the individual grains of metal so that they may be easily wiped away, thus exposing fresh metal to attack.

Just as rust can be wiped off a piece of rusty iron with a soft, dry cloth, so can oxides (in small enough amounts to be called tarnish) be wiped off by a passing piece of metal much more easily than by gouging out bits of the raw metal. In this

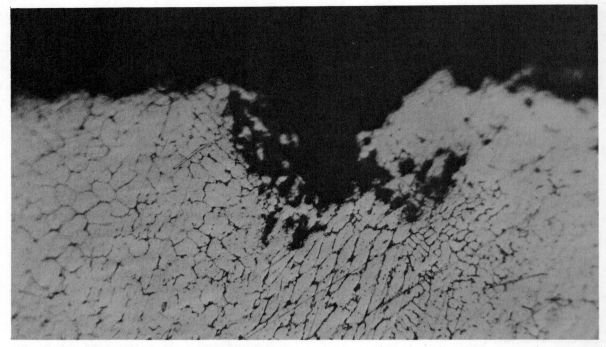

Figure 136. Photomicrograph showing corrosion by direct chemical attack on a metal surface. Note the broad pitting effect and intergranular nature of the attack. It penetrates the spaces between grains of metal rather than cutting through them, and progresses at random as the granular structure dictates.

instance, corrosion occurs because there is a bright, shiny surface of raw metal already exposed to the corrosive agent without that agent having to work its way down through a layer of oxides. This type of corrosion always presents a shiny, smooth *uncorroded* look simply because the protective film of tarnish is rubbed away as quickly as it is formed. This is what we usually look for, and we refer to it as "worn smooth."

We have used the example of wiping away rust with a soft, dry cloth to illustrate that an oxide film wipes away much more easily than raw metal. You may demonstrate to your own satisfaction that corrosion is a continuing process and that oxides are always in the process of formation. Take a piece of aluminum and shine it as brightly as you can, using any method you prefer; wash it well to remove all dirt, oil, or other foreign matter; rinse well and dry. Put it aside for a week, then repeat the process on just half of the surface previously shined. The newly shined surface will appear brighter than the rest of the surface.

This type of corrosion frequently shows up in a mortise lock at the spot where the tail of the latch bolt rubs the lock case. Although fretting corrosion is not limited to that particular location, it is limited to that type of circumstance (where parts rub together).

Stress Corrosion

The term *stress corrosion* refers to the fact that chemical attack on metals is most intense at the point at which the molecules which make up the metal are under the greatest mechanical strain. The situation may be likened to sneaking up on a weight lifter with a 200-pound weight hoisted over his head and jabbing him with a pin. The results are apt to be somewhat catastrophic.

The exact mechanico-chemical process by which stress corrosion occurs cannot be demonstrated by step-by-step observation, because it takes place in the internal structure of the metal, in the intermolecular spaces. It seems a fair assumption, however, that in the process of putting a strain on metal, the side which must elongate itself in order to prevent breaking must increase its intermolecular space.* Stated in another way, since the size of a molecule remains constant, the only way in which

---

* Intermolecular space, not intergranular space. Granules are clumps and groupings of molecules. If a stressline goes through a granule rather than around or beside it, that is the path which stress corrosion follows.

stretching can occur is by changing the distance between the molecules.

This increase in the intermolecular distance will allow corrosive agents to creep in between the molecules themselves. A wedge of corrosion is formed, with the corrosion product (or oxides) expanding in volume, forcing adjoining molecules yet farther apart, and exposing more fresh molecules to corrosive attack. The final result is a crack or break following stress lines and showing an oxide film all the way through the crack (except for a comparatively small portion which breaks clean and bright when the strain has exceeded the strength of the metal to resist).

This type of break is often referred to as *crystallization,* but actually it is a form of corrosion. A true crystallization break is really a comparatively rare thing and the entire broken surface is bright and lustrous, as one would expect a crystal to be.

The points in a lock where one might expect stress corrosion to occur are primarily in the cast iron case where the stresses are of the thermally-induced variety (the result of imperfect cooling of the hot casting); or in the springs, where the stresses are mechanically induced by the bending process in forming them or by the strain they are under in performing their function. In any case, however, the break will always appear at the point of greatest stress.

Electrolytic Corrosion

Electrolytic corrosion is a relatively uncommon type of corrosive attack because of the specialized conditions which are required for it to occur. All of the following conditions must be present:

1. Two or more different metals.
2. Moisture to provide the electrolysis medium.
3. A chemical compound in the water to convert it into an electrolyte.

Electrolytic corrosion occurs primarily at the junction, or point of contact, of two dissimilar metals. This is the most important clue to the recognition of its occurrence. Only one of the two metals involved will present a rough, pitted appearance as if dipped in a strong acid at the junction. This appearance diminishes rapidly as the distance from the junction increases. The other metal involved will present an almost untouched appearance; one piece may become locally plated with metal from the other. The plating phenomenon is more common when one of the metals is copper or a copper-bearing alloy. To obtain a battery effect, one metal serves as an anode; the other as a cathode.

A common misconception about electrolytic corrosion is the generally prevalent idea that a large amount of water is required. Some water is needed, to be sure, but the amount is much less than commonly believed. The electrolyte of a common flashlight battery or dry cell has only enough water to form a paste. It should be remembered that the less water present, the greater the concentration and strength of the solution which is the electrolyte. This will cause the activity of the electrolyte to become intensified and somewhat more localized.

The chemical agency which converts the moisture (water) present into an electrolyte may be almost

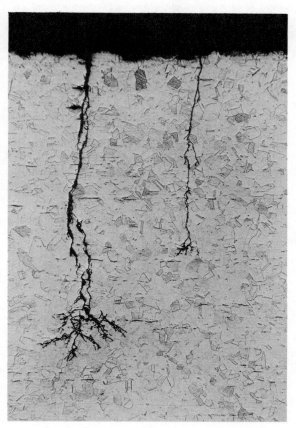

Figure 137. Photomicrograph showing stress corrosion cracking. Note that the cracking rarely bypasses the grain structure of the metal, and cuts through in a roughly straight line. The 'tree root', at the bottom of the cracks, is the characteristic form taken at a given point when maximum cracking occurs. The metal would have cracked in two if it had been thinner. Every locksmith has seen similar cracking in flat lock springs. This particular photo of stress corrosion cracking was chosen to illustrate how all stainless steel metals are subject to corrosion of this kind.

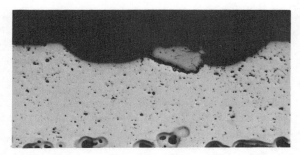

Figure 138. Electrolytic corrosion in aluminum alloy sheet. The irregularly shaped particle at top center is an iron particle inadvertently rolled into the sheet during fabrication. The iron is virtually untouched while the aluminum surrounding it is severely corroded by electrolysis.

anything; an acid, such as hydrochloric (muriatic), acetic (vinegar), or a mixture of two or more acids; a base (or alkali), such as sodium hydroxide (lye), or ammonium hydroxide (ammonia); a salt, such as sodium chloride (table salt) or sodium bicarbonate (baking soda). These are inorganic salts. Another class of salts which may form an electrolyte are the metallo-organic salts such as sodium stearate (soap) or the sodium aryl sulfonates (detergents).

Electrolytic corrosion is usually associated with the other types of corrosion, and though covering a small area and not too common in occurrence, it may have a locally severe effect. It is often of consequence in the case of small parts to which springs are attached, or where contrasting types of metal are in contact. It may also be important where extremely close-fitting parts come together.

Electrolytic corrosion is governed by the following factors:

(1) The composition of the contrasting metals; (2) the amount of water present; (3) the amount and nature of the electrolytic agent present; (4) the quality of the electrical contact between the dissimilar metals.

It is readily apparent that much of what we have referred to as wear, crystallization, and various other malfunctions which beset locks, are actually various types of corrosion. It is a rather sobering fact that corrosion, even in "normal" environments, does more damage to locks than wear. The lock manufacturer could greatly reduce this factor by the judicious use of appropriate plastics to replace many of the lock parts which are presently made of metal. But the locksmith can also do much to reduce the problem to more acceptable limits.

Diagnosis, Cleaning, and Preparation

T‌HE first step in the diagnosis of the ills with which a lock may be afflicted should be taken while the lock is still installed in its "in service" position. The following tests have been found extremely useful in avoiding guesswork and saving time:

1. Is the hardware held securely in place on the door?

2. Do knobs retract the latch properly and return to the neutral position when turned and released? In the case of locks having swivel spindles, can the knobs be turned in opposite directions simultaneously, released, and return to the neutral position?

3. Does the latch bolt retract and extend easily and fully?

4. Does the auxiliary latch retract and extend easily?

5. Is the deadlocking device properly timed?

6. Can the deadbolt be extended and remain extended even though a light end pressure be applied to the bolt while turning the key?

7. Does the key begin to turn easily, without jumping or excessive tightness?

8. In closing the door, do both latch bolt and dead bolt enter their respective pockets in the strike properly without undue force and without either of them showing signs of dragging on the top or bottom of its pocket in the strike plate?

If the answer to a question is no, you have an excellent clue to the location of the trouble. At least it will indicate the portion of the lock in which the difficulty may be found.

Having made these tests and noted the results, the lock may be removed from the door and its cover removed. The interior of the lock should then be examined before any internal parts are removed or disturbed. This examination may be rather superficial, seeking only broken or bent parts and signs of wear and corrosion which might not be so readily apparent even after the cleaning process has been completed. After this examination, the lock may be disassembled for cleaning. A very effective and speedy cleaning set-up consists of three solutions: (1) carburetor cleaner; (2) solvent; (3) methyl chloroform (preferred) or white gasoline to which sufficient carbon tetrachloride has been added to make it noninflammable.

If there are heavy deposits of congealed grease, it may be desirable to remove some of it by brushing with a solvent before immersing in the carburetor cleaner. This will speed up the cleaning process and prolong the life of the solution. Other than this, little or no brushing is required for the duration of the process.

The usual cleaning process requires only a few minutes in the carburetor cleaner, but longer immersion may be used to remove paint, lacquer, and extremely stubborn dirt. Next, immersion and agitation in the solvent is necessary in order to remove the carburetor cleaner. The parts will come out of this solution reasonably clean, but covered with the oily film which the solvent and traces of the carburetor cleaner leave. The parts are then immersed for a few minutes in the methyl chloroform to remove the oily film and provide quick drying.

The lock cylinder should be held under the

surface of the methyl chloroform and the key run in and out several times to make sure that all the previous solutions have been removed. The methyl chloroform will evaporate quickly and cleanly, even from the spaces over the pins where the springs are located.

Observe These Precautions

1. This process should be accomplished in a well ventilated area. The fumes from these cleaning agents are quite strong and the operator may become nauseated or unconscious unless adequate ventilation is provided. Even fatalities could result in extreme cases, but there is no danger if this simple precaution is observed. One should also be sure that the containers are covered when the operator is not actually working in the solutions.

2. Do not allow any of the solutions to contact eyes, mouth, or skin. These solutions, especially the carburetor cleaner, can produce serious effects if handled carelessly. The carburetor cleaner should not come into contact with your hands at any time, nor should parts with the cleaner remaining on them be handled, even with rubber gloves. The rubber gloves would be quickly ruined. A cleaning basket is provided with the carburetor cleaner and it should be used all the way through the cleaning process. Eye protection (or face protection such as a grinding shield) is absolutely essential to using this process safely.

3. Do not use this method for cleaning plastics unless the plastic has been tested to determine if it will withstand all the solutions. In case of doubt, clean all plastics with one of the dishwashing liquid detergents, and rinse carefully with clean, warm (not hot) water.

4. Remember that cleaning solutions are chemicals and may react, even to the point of explosion, if the wrong ones are mixed. Those listed in this process are safe in this respect (although not fire proof), but this is something to remember if other cleaners or solutions are tried.

Observe these simple precautions and you will find that this is a quick and easy way to perform all cleaning operations.

With the cleaning completed, a more detailed inspection of the parts may be made. Rough surfaces causing any abrasion may be smoothened; broken, bent, or unserviceable parts may be welded, repaired, or replaced; hubs may be smoothened; parts may be deburred and buffed if necessary; case and cap may be repainted and trim re-lacquered if needed. It is at this stage that all parts should be made operational and as nearly in their final form as possible before beginning assembly or lubrication. Sharp corners should be dressed off and all bearing surfaces made smooth. If areas of corrosion are present, all traces of corrosion should be removed by vapor blasting (if this method is available), by sand blasting with very fine grit sand, by wire brushing (very fine wire), or by buffing.

After all of these things have been done where needed, and the lock is clean, dry, and free from grit or oil and grease, lubrication and corrosion protection may proceed in accordance with the conditions to which the lock is exposed and the construction of the lock.

LESSON 20

Locks in Service

THE term *normal indoor service* may be defined for the purposes of this text as referring to those locks which are installed in locations that are not exposed to the elements, extreme humidity, or traffic requiring an extreme amount of opening and closing of the door. The temperature range should always be within reasonable limits on both sides of the door; hard slamming should not be required to close the door; and the stresses imposed on the unit should not be in excess of those which one would expect that particular make and model to withstand over a considerable period.

NORMAL INDOOR SERVICE

Examples of this category would be most locks in residential service other than entrance locks, office locks (except where extremely heavy traffic is involved), storage room and janitor closet locks, apartment door locks and hotel room locks. Not included in this category would be locks on doors subject to heavy slamming or heavy traffic, and doors in the path of a pressurized air system.

Key Locks

Lock cylinders, whether disc or pin tumbler, and lever locks, may be lubricated for this class of service in a satisfactory manner with a mixture of three parts of graphite and one part molybdenum disulfide (by volume). The lubrication should be carried out while the cylinder is removed from the lock, so that it can be turned upside down and the lubricant worked in around the tumblers with the key. If the cylinder does not work consistently while

in the inverted position, it is usually indicative of a weak or broken spring, or of grit remaining in the tumblers.

If it is necessary to open the cylinder or remove the tumblers, it is much better that the lubricant be used on top of the pins (if a pin tumbler cylinder). This will allow the lubricant to trickle down around the pins slowly as needed for a long time. Care must be used not to fill the space above the pins so full of lubricant that compaction occurs when a key blank is inserted in the cylinder.

Mortise and Rim Locks

Mortise and rim locks of the springlatch, deadlatch, or deadbolt varieties may be lubricated in most cases with Wonder Mist. If the lock case or working parts are die cast, the preferred lubricant would be moly spray, because of the peculiar ability of moly to cling to that type of surface. If the case is a particularly rough casting which cannot be readily smoothened, the preferred lubricant would be a silicone grease such as the K-2 spray lubricant, which may be dusted with the graphite and moly mix used for cylinder lubrication. Silicone is preferred to petroleum jelly, because it has much less of a tendency to run when the weather turns hot.

Cylindrical and Tubular Locks

Cylindrical and tubular locks in normal indoor service may usually be lubricated with a spray type silicone grease, such as the K-2 silicone lubricant, without disassembling beyond the extent required in cleaning. Good penetration may be obtained by

181

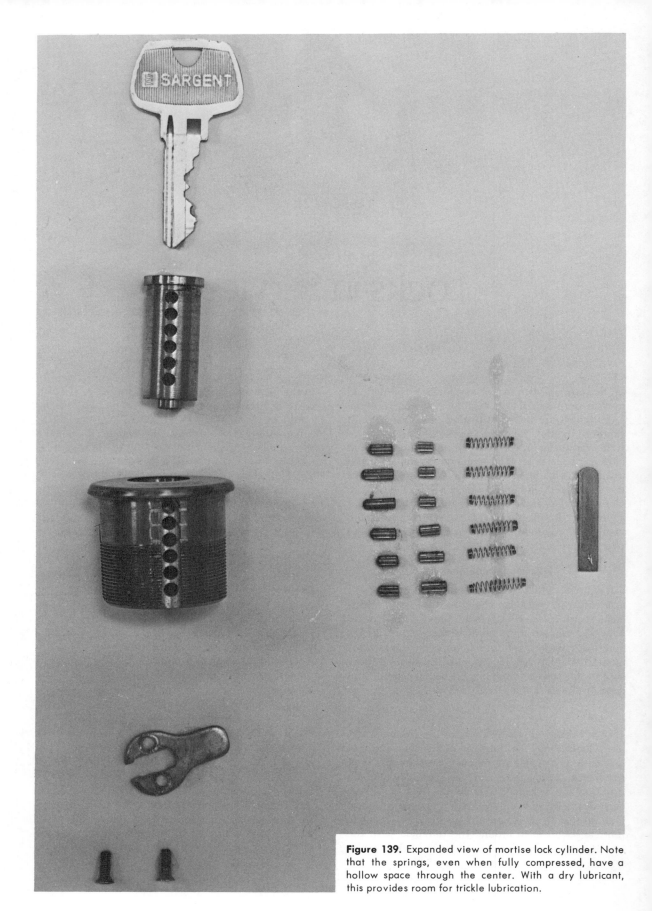

Figure 139. Expanded view of mortise lock cylinder. Note that the springs, even when fully compressed, have a hollow space through the center. With a dry lubricant, this provides room for trickle lubrication.

this method. It is particularly valuable in the lubrication of those latch and bolt assemblies which are sealed together at the factory.

Mortise and Rim Panic Exit Devices

Panic exit devices in both major categories may be lubricated with an overall coating of Wonder Mist or moly spray for the case and all parts, with an additional coating of silicone grease (K-2 or bulk) *except* at the points where a part is driven only by a small or comparatively weak spring, or by gravity drop, or by a combination of these.

No wet lubricant of any type should be used on any part so lightly driven. Coagulation of the lubricant or collection of a small amount of dirt or lint would render such parts inoperable. Wonder Mist or Slipspray may be used on exposed parts to avoid a greasy film or the extreme dirtiness of graphite or moly.

Disc and Pin Tumbler Utility Locks

Utility locks used on desks, files, storage cabinets, and the like, may be lubricated by dusting the graphite-moly mixture into the keyhole and working it into the mechanism with the key or a key blank. The lock usually does not need to be disassembled for this treatment.

Lever Locks

Lubrication of lever locks may frequently be done without disassembly by removing them from the door, dusting the graphite-moly mix into them, working the key thoroughly with the lock held in a number of positions to obtain maxium coverage, then blowing out the excess.

As an alternative, white gasoline or methyl chloroform may be used both to remove the excess lubricant and to carry it into the more-difficult-to-reach parts. If a liquid is used, it should be of a type which will evaporate completely. Jail type lever locks should always be completely disassembled and all parts lubricated with either Wonder Mist or Slipspray, with an additional coating on all working parts and the case interior of moly spray.

Combination Locks

The time-honored method for the lubrication of combination locks consists of wiping all parts with a light coating of petroleum jelly, in the hope of achieving some lubrication without piling up too much jelly in any one place and cause parts, which are supposed to be independent, to drive together instead of separately. To achieve the proper lubrica-tion of a combination lock is actually a much more complex, although less time consuming, matter. It is necessary to consider each part separately, according to its own requirements.

The following represents a new approach to this problem. It has been carefully tested over a period of years and has been found superior to both the oldtime method and all other methods tested. A lock in good condition so lubricated may reasonably be expected to stand up well for years of heavy service. In the interest of simplicity, the various parts of a combination lock will be discussed without regard to their placement in the lock, but with regard to the method by which they are lubricated.

Group A Combination Lock Parts

This group of parts may be lubricated by spraying with the K-2 lubricant in a sufficient quantity so that, when it is subsequently dusted with graphite-moly mix, the resulting mixture will retain good viscosity characteristics. Ideally, this is a little more viscous than the K-2 lubricant alone. The mixture should be carefully applied in order to avoid excessive spreading over the non-contact surfaces of adjoining operating parts yet to be lubricated. The points to be lubricated are:

1. Tumbler post and the adjacent surface on the case or cap which comprises the bearing surface for the bottom tumbler.
2. Bottom tension washers (between bottom tumbler and case or cap).
3. Bearing surface in dial ring or tube, and bearing surface for the driver. (Note that this does not include the dial bearing or the bearing portion of the driver.)

Group B Combination Lock Parts

This group of parts may be lubricated by spraying the bearing or working surfaces of the parts with Slip.* A light coating is all that should be used and no harm will be done if the entire part is coated. The spray should be kept off painted surfaces, to avoid softening paint which might be vulnerable to this sort of effect by silicones. Lubricate:
1. Wheel.
2. Fly.
3. Stationary spacer washers.
4. Tumbler retainer.
5. Driver.
6. Lever or fence and its retaining screw.

* Not to be confused with Slipspray, a duPont product. Slip is a light silicone oil in a spray can, and is the product of Certified Laboratories Inc., Fort Worth, Texas.

7. Butterfly lever in center of dial. (Used to control cam slide in certain types of manipulation resistant locks.)

Group C Combination Lock Parts

The bolt and the area of the case in which it slides should be lubricated with Slipspray, followed with a light spray of Wonder Mist, in order that these parts may have some lubrication and yet not collect dust and dirt as they would with wet lubricants.

The relocking device is intended to function only one time and hence requires little or no lubrication. At the most, a light spray of Wonder Mist might be used if the device does not seem to function properly without it.

The dial shaft and dial bearings may be lubricated with Slipspray. This contributes a desirable silky feel to the turning of the dial.

HEAVY INDOOR SERVICE

The term *heavy indoor service* may be defined for our purpose as "neither exposed to the elements nor in extreme humidity, and having a reasonable range of temperature limits" on both sides of the door. It would, however, include units which might be subject to one or a combination of the following conditions: harder than normal slamming required to close the door; heavy traffic which requires the lock to be operated with excessive frequency; a pressurized air system which exerts undue pressure, causing the lock to operate as if excessively tight, or causing the door to slam. Examples of this category would be heavy traffic office doors, public rest room doors, corridor doors in apartment houses or public buildings, photographic dark room doors, or heating and ventilating system doors.

Disc or pin tumbler lock cylinders and lever locks may be lubricated for this class of service with the graphite-moly mix, exactly as for normal indoor service. (See also the section dealing with pin tumbler utility locks in this class of service for the technique to be used in the event key use is extremely frequent.)

Mortise and Rim Locks

Mortise and rim locks in heavy indoor service may be lubricated for maximum service and dependability with the K-2 silicone spray lubricant and a dusting of graphite-moly mix, except for the exposed parts, which may be lubricated either with Slipspray or Wonder Mist. It is important in this class of service that the bolt and latch both be lubricated with a nongreasy, clean lubricant in order to ease the strain on these parts and the

internal mechanism, and to serve the additional purpose of reducing the fretting corrosion so commonly found at these points.

It is also important to remember to lubricate the strike plate and the contact surface of the latch and bolt pockets in the same way as the latch and bolt themselves. This will contribute greatly to the smoothness and ease of operation of the entire unit. If the lock contains parts of the gravity drop or lightly spring driven mode of operation, those parts should be lubricated with Wonder Mist or Slip in order to assure proper and positive functioning.

If the lock case is of a particularly rough casting, it is well to give it an initial coating with Slipspray before the other lubricants are applied. This helps to fill the small pockets which cannot be smoothened out, and contributes a silky feel to the operation of the lock.

Cylindrical and Tubular Locks

Cylindrical and tubular locks in heavy indoor service may be lubricated with the K-2 lubricant, followed by a dusting with graphite-moly mix of the case and the parts which it houses. Latch tube internal parts may be lubricated without disassembly with Slip by using the spray extension which can be obtained with this product; and the same treatment may be used for knob bearings, making disassembly unnecessary except to repair or replace broken or damaged parts. Lubrication of latches, bolts, and strike pockets with Slipspray or Wonder Mist is, as in the case of mortise and rim locks, important both to ease of functioning and to durability.

Mortise and Rim Panic Exit Devices

Lubrication of panic exit devices in heavy indoor service should begin with the careful smoothing of case and parts at all friction points, followed by an initial coating of Slipspray at all points where extremes of pressure will occur, whether torsional or sliding in nature. The next step is an overall coating with moly spray for case and all parts (except those which are exposed) and the point where the deadlocking lever comes in contact with the latch bolt.

The final application is a fairly generous one of K-2 lubricant covering the case interior and all working parts, except the point where the deadlocking lever contacts the latch bolt; and with the additional exception that none of this lubricant should be used on lightly driven or gravity operated parts, or on the points of the case with which they make contact in their course of operation. Latch

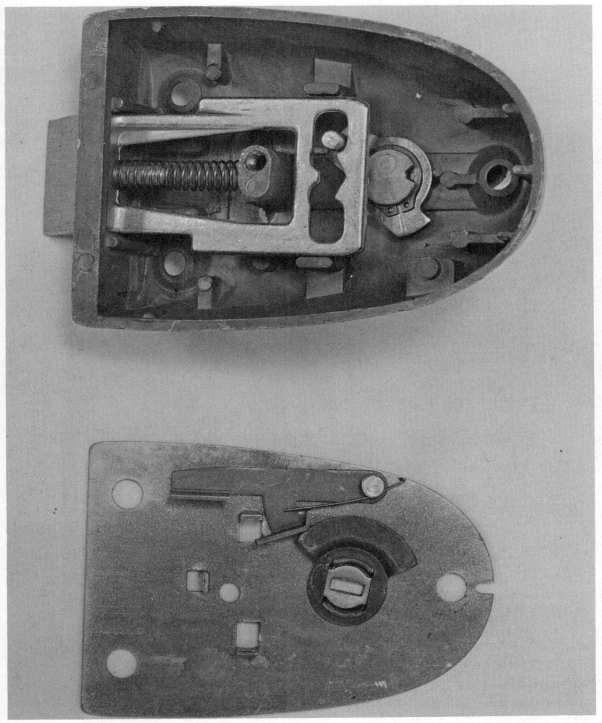

Figure 140. Rim deadlatch.

Figure 141. Mortise panic exit lock. Note that deadlocking lever is of the gravity drop type with light spring supplement.

bolts, auxiliary latches, strike plates, and strike plate pockets should be lubricated with Slipspray.

Disc Tumbler Utility Locks

Disc tumbler utility locks may be lubricated by first cleaning carefully with a volatile solvent, then spraying Slipspray into the keyhole and quickly working the key in and out of the lock for maximum distribution. Next, turn the key to the opposite position and repeat the Slipspray application. A final dusting with graphite-moly mix, with the lock in first one position and then the other, completes the process.

Pin Tumbler Utility Locks

Pin tumbler locks in heavy indoor service in those situations where normal lubrication by working graphite-moly mix in through the keyhole has proven inadequate may usually be lubricated by using the following method: The first step is complete disassembly of the cylinder and careful cleaning of all parts. The pins in a pin tumbler lock tend to wear, through fretting corrosion, at the point of contact with the key into a chisel-shaped point rather than a round one. These pins, where found, should be discarded as they tend to come up to a different level each time a key is inserted.

Next, check the level of all key pins in the plug while the key is inserted. If short ones are found, check the hole they fit into to be sure it is not egg shaped around the diameter of the pin. If it is egg shaped, the plug should be replaced if possible. If no replacement is available, be sure to use a long pin in that hole in order to minimize the tipping effect, even though it means a combination change.

All springs should have good tension; if existing springs do not have adequate tension, they should be replaced. Driver pins should, when dropped into the plug on top of the key pin with the key withdrawn, extend about .020 in. into the upper portion of the cylinder. Normally, this would be about .025 in. above the top of the plug. In cylinders where the drivers are the same length, regardless of key pin lengths, this would mean replacing all the drivers. This should be done regardless of factory method, in order to make the spring pressure uniform and obtain maximum performance of the key.

In some cylinders, where the spring chamber is deeper than ordinary, it may be necessary to make the drivers level out somewhat higher above the surface of the plug so that the key pins cannot be forced all the way up into the spring chamber.

Lubrication consists of spraying plug, cylinder, pins (driver and key), and pin holes in both plug and cylinder lightly with Wonder Mist and dusting lightly with graphite-moly mix when thoroughly dry. Add a small amount of the graphite-moly mix above the driver in the spring pocket if this is accessible, otherwise, the mix may be added on top of the key pins with the key withdrawn before the plug is inserted into the cylinder (with the key out).

Other parts of the lock may be lubricated with Wonder Mist and graphite-moly mix in the same manner as the plug. A small amount of graphite-moly mix in the keyhole after final assembly will help reduce key wear, and this is often an important factor in the overall serviceability of locks in this class of service.

Lever Locks

Lever locks in heavy indoor service should, whenever possible, be completely disassembled and cleaned. The levers should then be checked for rolled edges at the point of contact with the key, and any rolled edges that are found should be removed. Many lever locks have an inspection hole at the point where the fence enters the gate in the tumblers. If the lock being worked does not have one, it is well to drill one at this point, provided that this can be done without compromising the protection offered by the lock.

The lock case, cap, tube, and plug may be lubricated by spraying with Wonder Mist, followed by a light dusting of graphite-moly mix. Tumblers should be sprayed on both sides with Wonder Mist. Then place a small amount of the graphite-moly mix on a piece of paper, and rub the tumblers in it until both sides are shiny. The lock may then be reassembled without further lubrication. The fit of the key should next be observed through the inspection hole and a new key cut if tumbler or key wear is sufficient to cause the old key to fit improperly.

Normal Outdoor Service

The term *normal outdoor service* as used here may be defined as referring to locks which are installed in entry doors and which are not exposed to an unusual amount of traffic or hard slamming, and the stresses imposed on the unit are not in excess of those with which that particular type of unit is intended to cope. Extremes of weather conditions either do not exist, or some degree of protection is provided such as a porch, overhanging eaves, or a screen or storm door. If the lock has a cylinder with a keyhole which has no blind end at the back, and no storm door or other protection from the elements, it should be serviced as noted in the climatic extremes section.

Examples of locks in this class of service would include most locks on residential entry doors, some low traffic business entry doors, and sheltered entry doors of low to medium traffic warehouses, industrial plants, and public buildings. Not included in this category would be locks subject to other than occasional heavy slamming, heavy traffic, prolonged subzero temperatures, or temperatures colder than 20° below zero during any routine operating period.

It is especially important that lock housings and all associated parts be clean and free of oil and grease. Oil or grease may tend to either run or congeal with the temperature variants commonly found in this class of service unless the oil is of the multiple viscosity type or the grease is of the proper weight of silicone. If a silicone grease of a new type

Figure 142. Mortise deadlatch.

is used, it should be one which is labeled as neither melting nor freezing within the *extremes* of the anticipated temperature range.

Pin or disc tumbler cylinders may be lubricated for normal outdoor service with the graphite-moly mix in exactly the same manner as for cylinders in normal indoor service, except that if the cylinder must be disassembled for any reason, the tumblers, *and only the tumblers*, should be lightly sprayed with Wonder Mist prior to both reassembly and the graphite-moly lubrication.

Mortise and Rim Locks

Mortise and rim locks in normal outdoor service may, in most cases, be lubricated with an initial application of Slipspray followed by the K-2 lubricant. A light application of Wonder Mist lightly dusted with graphite-moly mix should be substituted for the K-2 silicone lubricant to lubricate those parts which are lightly driven or of the gravity drop type. Exposed parts such as latch bolts, bolts, strike plates, and bolt pockets may be lubricated with Slipspray. If the white residue which the Slipspray leaves is objectionable, it can be made invisible by an application of a light spray of Wonder Mist.

Cylindrical and Tubular Locks

Cylindrical and tubular locks in normal outdoor service may usually by lubricated with the K-2 lubricant without disassembling beyond the extent required in cleaning, as far as the case and the parts which it houses are concerned. The knob bearing and the surface of the lock housing upon which it runs should be lubricated with Slip, as should the parts inside of the tube. Latch bolts, bolts, strike plate, and strike pockets should be lubricated with Slipspray, Wonder Mist, or a combination of both.

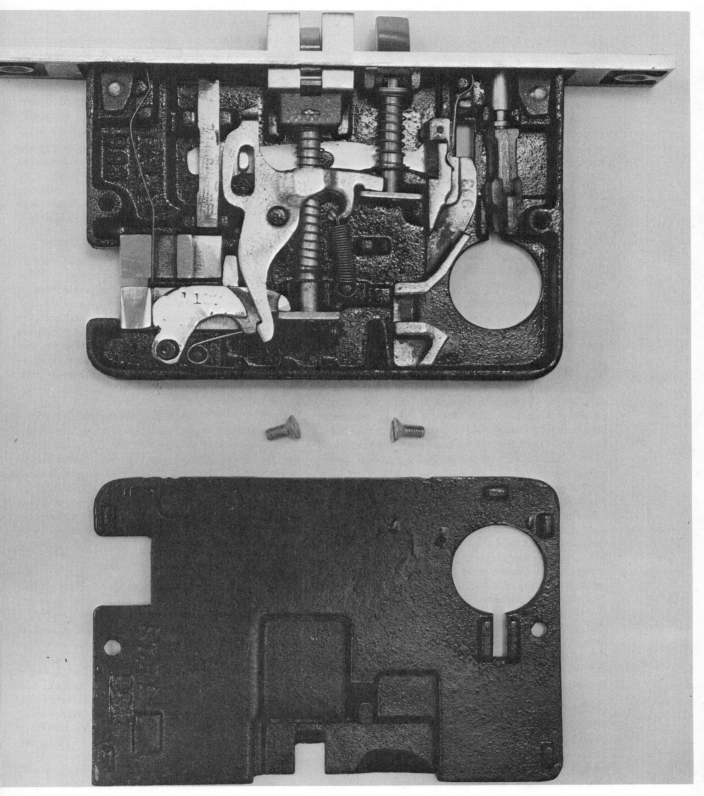

Figure 143. Mortise panic exit lock. Note areas of sliding friction at the latch lever lift blocks. Thumb latch tang and panic bar tang must both be well aligned on their respective blocks

Mortise and Rim Panic Exit Devices

Lubrication of panic exit devices in normal outdoor service should begin with the careful smoothening of case and parts at all friction points, followed by an application of Slipspray, then a coating of K-2 lubricant (Wonder Mist on gravity drop or lightly driven parts), and finally an overall dusting with graphite-moly mix. In the case of mortise units with lift type latch retractors, (see Figure 143) remember that lubrication at the point of contact between the latch retractor block or lever and the latch lift tang on both sides of the door will contribute to the smoothness and ease of operation of the entire unit.

The recommended lubricant for this point is a coating of spray-on acrylic lacquer, followed with moly spray. The moly spray will slightly dissolve the acrylic lacquer, and hence will cling firmly when dry. No K-2 lubricant should be used in contact with the acrylic lacquered parts as it may cause the lacquer to soften and gum up.

Disc Tumbler Utility Locks

Disc tumbler utility locks may be lubricated for all classes of outdoor service by first cleaning carefully with a volatile solvent, then spraying Slipspray into the keyhole and quickly working the key in and out of the lock for maximum distribution. Then turn the key to the opposite position and repeat the Slipspray application. A final dusting with graphite-moly mix, with the lock first in one position and then the other, completes the process.

In the case of cabinet locks, padlocks, and other related types, all working parts may be treated in the same manner: an initial coating of Slipspray with a subsequent dusting of graphite-moly mix.

Disc tumbler automotive lock cylinders, with the exception of the side-bar type, may be effectively lubricated and protected in the same way. The side bar type has such weak springing on the side bar that no Slipspray should be used. The automotive lock-latch unit may be serviced by the method discussed under the mortise lock group (in the section on climatic extremes) whenever disassembly is required for any reason.

Pin Tumbler Utility Locks

Pin tumbler utility locks in outdoor service are usually in the form of padlocks. However, the servicing method described here is extensible to other types of locks as well, since the principle is the same. As in the case of disc tumbler utility locks, the first step in servicing is careful cleaning to remove particles of dirt, dust, sand, etc., from the mechanism.

All working parts except the cylinder may then be sprayed with a moderate coating of Slipspray, which acts as a water repellant and lubricant. Lubrication is completed with an application of graphite-moly mix on working parts, including the cylinder mechanism. Moly spray may be substituted for the graphite-moly mix on working parts other than the cylinder. Its superior sticking qualities make it preferable for this purpose in those cases where the working parts are sufficiently accessible.

Lever Locks

Lever locks in normal outdoor service may be serviced by cleaning in the usual manner, disassembling only if necessary to clean properly or to make repairs. An application of Slipspray through the inspection hole into the tumblers and onto the working parts may be made, directing the spray into the tumbler mechanism by operating the key while the spray application is still liquid. Following this, an application of graphite-moly mix, worked in well, is all that is required.

Combination Locks

Combination locks in normal outdoor service are almost invariably padlocks of the cheap type with factory pre-set, non-changeable combinations in a sealed unit. Therefore, cleaning should be limited to a volatile solvent. A generous application of Wonder Mist should then be worked in by operating the dial in both directions, followed by a moderate amount of the graphite-moly mix applied in a similar manner.

Heavy Outdoor Service

The term *heavy outdoor service* as used here refers to those locks which are installed in entry doors, and which may be exposed to very heavy traffic, a moderate amount of hard slamming, or abnormal stresses as an inescapable part of their function. Extremes of weather conditions either do not exist or some degree of protection is provided so that rain or snow does not beat directly into the lock, and blowing dirt or sand is not a problem. If a lock has a cylinder with a keyhole which has no blind end at the back and no storm door or other protection from the elements, it should be serviced as noted in the climatic extremes section.

Examples of locks in this class of service would include high traffic entry doors to business firms, warehouses, industrial plants, and public buildings. Not included in this category would be locks subject to climatic extremes, such as frequent or excessive quantities of water, prolonged subzero temperatures, temperatures 20° or more below zero during any routine operating period, or blowing dirt or sand.

It is essential in servicing locks in this category that lock housings and all associated parts be clean and free of oil and grease before the application of any of the treatments. Presence of oil or grease may prevent some of the applied treatments from adhering properly to the treated surface, and in addition, may tend to either run or congeal with the temperature variants to which these locks are frequently exposed. Any silicone greases used in the lubrication of locks of this kind should be those which will neither melt nor freeze within the *extremes* of the anticipated operating range.

Disc Tumbler Lock Cylinders

Disc tumbler lock cylinders for locks in heavy outdoor service may be serviced by first cleaning with a volatile solvent, followed with Slipspray, then the graphite-moly mix. Both of these lubricants should be thoroughly worked into the mechanism with the key.

Pin Tumbler Lock Cylinders

The manner of servicing pin tumbler lock cylinders in heavy outdoor use depends on the amount of usage of the cylinder and not on the usage of the rest of the lock. If the cylinder usage is light, it may be lubricated with graphite-moly mix, just as for normal indoor service.

Cylinders which have only a moderate amount of usage may be lubricated in the manner indicated as appropriate for locks in normal outdoor service.

Cylinders having extremely heavy use (to the point where the key pins become wedge-shaped, rather than symmetrical, on the point) should always be disassembled, and worn tumblers replaced, as described in the section on pin tumbler utility locks in heavy indoor service.

In the case of door type cylinders where this problem is evident, there are two additional possibilities for accomplishing a reduction in pin wear. Some manufacturers make available a hardened key pin to help reduce this problem. In other cases, a possible solution is to replace the front three or four key pins, where the deformation is the most severe, with upper pins having a small ball bearing beneath them for the key to bear against, thus reducing both

pin and key wear. The ball bearing technique must not be used on all of the pins in a given lock, as this tends to compromise the security of the lock. At least two pins should be left without the ball bearings in order to retain adequate security.

Lubrication of cylinders after these pins have been replaced may be accomplished by lightly spraying the plug and all pins with Wonder Mist; dusting the plug and interior of the cylinder barrel with a light coating of graphite-moly mix; and adding a small amount (not enough to cause compaction) of mix to the top of the driver pin. If the top of the driver pin is not readily accessible, the mix may be added between the driver and the key pin. Adding the mix in either of these locations will provide long term trickle lubrication, but the above driver location is somewhat better in this respect.

Mortise and Rim Locks

Mortise and rim locks in heavy outdoor service may be lubricated with an application of Slipspray, followed with a moderate coating of K-2 lubricant, and a moderate application of the graphite-moly mix. A light application of Wonder Mist should be substituted for the K-2 silicone lubricant on those parts which are of the gravity drop type or are lightly driven.

In the case of a thumb latch type of lock, both the thumb latch tang and the surface which it contacts should be lubricated with a coating of moly spray, followed with a coating of a spray-can type of acrylic lacquer (clear) and a second light coating of moly spray. This coating should be confined to those portions which are not lubricated as much as other parts, because the K-2 lubricant or the Wonder Mist will eventually cause the lacquer to soften and become gummy, like old grease.

Cylindrical and Tubular Locks

Cylindrical and tubular locks in heavy outdoor service should normally be disassembled in order that the prescribed lubrication will reach all working parts. Before beginning any servicing of cylindrical locks for this class of service, it should be realized that many are simply not suitable for this type of service, regardless of anything which can be done with them.

In some cases, this is due to a design defect which causes the inside knob to come off in the hand as the door is opened or closed; in other cases, it is due to the use of die-cast metal in functional parts. However, the Schlage and some similar designs

Figure 144. Lock trims, knobs, and spindles need lubrication at friction points.

have proven acceptable in this class of service if properly lubricated and installed.

After disassembly and cleaning, the bearing surfaces of the knob and the housing surface upon which it bears should be lubricated with Wonder Mist, and the housing only should be lightly dusted with graphite-moly mix. Working parts and case should be lubricated with K-2 lubricant and moderately dusted with graphite-moly mix. The sealed tube and its working parts may be lubricated with Slip, followed by dusting the graphite-moly mix, in small amounts, into the tube through the back side. Slipspray lubrication for the strike and its pocket is essential for all locks in this class of service.

Mortise and Rim Panic Exit Devices

Mortise lock units in heavy outdoor service in conjunction with panic exit device hardware may be lubricated in the same manner as ordinary mortise locks in the same service class. It is important, however, that all surfaces involving sliding contact between two parts, or one part and the case, be as smooth as they can be made before lubrication. This is necessary in order to relieve the intensity of the great stresses always present in this type of unit.

Rim panic exit devices and the crash bar assembly for use with mortise units in this service class may be lubricated in the same manner with the exception of the pivot pins for the two ends of the crash bar. These may be lubricated as follows: If unworn, lubricate with Slipspray, and follow with a coating of Wonder Mist. If moderately worn, lubricate with nitrocellulose lacquer to which has been added enough of the graphite-moly mix to act as a heavy pigment. This coating, when completely dry, should be lightly sprayed with Slipspray to prime (or begin) the lubricating action.

The nitrocellulose lacquer should be painted on with a small art brush after the graphite-moly mix has been thoroughly stirred in. If the pin or its bearing should be extremely worn, the hole should be reamed and an oilite bearing inserted or an oversize pin used.

Disc and Pin Tumbler Utility Locks

Disc and pin tumbler utility locks may be lubricated for this class of service in the same manner as that described for the same type of lock in the section on normal outdoor service, with the exception of cases where a cylinder take-apart is required. For cylinder lubrication in these cases, see the section on lock cylinders under the "heavy outdoor service" classification.

Lever Locks and Combination Locks

Lever and combination locks are not ordinarily found in this service class. If they should be used, refer to the directions for "lever locks in heavy indoor service" or "combination locks in normal indoor service." Use of these locks in this class of service is not recommended.

Locks In Climatic Extremes

The term *climatic extremes* as used here may be defined loosely as any climate where man can survive or function. Specifically, it includes conditions such as blowing dirt, sand, snow, water, high temperature differential between the two sides of a door, and atmospheric temperatures ranging from 60° below zero to 120° above, with winds ranging to 150 miles per hour. This environment does not include manmade conditions such as corrosive vapors, nuclear radiation, or high temperature steam.

Locks in this class of service will not be broken down into specific types, but rather into the lock and cylinder portions thereof, since lubrication of all types is substantially the same. Before breaking them down even this far, there are some important principles involved in the servicing which merit discussion.

First, and most important, is that no wet lubri-

cant of any type should ever be used in any part of a lock subject to any of the conditions previously described. All lubricants must be dry to avoid collecting dust and sand, and must possess water repellant characteristics.

Second, all parts must be protected to the greatest possible extent from exposure of raw metal to water, whether blown in or of a condensate nature.

Third, all parts must be smooth with sharp edges removed in order to prevent the scraping away of protective coatings.

Fourth, all protection possible must be provided for the locks to keep direct contact with the elements as minimal as possible.

Lock Cylinders

Disc tumbler cylinders may be lubricated for this class of service by first spraying lightly with Slipspray and then dusting with the graphite-moly mix. Whenever practicable, it is preferable that the tumblers be removed for the Slipspray application in order to assure complete coverage.

Pin tumbler cylinders should be disassembled for lubrication. The plug only should be lubricated with Slipspray, and do not neglect the keyway. Graphite-moly mix should be used above the drivers or between the drivers and the key pin in order to provide trickle lubrication. The plug should be dusted lightly with the graphite-moly mix just prior to insertion in the cylinder. Cylinder connecting bars of rim cylinders and mortise cylinder cams should be sprayed with acrylic lacquer, with an additional coating of moly spray for the mortise cylinder cams.

This same treatment may be applied to automotive pin tumbler cylinders any time there is occasion to disassemble one.

Locking Units

As has been previously indicated, all parts must be smooth and free of sharp edges before any application is made. It is imperative that all parts be completely free from grease or oil. All parts, including case, knobs, internal parts, trim, spindle (except the pivot threads of swivel spindles) receive an initial coating of acrylic lacquer. This coating covers every part in its entirety, regardless of whether or not it serves a functional purpose.

The second coating consists of moly spray and covers only the interior of the case and the interior working parts. The third coating consists of Slipspray and covers all of the parts previously covered with moly spray, and additionally covers exposed working parts such as the bolt, latch bolt, strike plate, bearings of knobs, trim bearings, cylinder threads, and all screws used in assembly.

Swivel spindles are lubricated with Slipspray and a moderate dusting of graphite-moly mix. Cylindrical locks with sealed tubes may be lubricated with Slipspray followed with graphite-moly mix sifted in, or moly spray may be substituted for the mix.

This treatment has proven effective in actual service under a combination of all of the adverse conditions listed at the beginning of the climatic extremes section.

Locks In Corrosive Vapor Service

The term *corrosive vapor service* includes a wide range of corrosive agents, and a variety of conditions which affect the speed and extent of the corrosive process. For the purpose of this discussion, these corrosive vapors will be grouped according to the effectiveness of the techniques used in minimizing their destructiveness:

Group 1: Halogen vapors (chlorine, iodine, bromine, fluorine, and their derived acids, such as hydrochloric acid, etc.).

Group 2: Sulfur oxides and derived acids (e.g., sulfur dioxide, sulfurous and sulfuric acids).

Group 3: Nitric acid vapors.

Group 4: Organic corrosive agents. These include a wide variety of exotic chemicals as well as such prosaic ones as vinegar (acetic acid), carbolic acid (phenol), acetone, and many others with a wide range of strength and type of effect.

Other factors which affect the speed and extent of the corrosive process are:

1. *The nature of the metals which are being subjected to attack by a given corrosive agent.* Some metals react much more vigorously with a particular corrosive than do other, more inert metals. For this reason, locks having aluminum or magnesium parts or their alloys should never be used in any type of corrosive vapor service. Copper and its alloys withstand most corrosive vapors poorly, and in the presence of nitric acid, they are useless. Iron and steel are usually better. The proper grades of stainless steel are far better, and best of all in terms of corrosion resistance are certain plastics such as nylon, teflon, polyethylene, and some others, depending on the nature of the corrosive agent.

2. *The presence of humidity or condensed water.* These will greatly accelerate the corrosive process; therefore, a water repellant treatment is essential. Remember that the repellant must meet two conditions: (a) It must be inert with respect to the corrosive agent involved. (b) It must be dry, rather than wet like oil, or sticky like grease, because even though water and oil are not mutually soluble, tiny droplets of water may become trapped in oil or grease, forming what is known as a colloidal suspension. A colloidal suspension involving water, a corrosive agent, and a susceptible material is just as disastrous as suspending the susceptible material in a solution of the corrosive agent.

3. *Temperature of the part being attacked.* This is a varying and often confusing factor in the rate of corrosion of a lock. Normally, the warmer the temperature, the more rapidly a corrosion reaction takes place. The reverse is often true in the case of a lock in atmospheric corrosion conditions because the lower temperatures and the large surface of exposed metal tend to cause condensation where water would not otherwise be a major factor. The introduction of water in liquid form is more than enough to offset the accelerated corrosion rate of higher temperatures with the water remaining in vapor form.

4. *The presence of two or more different metals and their composition.* These are factors in varying the speed and type of corrosive attack due to the creation of a battery type of effect. It should be remembered that the essential components of a battery are two different metals in the presence of a corrosive solution (electrolyte). This type of situation creates the factor of electrolytic corrosion and will greatly accelerate the corrosive process in the immediate vicinity of the point of contact of the dissimilar metals.

5. *The presence of a catalyst or an inhibitor.* This will affect the rate of corrosive attack; a catalyst speeds up and an inhibitor slows the rate of a chemical reaction without being changed themselves. The existence of a catalyst should be suspected whenever the rate of corrosion exceeds by a wide margin what would otherwise be expected from the nature of the corrosive agent involved and the metal upon which it is acting.

Objectives of Protective Measures

Considerable progress has been made in the development of inhibitors for certain types of corrosive attack, principally with respect to corrosion in the presence of fresh and salt water. The applicability of these inhibitors to corrosive atmosphere service is not yet established and their use here should be approached with care, and only experimentally, until sufficient data is available to evaluate their usefulness in a specific situation.

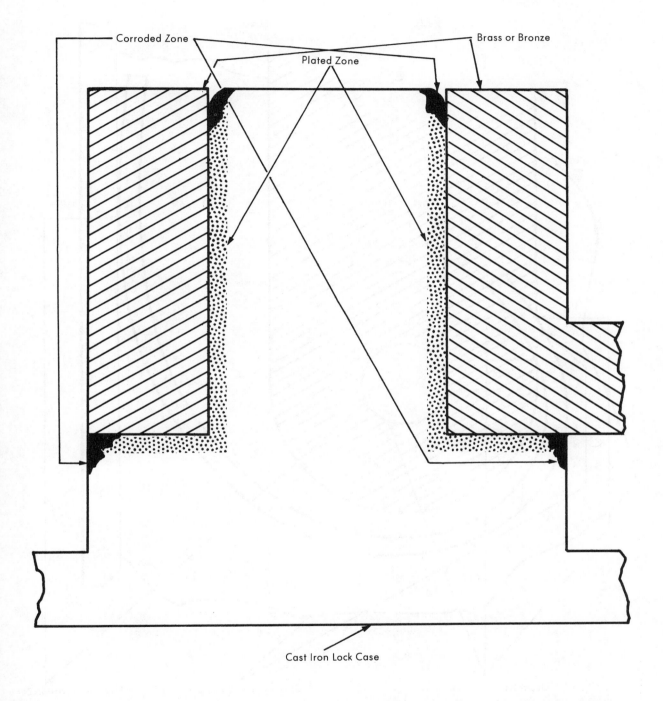

Figure 145. Typical pattern of electrolytic corrosion in a lock. Although the plating action is common enough to merit illustration, it is by no means universal to this type of corrosive action.

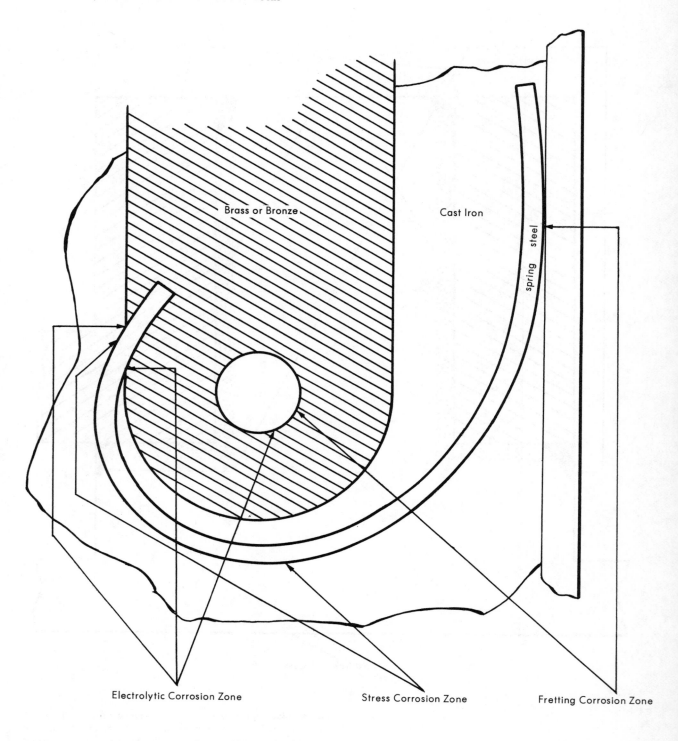

Figure 146. Distribution pattern of typical corrosive attacks. This illustration shows the areas of attack, in a single mortise lock function, of the more specialized forms of corrosion. These selective forms of attack may or may not occur in association with one another. Likewise, they may or may not occur in association with the overall corrosion of a direct chemical attack.

Protective measures taken to minimize corrosive attack on a lock must meet a wide variety of criteria. Protection must be furnished to all parts, both operating and nonfunctional, otherwise the corrosive agent may work its way under the protective coating and into a vital area of the lock. Protective coatings must provide a satisfactory resistance to all of the adverse conditions to which the unit may be subjected.

Measures must be taken to exclude water from the metal surfaces to be protected. Mechanical abrasion which might damage the protective coating must be avoided by careful smoothening of all surfaces where there is any possibility of motion between two parts or a part and the case. Lubrication must be supplied which is of a type compatible with both the corrosion factor and the other conditions to which the lock is exposed.

Adequate clearance must be provided between parts in order that the applied protective coatings do not change the tolerances to the point of impeding their function. The various coatings applied to achieve these criteria must be mutually compatible; that is, they must not interfere in any way with one another in meeting the specific purpose for which each is intended.

The solution to any lock problem involving corrosive vapors is likely to wind up being a compromise between cost and results. For example, the optimum solution to one problem in corrosion protection appears to be to plate all parts with ruthenium. According to available information, ruthenium would provide a protective coating for that particular environment which would provide optimum corrosion protection as well as excellent wear characteristics. It developed, however, that the price of such a plating job would be about $50,000 to $100,000 per lock.

Although some lock manufacturers at present use some stainless steel in the manufacture of certain locks, the use of this fine, corrosion resistant metal is largely confined to those places where it shows, the theory apparently being that it should look pretty whether it works or not.

Lock manufacturers seem quite reluctant to discuss seriously the possibility of building mortise locks or panic exit devices of the mortise type entirely of stainless steel; however, a few firms are now producing high quality cylindrical locks of stainless. The lock cylinder itself is not stainless in any of them, though, to the best of the author's knowledge.

The chart which follows and the elaborating remarks which expand upon it may be used to effect whatever degree of protection may be both desirable and economically realistic. It should be remembered that even though the replacement cost of the original unit may be less than the cost of the protective measures taken, such measures may increase the life expectancy of the unit to anywhere from 3 to 10 times that of an untreated unit. This alters the picture of economic feasability to favor treating over replacement.

The table of corrosion resistance accompanying this section is intended as a guide showing the approximate relative effectiveness of various protective measures which may be taken. Most of the values are based on field trials by the author. However, some (mostly with respect to sulfur oxides) must be considered as theoretical, even though based upon the best data available. In any case, wide variations in the concentrations of the corrosive agent may cause wide changes in the length of time for which adequate protection may be furnished. However, the *relative* effectiveness of the various protective measures taken should change but little.

It is well to remember that the treatments listed in the table represent only a small portion of the possible treatments available today, and more are being developed all the time. Each corrosion situation is unique and must be considered on its own merits. Although the table covers a wide range of situations, it is possible that there may be a better solution to a specific problem.

When choosing a treatment from the table, or from other sources, a number of criteria must be observed:

1. Will the treatment offer resistance to the corrosive agent under the conditions at hand?

2. Are changes in the working tolerances of the lock necessary in various places to accommodate the physical bulk of the treatment selected?

3. Does the treatment selected provide the proper degree of lubricity for adequate functioning of all parts, or must an additional treatment be provided to obtain it?

4. Are all of the treatments indicated completely compatible with one another?

Combinations of certain treatments will result in better protection than any one will furnish alone, because of specific properties inherent in each which are not supplied by the other. For example, adding graphite to the clear nitrocellulose lacquer as a pigment increases its corrosion resistance by about one point and its lubricity by about three points (it is necessary to prime this lubricity with either a graphite rub or a coat of moly spray until some of the graphite pigmentation can work its way to the surface).

A further coating of Slipspray would provide valuable additional water repellant qualities and reduce the lubricity very little. This particular combination is quite effective in preventing corrosive attack by halogen acid vapors and provides adequate lubricity for most purposes. It is frequently necessary to relieve close tolerances of working parts in order to accommodate the physical bulk of the nitrocellulose graphitic lacquer.

A clear acrylic lacquer spray followed immediately with moly spray (to obtain penetration of the moly into the lacquer), dried and followed with Slipspray and a light second coat of moly spray, should accomplish about the same results, unless some type of organic solvent is encountered. Acrylic lacquer has virtually no resistance to many liquid solvents, and this factor may be useful in cleaning and servicing at later dates, but could be harmful if the lock is exposed to these solvents while in service. A few comments on the usage of some of the treatments listed in the chart may be in order at this point.

Clearance relief must be provided for closely fitting parts whenever the treatment involves a solid substance of appreciable film thickness. Those treatments listed in the chart which frequently require this relief are: all of the lacquers, but especially the nitrocellulose type; chrome plating; and teflon coating. Relief for lacquer coating is not as critical, and shop rule-of-thumb estimates can be made with an acceptable degree of accuracy, and can be increased if necessary with but little inconvenience.

Clearance relief for chrome plating or teflon coating must, however, be carefully planned. Prior consultation with the shop which will do the actual plating or coating is absolutely essential in order to determine the finished thickness of the material to be applied and the accuracy with which that thickness can be maintained. A shaft or part enclosed on two opposite sides will require an *increase in clearance* equal to four times the thickness of the coating.

It is good practice to allow an extra 20% increase for variations in the coating thickness. For an applied coating of .002 in., this would be an increase in required precoating clearance of approximately .009 in. The formula for obtaining these figures, letting I be the increase in required clearance and T be the thickness of the coating to be applied, would become:

$$I = 4T + 1/5T \text{ or } I = 4.2T$$

In the event of multiple coatings of this type being envisioned, the formula would evolve somewhat differently:

If I = increase in clearance
If T_m = coating of greatest thickness
If T_s = sum of the thicknesses of all other coatings
Then $I = 4.2 T_m + 4 T_s$

Proper order of application is of considerable importance if maximum effectiveness is to be obtained from any given series of treatments. The first application must necessarily consist of the basic, more solid corrosive protection in order that it will adhere well to the metal, unless the initial coating is fully soluble or suspendible in the second. This may be followed by dry spray lubricants, then wet lubricants (as in noncorrosive indoor atmospheres), and finally, the dry, powdered lubricants. This is not to imply that all, or any particular one, would be used in any given situation, but merely states the order of their usage when and if used.

CORROSION PROTECTION

Treatment	Corrosion Resistance To:				Radiation Resistance	Lubricity	Treatment Compatability
	Halogen	Sulfur Oxides	Nitric	Organic Agents			
1. Clear nitro-cellulose lacquer	3	3*	5	4-5	2	5	2, 4, 5, 10, 12, 13, 14
2. Graphited nitro-cellulose lacquer	2	2*	4	4-5	2	2	1, 4, 5, 10, 12, 13, 14
3. Clear acrylic lacquer	3*	3*	5*	5	2*	5	4, 5, 10, 12, 13, 14
4. Moly spray	2	1*	3	1	1	1	ALL
5. Slipspray	2	2*	2	1-4	1	4	ALL
6. Wonder Mist	4	4*	5*	5	3*	3	4, 5, 7, 8, 9, 10, 11, 12, 13, 14, 15
7. Petroleum jelly	5	4*	5	5	5	2	4, 5, 6, 8, 9, 10, 11, 12, 13, 14, 15
8. Cup grease	5	5	5	5	5	5	4, 5, 6, 7, 9, 10, 11, 12, 13, 14, 15
9. Oils	4	5	5	5	5	2	4, 5, 6, 7, 8, 10, 11, 12, 13, 14, 15
10. Chrome plating	2	3*	2	1	1	5	ALL
11. Teflon coating	1	1*	1	1	1	3	4, 5, 6, 7, 8, 9, 12, 13, 14, 15
12. Graphite	2	2	3	1-3	1+	1	ALL
13. Moly	2	2*	3	2-3	1	1	ALL
14 Graphite-moly	2	2	3	1-3	1+	1	ALL
15. Silicones	4	4*	5	5	4*	2	4, 5, 6, 7, 8, 9, 10, 11, 12, 13, 14

* According to best data available to the author, field trials incomplete.
+Do not use in a neutron field or near an atomic accelerator! (Otherwise rating of 5)

Ratings are from 1 for superior to 5 for unsatisfactory, but may be subject to considerable variation as conditions vary. For example, some of the dry powders may furnish excellent short term protection, then slough off. Some of the treatments may be penetrated through their own porosity with respect to the agent, or by colloidal suspension. Careful, intelligent selection from the table can help greatly in combating corrosion. The author has used these treatments to extend the life of a lock to about 20 times its untreated expectancy.

AVAILABILITY OF MATERIALS

LACQUER: Clear, both acrylic spray cans and nitro-cellulose, from the local automotive wholesaler or body shop.

GRAPHITED NITRO-CELLULOSE LACQUER: Mix thoroughly equal volumes of clear nitro-cellulose lacquer and powdered graphite. Specific applications may require adding more of one or the other. (Graphite-moly mix may be substituted for the graphite.)

MOLY: (Molybdenum disulfide) — from the local automotive wholesaler.

SLIPSPRAY: (Powdered teflon spray lubricant) — from the local du Pont distributor, usually one of the automotive wholesalers.

SLIP: (Light silicone spray) — Certified Laboratories, P.O. Box 2493, Fort Worth, Texas.

WONDER MIST: (A light silicone grease) — from Cherry Creek Locksmith Service, Denver, Colorado.

K-2 LUBRICANT: (A light silicone grease) — from Curtis Industries, Eastlake, Ohio.

CHROME PLATING: Custom chrome plating is available in most large cities. Care should be taken to select a shop which can maintain the desired coating thickness with reasonable accuracy and apply it tightly, as corrosive vapors tend to lift it even at best.

TEFLON COATING: Contact E.I. du Pont Co., Wilmington, Delaware, for the name and address of the custom coating shop nearest your city. Check with the shop about the thickness of the coating. Do not attempt to do this work yourself — it can be dangerous!

GRAPHITE: Dixon's Microfyne — available from a locksmith supply house.

GRAPHITE-MOLY MIX: Three parts graphite and one part moly (by volume).

PETROLEUM JELLY: From the local drug store.

OILS AND GREASES: From the corner service station, or from the bulk distributor if something special is required.

NEW LUBRICANTS & CORROSION INHIBITORS: Like gold, hard to find!

Reminder: All spray-on coatings and lubricants must be applied in a well ventilated area. As a safety measure, all spray materials and solvents of any kind should be considered highly flammable until proven otherwise. Observe the manufacturer's precautions regarding use of all treatment materials, lubricants, and solvents. Be especially careful when mixing unknown products which might react chemically; if you have any doubt at all about mixing them — don't.

Installation Criteria

Rim lock cylinders in wood doors may be located and drilled by a template, and the cylinder to be used is the one furnished for that particular lock, or by careful measurement. If the cylinder is to be located by measurement, it is important to remember that it is the cylinder connecting bar, rather than the cylinder center, which must be exactly opposite the slot in which the cylinder connecting bar fits to operate the lock mechanism. The hole which is to accept the cylinder is drilled on the cylinder center.

The cylinder connecting bar of a pin tumbler cylinder is always below the center of the cylinder; therefore, in order that the cylinder connecting bar align properly with the lock part which it actuates, the hole for the cylinder must be drilled above the desired position of the cylinder connecting bar the exact distance that the connecting bar is below the center of the cylinder.

In locating the horizontal position of the cylinder, it should be remembered that on a standard 1-3/4 in. door, the edge of the door is beveled so that the cylinder center as measured on the outside edge of the door would be 1/8 in. further from the edge than the desired position on the other side of the door. The bevel on the edge of the door should be verified with square and rule before any cutting is done, and any adjustments necessary made before drilling the hole for the cylinder.

When the cylinder hole has been drilled in the door, the cylinder may be inserted and the cylinder clamp plate attached. The position of the clamp plate should then be carefully delineated by going around the edge of it with a sharp knife or chisel as deep as the clamp plate is thick. The clamp plate should then be removed, and also the wood beneath it, to a depth equal to the thickness of the plate, so as to allow the cylinder clamp plate to set flush with the inside surface of the door, rather than riding on top of the door surface. This produces a firm installation which will not twist or loosen readily.

RIM LOCK CYLINDER INSTALLATIONS IN METAL DOORS

The cylinder hole in the outside surface of a hollow metal door may be located in exactly the same manner as for a wood door. The hole must not, however, be cut through the inside surface of the door. The inside surface should be cut with three holes which correspond to, and replace, the holes in the cylinder clamp plate, one being for the cylinder connecting bar and the other two for the cylinder clamp screws.

Some cylinders have the cylinder clamp screws in the same horizontal plane as the cylinder center, and this makes it relatively easy to use the cylinder clamp plate as a template for locating the position of the three holes. It need only be placed on the horizontal and vertical lines marking the cylinder center's position on the door in such a way that the lines are centered in all of the three holes and their position marked.

The holes may then be drilled in the same size as the corresponding holes on the cylinder clamp plate. The clamp plate may be discarded and the cylinder installed in the door, using the three new holes in

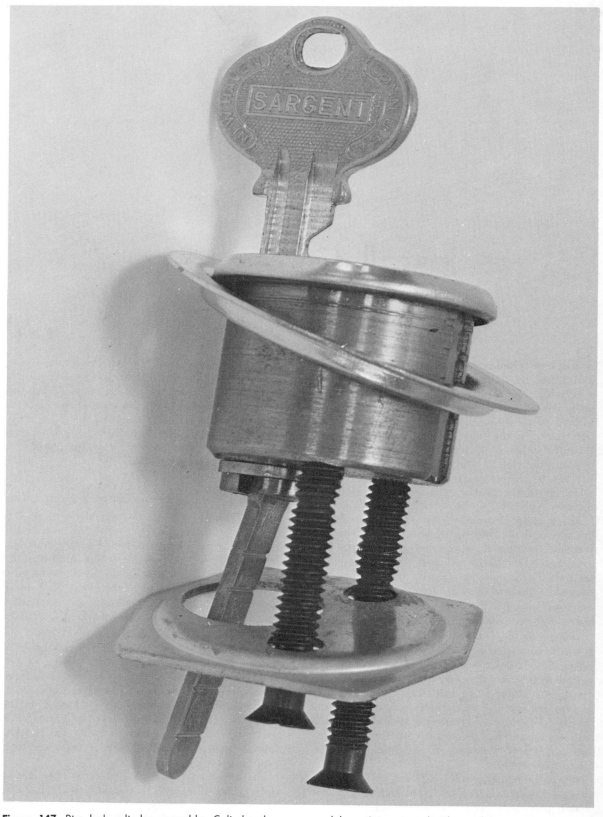

Figure 147. Rim lock cylinder assembly. Cylinder clamp plate is mortised flush with door surface when used with wood doors. It is not used with metal doors, except as template for cutting the inside of the door to a similar pattern.

the inner surface of the door to replace the clamp plate. If the cylinder clamp screws are not on the same horizontal plane as the center of the cylinder, measure down on the inside surface of the door, below the marked position of the cylinder center, the same distance that the cylinder connecting bar is below the center of the lock cylinder. Use the connecting bar hole in the clamp plate for a locating reference for the clamp screw holes.

Remember that the clamp screw holes must be in a horizontal plane for the lock to function properly.

This procedure, when properly carried out, will provide the same firm and rigid mounting as that described for wood doors.

Rim Lock Installation

Once the cylinder installation has been completed, the outside trim may be installed (if any is used, as in the case of a rim panic exit device with an outside function). Those holes which go through the door may be located either by template or by careful measurement. The holes almost universally lie in the same vertical plane, that is, they are nearly always along the same vertical center line as the cylinder hole. The attaching screw holes (for screws which go through the outside trim plate and into the door) may be located either by template or by setting the trim in place and marking the centers of the holes with a machinist's centering punch (not a common center punch).

The outside trim having been firmly anchored in place, the next step is to cut the cylinder connecting bar and thumb latch tang (or knob shaft) to length. The length of both of these is often very critical. If the cylinder connecting bar or the knob shaft is too long, it may protrude into the lock case far enough to obstruct the travel of another functional part. If too short, it may work properly for a time and then fail to operate due to the slight springing of the connecting bar or due to the slamming of the door having caused a slight readjustment of some part.

If a thumb latch tang is too long, it may bind against the case of the lock and jam in the retracting position, or strike against some obstruction and fail to retract the latch completely, because its normal arc of travel has been restricted. However, this latter symptom may be caused by improper positioning of the thumb latch relative to the lock or it may be hitting the cut hole in the door. It is up to the locksmith to determine which is the case and to correct the problem.

The proper length of these parts may be determined by holding the lock case up to the inside of

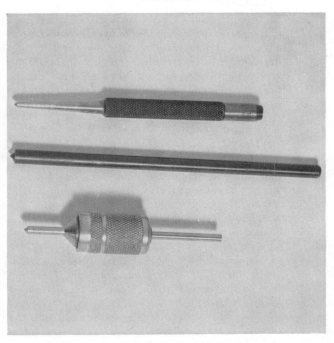

Figure 148. Top - Common center punch. Center - Machinist's centering punch for straight through holes. Bottom - Machinist's centering punch for both countersunk and straight through holes.

the door, just as if it were to be installed, and then fit it by cutting off just a little at a time until the lock mechanism functions perfectly.

The lock case is now ready to be installed on the inside of the door. It may be positioned by inserting the cylinder connecting bar into the cam which it operates, holding the lock case to the door, sliding the lock case up and down and from side to side to the limits imposed by the cylinder connecting bar, and marking the limit of travel in each position. A mark centered between each of the limit of travel positions will usually give the optimal positioning of the lock case on the inside of the door. It is well, however, to test the function of the lock with it held in this position before drilling or marking any locations for mounting screws.

A small adjustment in the positioning of the lock on the door is occasionally necessary. It is important, especially in the case of a panic device, that the lock case be square with the door. If a panic bar is not perfectly square with the door, it is usually possible to bend the arm on the lock case slightly to bring it square. This must be done before the inactive arm support is attached to the door, or there will be a bind in the mechanism, or the bar will have a permanent droop.

If a panic device should have a bind, a few taps with a plastic faced hammer is usually sufficient to relieve it. This tapping must be done in the right place and direction to relieve the bind. It should not be attacked blindly, but with the idea of moving the binding part away from the part it is binding on just enough to permit proper functioning of the parts.

Mortise Lock Installations in Wood Doors

The first step in the installation of a mortise lock in a wood door is the inletting of the lock case and front into the edge of the door. This may be done either by laying out the mortise with a template, or by careful measurement. Either method produces satisfactory results if done with care. However, the finished pocket in the door should meet the following standards:

1. It should be smooth on the inside, without bumps, lumps, or other protrusions, and free from chips or shavings, whether loose in the bottom or clinging to the sides.

2. The case pocket must be set straight into the door and be well centered.

3. The lock case and front should meet no resistance when you insert it into the pocket. If the case is squeezed, the functions of vital parts of the lock may be impaired, either at the time of installation or when a change of humidity causes the wood of the door to swell.

4. The case pocket should not be so wide that the lock is without support from the wood if a pulling force operates on it, either from the knobs or from a pressure on the cylinder. Ideally, there should be about 1/32 in. clearance on either side of the lock case to the wood; in other words, the mortise should be about 1/16 in. wider than the thickness of the lock case.

5. No lock should be mortised into a wooden door unless the stile is wide and thick enough to retain adequate strength in the door when the mortise is cut for the lock case.

After you have cut the mortise for the lock case and front, the holes may be laid out and cut for the cylinder, knob spindles, and turn knob (if any). Layout may be accomplished by template, measurement, or by a combination of these methods. Cutting of cylinder and knob spindle holes may best be done with a hole saw used in a power drill because it is not only faster, but has less tendency to splinter the veneer (or surface) of the door than an expansion bit. An auger bit is usually preferable for cutting the hole for a turn knob, as this hole is considerably smaller, as a rule, than the others, and

the splintering tendency is substantially less than with an ordinary metal drill.

If the lock trim is in the form of escutcheon plates rather than rosettes, it is not only possible, but usually desirable to make the accessory holes in the wood somewhat larger than most manufacturers' recommendations. An extra 1/8 in. to 3/16 in. helps to avoid many problems of alignment and binding; escutcheon plates cover them, so there is no problem so far as appearance is concerned. If the trim is in the form of rosettes, it is necessary to be more cautious because of the closeness of the mounting screws to the edge of the hole. The cylinder hole should be kept smaller than the diameter of the threaded portion of the cylinder plus the largest (outside) diameter of the cylinder ring and the sum of these figures divided by two, thus:

C = cylinder diameter
R = ring diameter (outside)
M = largest permissible hole

$$M = \frac{C + R}{2}$$

Keep the hole cut just a little smaller than M.

After the accessory holes have been cut, the case cover (cap) attaching screws and the front attaching screws should be checked to make sure they are firmly tightened. The lock case assembly may then be placed in the door and screwed down. This is the permanent installation for this assembly and there should be no need to remove it from the door again.

The attached positioning of the trim is of critical importance if maximum durability is to be obtained from a mortise lock. The following method gives good results and it is quick and relatively easy to follow:

1. Insert the cylinders through the escutcheon and screw them loosely into the lock, omitting cylinder rings.

2. Insert the spindle through the escutcheons and the lock and screw on both knobs until tight enough to hold escutcheons in place.

3. Center the cylinder in the hole in the escutcheon, then wobble (not rotate) the knob right and left until a median is established. Two or three small pencil marks adjacent to the knob at the edge of the escutcheon are of assistance in establishing this. Recheck to be sure the cylinder is still centered in its hole and tighten the knobs to hold the escutcheon *firmly* in place while the escutcheon

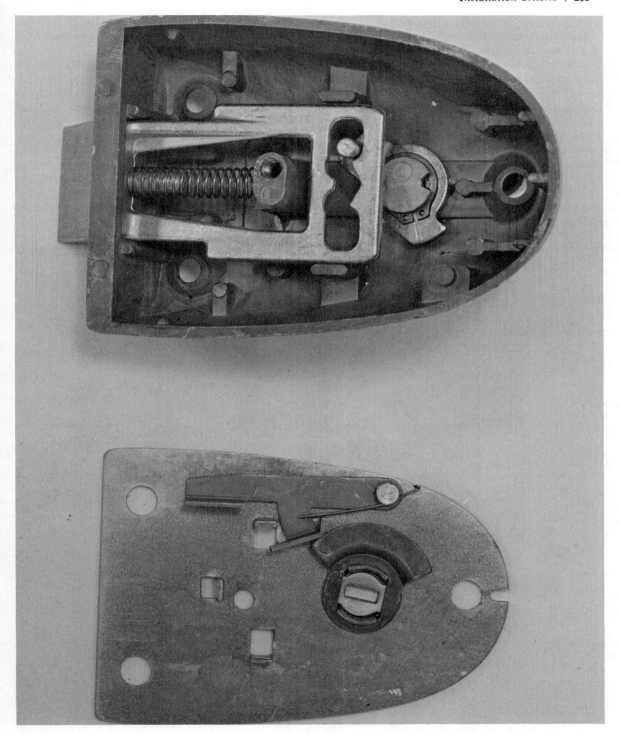

Figure 149. Rim deadlatch. Note that a cylinder connecting bar which protrudes too far into the lock case may cause the mechanism to jam.

attaching screws are located and installed.

The remaining escutcheon may be positioned in exactly the same manner if it, too, has a cylinder. If it does not have a cylinder, the knobs should be just tight enough to hold a position in which the escutcheon is set; the escutcheon may be set by measurement to be parallel to the edge of the door. The knob may then be wobbled both horizontally and vertically, in a manner similar to the one previously outlined. The escutcheon is then re-checked to insure that it is still parallel to the edge of the door, the knob screwed down to hold it tightly, and the attaching screws located and installed. The knobs may next be suitably loosened and the knob set screws tightened.

The knob set screws must be positioned to bear on the flat of the knob spindle. The spindle itself, if of the screw coupled swivel variety, must be adjusted before installation so as not to bind when the knobs are turned independently. This is done by screwing the parts together until they just start to bind, then backing off three-quarters of a turn.

If thumb latches are used instead of knobs, the tang should be centered as precisely as possible under the piece it is to lift. Under no circumstances should it either lap over onto the piece the other thumb latch is supposed to operate, or bind on the lock case on the other side, nor should it come into contact with the tang of the opposing thumb latch.

The thumb latch tangs should be cut off so that they just reach under the lock and even with the further outside wall of the lock case when in the latch retracting position. Then the tip should be rounded on the top side of the tang so that the further case wall will not obstruct its full travel, and leave no sharp edge or corner on the tip to hang up between the lift block and the lock case.

4. The striker plate should be positioned so that it

Figure 150. Mortise deadbolt. Both excessive squeezing of the case, or chips and shavings from mortising, may impede the function of the operating parts.

Figure 151. Mortise deadlatch. Excessive pressure on a lock case may cause the latch or the deadlocking lever to bind and malfunction. Misalignment within the door may cause the latch to bind. Other parts are also susceptible to malfunction from excessive pressure on the case, but to a lesser extent.

is not necessary to slam the door, even in the least degree, in order to engage the latch bolt. Dead bolts should also throw freely and never touch the strike plate unless the knob is first turned. The wood behind the bolt and latch bolt pockets should be cut out deeply enough so that latch bolt and dead bolt may both be extended fully and have some clearance without touching wood.

The wood should also be undercut slightly under the leading edge of the pockets to avoid any possibility that auxiliary latch bolts on deadlatches are supposed to ride the surface of the striker plate, rather than enter it.

Mortise Lock Installations
in Hollow Metal Doors

Hollow metal doors are made to accept a specific make and model of mortise lock, with the mortise built in to accept the lock. Usually the cylinder and knob holes are precut, and frequently the holes for the escutcheon plates are drilled and tapped. Such a door presents few problems as long as the lock being installed is the same as the door was built to accept. If such a lock is not available, one should be selected which will fit the mortise in the edge of the door, and whatever changes are necessary should be made in the size and location of the cylinder and knob holes and the holes for the trim.

Locating, or relocating, of these holes may be done exactly as described for wooden door installations; however, all screws should be of the machine screw type rather than wood or sheet metal screws. Screws having a fine thread usually hold better in thin metal than screws with coarser threads. If the metal is extremely thin, punch the holes with a very fine, sharp center punch instead of drilling them. This dimples the metal inward and provides more surface for the threads.

The better grades of hollow metal doors are made with two spring clips, one on either side of the recess for the lock. The clips hold the lock firmly in position in the door, and they should be left intact unless they put such a squeeze on the lock case as to impair the functioning of the lock. If they are so

tight as to prevent the proper functioning of the lock, they may be pressed away from the lock case enough to relieve the pressure. In an extreme case, they may be cut entirely away, using a cold chisel or hacksaw blade. This should not be done unless there is no other way to relieve the excess pressure.

Cylindrical Lock Installations

Cylindrical lock sets, because they are made in two parts which become a functional whole only in the door, have somewhat different and more rigid installation requirements than either rim or regular mortise lock sets. Metal doors which are precut for the type of cylindrical set being used present few problems if the doors are well made. An installation in a wood door presents more problems, however.

1. The hole for the lock housing must be cut with exactly the right amount of backset from the edge of the door for the housing slots to engage the tangs of the latch tube. The wood of the door should provide support for the opposite side of the lock housing so that the two sections cannot become disengaged. In addition to this, the hole must be cut squarely

through the door or a strain will be placed upon the mechanism which will cause the lock to bind.

2. The hole for the latch tube assembly must be bored true with the door in every direction, or the latch end will be forced to slide across the latch retracting yoke or the lock housing, with the result that the latch will bind and possibly be damaged.

3. The lock housing depth adjusting collar (on the keyed knob side) must be adjusted so that the latch end enters fully into the latch retracting yoke, engages it completely, and yet does not enter far enough to cause the latch end to bind against the housing on the far side when the housing is securely tightened onto the door.

4. The strike plate must be set so that the latch will enter it freely, without hesitation and without having to slam the door. The wood must be slightly undercut from the edge of the strike plate pocket which the flat face of the latch contacts. This is to insure that the latch will extend fully into the pocket of the strike. In addition to this, the depth of the pocket must be such as to make sure that the latch will not strike the bottom of the pocket before the latch is fully extended, even though the door may swell in wet weather.

If the lock is of the deadlatch type, it should also be remembered that the deadlocking plunger (along the flat side of the latch) must not be permitted to enter the strike pocket at any time or under any conditions, or its function will be negated.

Figure 152. Cylindrical deadlatch. Note that the latch tang must be drawn straight into lock housing by the yoke. For maximum performance, the tang should be carefully centered in the yoke to avoid thrusting the latter sideways and causing increased friction. The deadlocking plunger beside the latch must not enter the pocket of the strike; yet the latch must enter fully into the strike in order to engage the deadlocking mechanism built into the latch tube.

LESSON 22

Servicing Locks in a Radiation Field

With the advent of nuclear generation of electrical power in significant quantities there has arisen the need for knowledge of some of the do's and don'ts of servicing locks in areas where radiation is present. Many locksmiths now have a nuclear facility near them and in the future, many more will. Much is now known about the special problems which such an environment poses, and more is yet to be learned. The hazards involved are very real, to be sure, but the control of the hazards is rigid. As long as the locksmith is alert, plans his work well, and takes full advantage of the hazard controls available, the risk is small. Carelessness is the principal danger in a radiation environment.

The pages which follow will show as objectively as possible what some of the principal hazards and conditions are, and will offer broad suggestions on how to deal with them. These methods have been in use for several years without causing difficulty. In the interest of simplicity, the author has presented many broad statements without elaborating, but in reading these pages, one should realize that the number of variables encompassed is tremendous and would fill many volumes if presented in detail.

It should be understood that the author is not a nuclear physicist, but a locksmith, and he offers a locksmith's viewpoint. The suggestions are intended to help others in the profession. The assistance of qualifed people who understand the radiological conditions present will also be available to help locksmiths cope with lock problems in such an environment. There are types of radiation other than those discussed here, and other kinds of situations (principally around nuclear accelerators), but the following are the most likely to be encountered by the locksmith; therefore, the discussion is limited to them.

Classes and Types of Radiation

To those interested in working effectively in a radiation environment, radiation is considered as belonging to one of two classes and one of three types. The two classes of radiation are:

1. *Direct* or emanating from a specific and identifiable source.

2. *Contamination* or emanating from multiple sources which have no specific location other than as originating from a certain object or area. An example of this would be dust which contains radioactive particles of minuscule size, or a liquid containing a radioactive material in solution.

Work being done in an environment of *direct radiation* is governed by the following factors:

1. *Strength of the source.*

2. *Distance from the source.* Field strength diminishes in proportion to the square of the distance from the source in the case of dimensionally small sources.

3. *Shielding.* Interposition of shielding (lead, concrete, water, etc.) between the worker and the source of the radiation will reduce the field intensity.

4. *Control of exposure time.* The human body can acquire a certain amount of radiation exposure without a significant impairment of any bodily function. The "Radiation Protection Guides" by

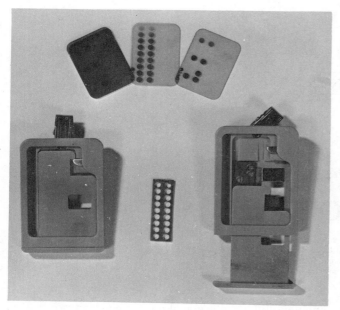

Figure 153. Film badge. Worn by every worker in a potential radiation field. It consists of a carrier housing containing special photographic films and various degrees of shielding to measure the intensity and energy level of all radiation striking it on an accumulative basis. Many other personal exposure metering and measuring devices are available, but this is the standard device used to measure the accumulated dosage which is part of all workers' exposure record. The assembled film badge is attached to the clothing and measures all types of radiation except alpha.

which government and industry have set specific standards for controlling the amount or dose of radiation which a person is permitted to accumulate, are set at a small fraction of the impairment level.* These levels are expressed primarily in terms of field intensity and time for each exposure and include an overall annual and lifetime limit.

Work being done in an environment of radioactive contamination is governed by the following factors:

1. Strength of field in the work area.
2. Control of exposure time.
3. Protective equipment.
4. Decontamination.**

Certain protective equipment is used to prevent radioactive particles from being carried on the person or clothing, or being inadvertently introduced into the body in eating, drinking, breathing, or through the eyes, or a scratch or cut.

*Further information on this subject may be obtained from a government publication: *Permissible Dose from External Sources of Ionizing Radiation*, Handbook 59, Department of Commerce, U.S. Bureau of Standards, available from the Superintendent of Documents, Washington, D.C., 20402.

Protective equipment includes a wide variety of items, such as rubber gloves, coveralls, rubber or plastic overshoes and boots, respirators, gas masks, hair covers, and external air lines. Their use is governed and the selection made in accord with the nature of the contaminant and other environmental factors.

Once the lock has been removed from the contamination area, it should be carefully checked for contamination. If none is found, work may proceed as usual. If, however, the lock is found to be contaminated, it should be decontaminated until no significant radiation is detected before proceeding. This, and every operation involving radiation, must be done under the supervision of qualified radiation protection personnel. The decontamination process itself may consist simply of a wash with solvent, or soap and water, or both. In particularly stubborn cases, it may be done by processes known as *electrolytic decontamination* or by *ultrasonic decontamination.*

Frequently all decontamination is done by specially trained plant personnel. Electrolytic or ultrasonic decontamination, where available, is always done by such personnel. All solutions and equipment used in the decontamination process must be considered to be contaminated, even after the slightest use or contact, and must be handled and disposed of according to the instructions of radiation protection personnel. The question of whether or not a lock may be buffed, welded, ground, or filed on, should always be referred to radiation protection personnel, as otherwise permissable levels of contamination may become excessive under those conditions.

In the matter of the protection of people who must work in a radiation field, three types of radiation are significant:

1. Alpha
2. Beta-Gamma
3. Neutron

Alpha radiation particles, although the heaviest in this group, move relatively slowly, are incapable of penetrating the skin, and hence are relatively harmless *as long as the source of emission is outside the body.* Once a source of emission is inside the body, however, they can be deadly. Under this condition, alpha radiation is perhaps the most dangerous of all, because it does not penetrate to

**Decontamination: The removal of part or all of a radioactive contaminant, thus reducing the radiation intensity, and resulting in a proportional extension of the exposure time permissible.

Figure 154. Protective clothing and equipment. Items shown represent only a small part of the protective measures which can be used to protect personnel against radiological contamination. Top - Respirator, plastic shoe cover, full face, remote air supply mask. Center - Full face, wide vision respirator; cotton lab coat; cotton coveralls; cotton hair protector. Bottom - Polyethylene bag for carrying contaminated articles; full face, anti-fogging respirator; light weight rubber boot; industrial type hard hat; surgical type rubber glove; standard weight rubber glove. Other items may include all plastic or rubber suits, self-contained air packs, etc.

the outside of the body to show on detection instruments.

Special instruments must be used to detect alpha radiation even under the most favorable conditions, as the wall of the ordinary Geiger counter tube is impervious to alpha particles and hence cannot detect them. If there is any doubt as to the possibility of *alpha contamination* in a work area, the locksmith should insist on an *alpha survey*. Radiation protection personnel make this survey with a special *alpha counter*. (See Figure 155.)

Beta and gamma radiation are the most common types of radiation and are usually associated, that is, they are frequently emitted by the same source. Gamma radiation, similar in many respects to the X ray, is light and fast, penetrates deeply, and is readily detectable with the Geiger counter, as is some (but not all) beta radiation. Because of the comparative ease of detection and metering of beta-gamma radiation, it is the easiest to cope with and to work around as long as the rules and the instructions of radiation protection personnel are scrupulously observed.

Neutron radiation particles are heavy with a wide range of speed and energy, and must be considered to be penetrating. The fact that this is the particle which is used to split the atom in a reactor is indicative of the havoc which the neutron is capable of causing in both man and materials. Some of the effects which neutron bombardment may cause in materials are:

1. "Growth" of metals (expansion in volume).
2. Radioactivity, artifically induced, in certain metals.
3. Embrittlement of many plastics.
4. Chemical changes.

Figure 155. Radiation "counters." These instruments are used to measure the various types of radiation in order to determine the strength of the radiation field.

5. Transmutation of elements.

Neutron radiation can be shielded, but shielding is vastly more difficult than with most other types of radiation. This type of radiation is usually found in impressive levels only in a nuclear reactor, an accelerator, or a neutron source (which is created by the careful and painstaking separation and combining of isotopes), or just possibly by the inadvertent combination of nuclear reactive material. Neutron sources, fissionable material, or their shielding must *never* be moved in any way from the position in which they are found, because to do so might cause a critical mass to be assembled. There are much more practical methods for suicide.

Neutron emission occurs in nature principally in radium, uranium, thorium, and some other naturally radioactive elements, and their ores and compounds. Neutron emission from these sources is usually of far less intensity and can be handled with much fewer precautions and with less shielding. In some cases, no shielding is required at all.

Suggested Rules for Working in a Radiation Area

1. Before entering an area where radioactive materials are handled or used, ask for a survey by qualified radiation protection personnel, and abide implicitly by their instructions and recommendations. If any of the instructions are not perfectly clear, do not hesitate to ask for an explanation.

2. Examine the job and the work area to determine what must be done and the facilities available for doing it. This examination should determine the basic tools required, how much room is available to work in, position and type of electrical outlet, and any unusual situation with regard to both lock and door or work area.

3. Leave the area and plan the job in every detail. Every move should be carefully planned in advance. What will be the first tool to use? Which tool will be used next and how many operations can it be used for? What additional tools will be needed for standby in case unexpected difficulties are encountered? Should the lock be removed from the area for work, or will less total radiation exposure time result if the work is done on the spot? Are new screws available for the lock and trim? If so, use them. Are new parts available for any defect which might be encountered?

Arrange primary tools in order of use, keeping them well separated so that each is instantly available. Remember that once a tool has been touched with contaminated gloves, it too must be

Figure 156. Turnstile monitor. Used to prevent egress from a given area by anyone carrying radiological contamination upon his person or clothing. As the person passes between an array of counter tubes, the monitor alarms and automatically locks the turnstile should any contaminant be present.

presumed to be contaminated. Standby tools need not be arranged in order of use, but should be arranged well separated so that they are available without delay and without danger of contamination through sorting while the job is in progress.

Arrange each screw in order of use and lubricate the threads. Have lubricant handy to lubricate the threads in the door or in the lock if that should be necessary. Keep used tools separated as far as practical from those which have not been used in order to avoid their contamination.

4. When leaving the work area, keep all possibly contaminated articles in a closed container or bag until they can be monitored for contamination.

5. When removing protective clothing, be sure that no external portion of the clothing is allowed to touch skin, hair, or personal clothing.

6. Use hand and foot counter. (See Figure 157).

7. Request a personal survey to make sure that there is no residue of contamination elsewhere on body or clothing.

Some of the points mentioned under item 3 regarding the planning and layout of work may not be necessary if the radiation exposure is mild, and may be varied if radiation protection personnel allow sufficient time to get the job done without the necessity of these precautions. It will depend largely upon the time available to do the job in comparison with the time required to do it without such detailed planning. In any case, it is desirable to keep the time spent in even a mild radiation area at a minimum.

EFFECTS ON LOCKS AND LUBRICANTS OF RADIATION

The neutron, with its atom-breaking qualities, has potential for inducing spectacular effects greater than those of any of the radiation particles which we have considered here. It is fortunate that the neutron is normally found in high intensity levels only in the vicinity of an atomic reactor or accelerator. Although neutrons may also be found in association with other materials, some of which occur in nature, they are normally of such an intensity level that such phenomenal effects as transmutation require time which, if not in terms of geological eras, is at least too great to be of significance in the servicing and handling of a lock.

Neutron Radiation

Some of the less spectacular effects which may result from exposure to low intensity neutron fields and are of significance to locksmiths are observable primarily as chemical changes. These changes in-

Figure 157. Hand and foot counter. This is a very sensitive instrument and is used to detect the presence of radiological contamination on the hands or feet. If limbs carry any contamination, an alarm is shown and the strength of the contamination registered.

clude such things as an accelerated corrosion rate (the radiation serving as a catalyst), and polymerization of greases and oils, changing them to thick, black, tar-like substances.

Higher intensity fields may produce additional changes, ranging from changes in hardness and embrittlement of metals to inducing intrinsic radioactivity in metals and lubricants. Of particular significance to locksmiths in this regard is the phenomenon known as the carbon-14 transmutation where ordinary carbon (as in graphite) is transformed into a radioactive form of carbon. Because of this effect, no graphite should be used in a neutron field of any appreciable intensity whatever unless its use in that specific circumstance is first approved by a qualified nuclear physicist or reactor engineer.

Graphite, being light and powdery, tends to fly about and to stick to objects with which it may come into contact; therefore, the locksmith has a direct obligation to avoid using it in a situation which might result in scattering a radioactive carbon isotope about an area where people might come into contact with it or even inhale it. Transmutation of metals in a lock is not likely to be observed, but if it should occur, the lock would have sufficient intrinsic radioactivity to require that it be handled with tongs, and no repair of such a lock, or other servicing, would be practical.

Beta-Gamma Radiation

Beta and gamma radiation are quite similar in their effect on locks, although the destructive potential of gamma radiation is consistently high and that of beta may vary rather widely. Gamma radiation, because of its similarity to X rays, is somewhat more likely to produce an observable effect deep within the mechanism of a mechanical device such as a lock.

Beta-gamma effect usually takes the form of an accelerated corrosion rate, the polymerization of petroleum lubricants, and the embrittlement of many plastics and lacquers. Direct damage to metals, such as embrittlement or transmutation are not significant, except for such damage as may be associated with an accelerated corrosion rate.

Alpha Radiation

Although the alpha particle is heavy, it lacks penetrating ability and cannot get through even the most common obstacles for more than a limited distance. Hence, any damage to a lock is confined to the immediate vicinity of an alpha emitting particle of contaminant, and is largely confined to a slightly accelerated corrosion rate. The intensity of field of alpha radiation is limited by its lack of penetrating ability, as the alpha particle is slowed and stopped in a comparatively short distance even in travel through a medium as tenuous as air.

The most significant aspect of alpha radiation is its difficulty of detection and the possibility of a lock becoming contaminated with an alpha producing source. Although alpha radiation is not particularly dangerous outside of the body, it is an entirely different matter if alpha-generating dust or particles are ingested or become lodged in eyes or mucous membranes. Extreme precautions are justified if a lock is suspected to be alpha-contaminated, both to protect the locksmith and to prevent dissemination in the area where he does his work and the area through which the lock must be transported (from the door from which the lock is removed to the place where the work is done).

Suggested Rules for the Protection of Locks in a Radiation Environment

1. Use no graphite in a neutron field. Moly (molybdenum disulfide) is suggested as an acceptable substitute. Graphite may be used, if desired, in other radiation fields where neutron radiation is not present.

2. Avoid the use of wet lubricants to the greatest possible extent. If wet lubricants are used, the thinnest possible coating should be applied, in order to avoid the extreme gumming-up which takes place with wet lubricants.

3. Teflon and nylon are two plastics which withstand radiation embrittlement much better than many other plastics; therefore nylon parts need not be replaced and a powdered teflon lubricant such as Slipspray may be used effectively in this environment. The powdered teflon lubricant has the additional advantage of inhibiting corrosive attack; however, it is frequently not as slick as the lubrication requirements for the particular lock function and in this case, another lubricant must be used with it (for example, moly spray, or a light rub with powdered moly).

4. In reassembling and reinstalling a lock in a radiation environment, each screw and the part it is screwed into should be lubricated with Slipspray or Wonder Mist. This procedure will facilitate reassembly and replacement, and help to prevent corrosion, thus facilitating future dismantling.

5. When servicing a lock in a radiation environment, the lock must be made as nearly perfect, both in function and in fact, as skill and ingenuity can make it. A lock failure in a radiation area could mean not merely an inconvenient delay, but the loss of life itself.

VI

Key Duplicating and Code Key Cutting

LESSON 23

Key Duplicating

A number of factors in key duplicating serve to distinguish a properly made key from the common dime store duplicate. Some of these are: the availability of the proper key blank, use of the proper machine for duplicating a particular type of key, use of the correct type of cutter, the accuracy inherent in the machine itself, and the skill of its operator.

Let any of these factors not be taken into full consideration, and the resulting key will be little or no better than the dime store key which may operate the lock — after a fashion, and cause speedy deterioration of the mechanism. This sort of slip-shod key duplicating seems to do little to tarnish the reputation of those places which simply duplicate keys; but if a locksmith turns out keys of this kind, his reputation is very quickly gone, not only with respect to key duplicating, but his other work suffers as well.

Much of this chapter is about the duplication of pin and disc tumbler keys because of their popularity; however, many of the things which are true of the duplication of pin tumbler keys are also applicable to keys for lever tumbler locks, disc tumbler locks, and warded locks.

Figure 158. Warding cuts on cylinder key blanks. Observe the way in which the configuration of these cuts shows up in an exaggerated form at the terminal end of the cuts. Illustration courtesy of Eaton Yale & Towne, Inc.

KEY BLANK SELECTION

The first step in key duplicating is the selection of the proper key blank. A number of factors go into the selection, including the shape and location of the grooves which allow a particular key blank to enter its proper keyway. These grooves are those which are milled lengthwise in the key bit to permit the key to pass the keyway wards and enter the lock; naturally, each of the grooves cut in the key blank must be in the proper location and of the proper shape and dimension to pass the wards.

Some of the common shapes of the warding grooves are shown in Figure 158, together with the configuration at the shank end which the warding cut creates at the point where the mill quits cutting. Often the terminal end of a warding cut will give a reliable indication of the shape of the groove in situations where there otherwise might be some doubt.

Keyway wards are often to be found in the same location as those on another key blank, but the shape of one or more of these warding grooves, or the difference in size, prevents the key blank from entering a keyhole which it is not intended to fit, or vice-versa. It is true that key blanks in so-called *master sections* are available, as shown in Figure 159, which will pass the keyway wards on two or more different sets of keyway wards or key sections,

as they are called in reference to the key blank.

Essentially, a locksmith is in business to sell protection to his customers. It is *not* good ethics to use one of these master sections to duplicate an ordinary operating key for a customer. In using such a master section, the key made for one person may also operate another customer's lock, if it should be keyed to the same combination, even though utilizing a slightly different key section. This is often done in dime store operations. There is no regard to the value of protection for other customers. Safety is lost through the use of this type of key blank to duplicate individual keys.

Certainly, the key will work in the customer's lock for which it was made, but it may also work in the lock of another customer, who depends upon it to protect his valuables. For this reason, the use of master sections in duplicating an individual key is not ethical. A key which is duplicated should be duplicated in *all respects* as identically as possible to insure your own customer's satisfaction, as well as to protect the general public against unplanned interchanges of keys.

Another factor which is often confusing, particularly in the duplication of worn keys, is the thickness of the key blank which should be used. A number of key blanks have the same section, but vary in thickness. If a blank which is too thick is

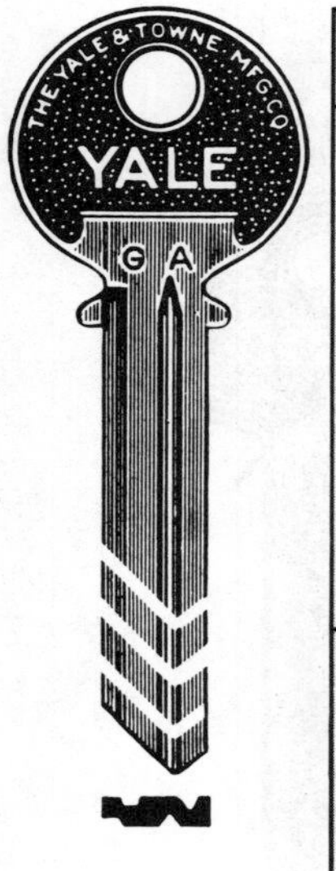

PINS	ROUND BOWS*	PRICE GROUP A* SURETY SECTION SUFFIXES									PRICE GROUP AS* MASTER-KEY SECTIONS	
											MASTER-KEY SECTION* SUFFIXES	CAT. NO. ROUND BOW MASTER-KEY BLANKS
4	9½	GA	GB	GC	GD	GE	GF	GG	GH	GK	GMK	109½
4	9½	SA	SB	SC	SD	SE	SF	SG	SH	SK	SMK	109½
5	8	GA	GB	GC	GD	GE	GF	GG	GH	GK	GMK	108
5	8	LA	LB	LC	LD	LE	LF	LG	LH	LK	LMK	108
5	8	SA	SB	SC	SD	SE	SF	SG	SH	SK	SMK	108
5	8	TA	TB	TC	TD	TE	TF	TG	TH	TK	TMK	108
6	11	GA	GB	GC	GD	GE	GF	GG	GH	GK	GMK	111
6	11	LA	LB	LC	LD	LE	LF	LG	LH	LK	LMK	111
6	11	SA	SB	SC	SD	SE	SF	SG	SH	SK	SMK	111
6	11	TA	TB	TC	TD	TE	TF	TG	TH	TK	TMK	111
6	11	VA	VB	VC	VD	VE	VF	VG	VH	VK	VMK	111

GA GB GC GD GE GF GG GH GK GMK

GMK shown.

Typical of LMK, SMK, TMK, VMK.

SURETY SECTIONS GA Series shown. Typical of LA, SA, TA, VA Series.

*COMBINATION SURETY SECTIONS and MASTER-KEY SECTIONS. Used for extensive pin tumbler master-keyed and grand master-keyed systems. The blanks will enter plugs of the several sections indicated by the symbols stamped on each blank.

Figure 159. Master section and the individual keyway sections usable with it. Illustration courtesy of Eaton Yale & Towne, Inc.

chosen, it will not enter the keyway. Conversely, if a blank is chosen which is too thin for a particular keyway, the blank will tip slightly when it is turned in the lock, often tipping the tumblers, or sliding the blade of the key partially out from under the tumblers, resulting in failure of the tumblers to align properly at the shear line.

Where various thicknesses of key blanks exist for any particular grooving of a key, one must judge the amount of wear on the key which is brought in to be duplicated, and select the key blank accordingly. For example, if a key is heavily worn, and yet the blank is about the same thickness of the thinner of two possible thicknesses of available blanks, then the thicker of the blanks would be the logical choice. If a key is slightly worn when it is brought in to be duplicated, and the thickness of the blank is about the same as the thinner available key blank, then the thinner blank is usually the better choice.

A great many key blanks are available with different lengths in the bit section of the key, as shown in Figure 160. The length of blank which should be used depends upon whether the lock is equipped with three, four, five, six, or seven tumblers. Suppose a key blank is chosen for, say a six-tumbler lock, when the lock is actually of only five-tumbler construction, and the keyway is blind or obstructed behind the fifth tumbler. The six-tumbler length of blank may not enter the keyway far enough, before striking against the obstruction, to position the cuts properly under the tumblers, which they should operate *before* striking against the obstruction. The result is a misfit or inoperative key.

Conversely, if a key blank is chosen in a four-tumbler length for use with the same five-tumbler lock previously mentioned, there will not be sufficient room on the tip of the blank for a cut to operate the fifth tumbler, resulting in an inoperable, or improperly operating key.

Another factor which can be of considerable importance in key duplicating, yet is often neglected, is the positioning and dimensioning of the stops. The stops are the small protrusions at the bow end of the key bit. They stop the foreward progress of the key into the lock at such a point as to position each cut on the key directly below the tumbler which it must operate.

Some keys are equipped with only one stop at the cut edge of the key. This type of stop (and most other stops) is designed to operate on the surface of the outer end of the lock plug, positioning the cuts relative to that surface. Some keys are equipped

Figure 160. Bit-length of cylinder key blanks. Key blanks are available with different lengths of bit, for use in locks with varying numbers of tumblers, and each must be of the correct length for the lock with which it is to be used. Illustration courtesy of Eaton Yale & Towne, Inc.

4 PIN—10½
5 PIN—13½
6 PIN—12½

with a second stop opposite the cut edge, that is, at the uncut edge of the key bit. In many instances this stop contributes to nothing more than the symmetry of the key and serves no useful purpose. On rather rare occasions, the back stop on a key is used for positioning a key in the keyhole. There are, however, a number of keys where both stops are used simultaneously to position the key, particularly in pin or disc tumbler locks or warded locks.

Keys for disc tumbler locks are an outstanding exception to the single stop alignment, since in most disc tumbler locks both stops are operative and must be correctly positioned. The dimensioning of the stops can be a very important item also, particularly in automotive keys and other disc and pin tumbler keys where the stops actually penetrate beneath the nominal surface of the plug, and come against a shoulder which is inside the keyway. Keys for locks of this type must have stops which are small enough to enter the space provided; yet they must be large enough to fulfill their function in properly positioning the key, while allowing full penetration of the key into the lock.

The shape of the stop is rarely important except in the case of those stamped keys where the surface of the stop which comes against the lock itself may not be precisely square with the bit of the key, but has an angular taper. If the angular taper is not the same as that of the original key, then the duplicate key will penetrate the lock either too far, or not far enough, resulting in a misalignment of the cuts with the tumblers they are supposed to operate.

Stops are rarely used on keys for lever locks; instead, the key is usually stopped by the end of the key coming in contact with the inside of the lock case, or some part of the lock. In those rare

Figure 161. Stop variations. This illustration shows some of the variations which may occur in the stops on a key. The stops help to position a key in a lock and can be very important in the selection of the proper key blank. Illustration courtesy of Eaton Yale & Towne, Inc.

instances where a stop is encountered on lever lock keys, the positioning, dimensioning, and style of the stops should be scrupulously observed.

The length of the shank of the key is usually of minor importance in key duplicating, with the exception of certain locks, primarily of the automotive type. In these, the stop is recessed rather deeply into the plug and the shank of the key must penetrate the plug for a considerable distance to permit the stops to come against their shoulder inside the plug. When this situation prevails, the shank must be of an adequate length to ensure the penetration of the stop into the plug to the required depth. Very few locks other than automotive locks have this requirement, but it is found on a few disc tumbler padlocks and filing cabinet locks, and one or two old-style push-key types of the pin tumbler padlock.

The push-key padlock key is readily recognizable by virtue of its being cut on a flat steel key blank, with the cuts made in the pin tumbler style of cut. This particular padlock has the keying built directly into the butt end of the shackle. Opening the

padlock is a matter of putting the key into the lock and then either pushing on the key or pulling on the shackle. Since this involves pulling the entire lock plug more deeply into the lock, the shank of the key must be sufficiently long to accommodate the amount of pull of the complete plug, key and all, into the lock.

Original keys for this type of lock may be made by the wiggle system. It is preferable, however, to use a homemade key blank for the original key, rather than the factory prepared blank, because the steel key blank is too hard to take a suitable impression from the tumbler. A homemade blank cut from a sheet of brass is usually preferable for this work. Tension must be applied by pulling on the shackle, rather than turning, as in most impression work, since the shackle will not turn. This brass first key may then be duplicated on the steel blank to provide the customer's key.

The shape of the tip of the key, particularly for disc and pin tumbler locks, serves a function in aligning the key to enter the keyway properly. Since there is a dimple in the surface of the lock plug at the position where the key enters the keyway, the point of the key in the bottom of the dimple in the plug face aligns the key with the keyway.

A tip which is too high or too low will place the grooving for the keyway wards out of alignment with the wards which the grooving must pass, resulting in difficulty in getting the key into the lock. In addition to this, if the extreme tip of the key is positioned too high up (too near the cut edge of the key), it will come against the tumblers, which then provide a firm stop for the key and prevent it from entering the lock at all. If the extreme tip of the key is too near the back of the key, the only obstacle presented will be faulty alignment of the key by the dimple in the face of the lock plug.

Lever tumbler key blanks usually have a small nose, or tip, which juts out from the true tip of the key at a right angle. The size and length of the nose is important in that it must be small enough to penetrate through the lock case as far as the length of the lock plug permits. Yet, it must fit the available hole snugly enough to prevent the tip of

the key from wobbling excessively if lateral pressure is placed on the key. The nose simply serves to maintain the correct tumbler alignment by preventing the key from wobbling in the lock.

Criteria for bit key blank selection are somewhat different, as shown in the accompanying illustration, Figure 162. In this illustration, dimension A represents the thickness of the lock case. A blank must be selected in which dimension A is no greater than the corresponding dimension of the original key. The key duplicating machine will trim away any excess of bit at this point which may be necessary to increase dimension A. It cannot help in any respect if dimension A on the key blank is already greater than the corresponding dimension A on the original key.

Dimension B represents the distance from the near outside of the lock case to the inside of the far lock case wall. In other words, B minus A equals approximately the inside depth of the lock case and must be correct if the finished key is to pick up all of the tumblers and throw the bolt properly. Dimen-

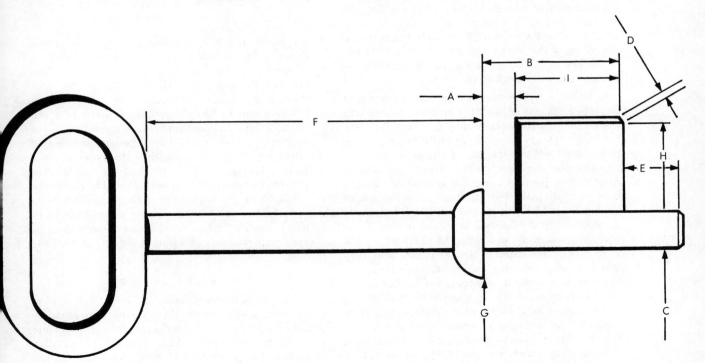

Figure 162. Points of comparison for the selection of a bit key blank. Dimensions C and D are critical and must be correct on the raw blank. Dimensions A, B, H, and I require only that there be enough stock to allow the key to be cut to the required dimension. Dimensions F and G usually will vary considerably without affecting the proper operation of the key. Dimension E is almost never of significance to the operation of the key and is quite sufficient if it supports the tip in the lock.

sion E is much less critical in that it simply represents the amount which the stem of the key protrudes on the opposite side of the lock. This type of lock is usually provided with a through keyhole so the key may be inserted from either side of the lock.

Dimension C represents the axis upon which the key rotates. It is a self-supporting key, with the support being derived from the diameter. For this reason, the diameter is quite critical in most lever lock keys. Too large a diameter at C will prevent the key from entering the lock. Too small a diameter will permit the key to drop too far down into the notch which allows the bit of the key to enter; the result is a misalignment of the tumblers, with consequent failure of the key to operate, no matter how carefully the actual cutting may be done.

Dimension H represents the height of the key bit above the surface of the pivot portion of the key bit. It is important only to the extent that it must not be less than that of the key to be duplicated. Dimension D represents the thickness of the key and is important in several respects on some keys. First, it must be narrow enough to permit the key to enter the lock. Second, it must not be narrower than the original or there will be a difference in the timing of the lifting of the lever tumblers, resulting in failure of the key to lift the tumblers high enough for the fence to enter the notch in the levers.

In addition to the overall thickness as shown at D, if the thickness of the bit is tapered, as shown in Figure 163, the same taper must be selected for the key blank. This is necessary in order to provide the proper thickness for the key at the point where it contacts the tumblers, as previously mentioned. Also, it must provide a suitable amount of material from which to make any warding cuts in the side of the key bit (which this type of key blank frequently incorporates in the finished key), and it must do so without excessively weakening the key bit. The ward cuts are not found in the key blank, since they may be made on the duplicating machine. However, the necessary amount of stock must be there, and correct in taper, in order to provide for the finished key to enter the keyhole which it is supposed to enter, to the exclusion of similar keyholes having merely different keyway wards. On this type of lock, these keyway wards are often used to provide a greater number of key changes than would otherwise be possible.

The collar, indicated in Figure 162 as dimension G, serves as a stop for the key. The diameter and length of the collar are both usually unimportant to

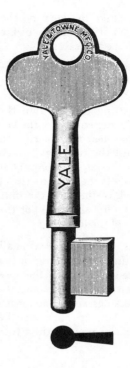

Figure 163. A variation of the bit key blank. It may have a tapered bit as shown. This type of key bit is commonly used for locks having keyway wards. Illustration courtesy of Eaton Yale & Towne, Inc.

the proper functioning of the key, as long as the diameter of the collar is sufficient to provide a positive stopping action to the penetration of the key into the lock while allowing the key to enter through any keyhole escutcheon.

Dimension F, like the length of shank of a pin tumbler key, is usually of small importance, the only criterion being that it must permit the insertion of the key to the stop. In some rim types of lever locks, this means that it must be capable of penetrating completely through the door, plus the thickness of any keyhole escutcheon present. Conversely, dimension F should not be so long as to make the key more unwieldy to carry than is necessitated by the penetration requirements of the door.

TYPE OF CUT ON THE KEY

Since keys for lever locks, bit keys, and warded lock keys have substantially the same type of cut in virtually all cases, these locks present only problems of dimension, rather than type. However, disc tumbler and pin tumbler keys may have one of three different basic types of cut, and these various types can pose a problem in key duplication.

Figure 164 shows the different types of cuts found on pin and disc tumbler keys and the ways in which the tumblers are set into the cuts in order to operate the lock. Notice that all of these types of cut have a slope on the sides to permit the key to lift each succeeding tumbler as the key is inserted and withdrawn from the lock. If the sides of the cut were square, or even nearly square, the key would not lift the tumblers out of its path, hence it could neither be inserted nor withdrawn.

A key duplicating machine must have a cutter, the angle of which is at least as steep as the angle of cut on any key which may need to be duplicated. Otherwise, the cutter of the machine will not penetrate sufficiently to make the bottom of the cut correspond to the original in either depth or configuration, however accurately a machine may be set. In the V bottom cut shown at A, the tumbler may not set on the bottom of the cut in the key, but rather, may rest upon the sides of the cut, making it essential that the angle of the sides be exactly that of the sides of the cut of the original key. Too wide an angle would mean that the tumbler penetrates too deeply into the cut key; too narrow an angle would mean that the tumbler is held further up out of the key. Either situation results in misalignment of the tumblers at the shear line.

It is worth noting at this point that the positioning of the tumbler cuts relative to the stop is much more critical with this type of cut than with others. Even a small misalignment will cause the tumbler to ride up the slope of the cut, again resulting in misalignment at the shear line. The point of the cutter used in cutting such a key must come almost to a sharp point in order to penetrate deeply enough

to make this type of cut. This factor requires that you carefully select the cutter or machine with which to duplicate the key.

The flat-bottomed cut, or *truncated V cut*, shown at B in the illustration, is probably the easiest of all to duplicate and is probably the most commonly found cut on original keys. It permits the use of a cutter with a slight flat on the tip. This helps to minimize the striations which are caused by the key blank traversing the cutter. This also makes the longitudinal positioning of the cuts less critical than is the situation with other types of cut.

The round-bottom cut shown at C allows the extreme tip of the tumbler to touch the bottom of the cut, as does the flat bottom cut shown at B. However, the positioning of the cut longitudinally in the key is as critical as it is in A, because any deviation from the normal positioning of that cut under the tumbler will cause the tumbler to rise above the shear line. This was the situation with A, although the condition is not so acute when the round bottom cut is used. The round bottom cut utilizes more space at the bottom than the V-bottom cut, and in some cases, more than the flat bottom cut, although the curvature of the cut bottom helps to minimize this more than a casual inspection would indicate.

The sides of the cut on these keys is often more steep than commonly found, on account of the extra space often used at the bottom, and the requirement to lift the tumbler sufficiently for high cuts in the next position. This means that the duplicating machine tends to turn out the duplicate key with a cut more like the V-bottom type than as it should be — with a cut of the round-bottom type. For this reason, it is essential to watch the configuration of the bottom of the cuts on the finished duplicate closely to ensure that the full curvature of the round bottom has been properly duplicated and has not been interfered with by the configuration of the cutter.

Key Duplicating Machines

Key duplicating machines are made up of three essential elements: a pair of vises, which are coupled together and move in unison while one vise holds the pattern key and the other holds the key blank; the cutter, which cuts the key blank to the proper shape and depth; and a guide, which works against the cut edge of the original key to control and limit the movement of the pair of vises holding the two keys, thus controlling the penetration of the cutter into the key blank. Other elements are, of

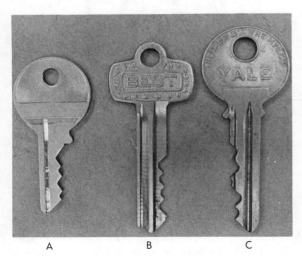

Figure 164. Principal types of cut used on pin and disc tumbler keys. A, V-bottom cut; B, flat-bottom cut; C, round-bottom cut.

course, present in key duplicating machines, and are important, but only these three are directly concerned with the cutting of the key. Other parts and factors in a key machine contribute to the maintenance of the accuracy in duplicating which the machine is capable of in a much less direct manner.

Key machines are made for a multiplicity of purposes. Some are limited to the duplication of cylinder keys only. Others are for duplicating keys for both cylinder and lever tumbler locks. Yet others are made to duplicate cylinder keys, lever tumbler keys, bit keys, and warded keys. Other special purpose machines are made for duplicating keys which have cuts on the sides of the key blanks, rather than on the edge. Other special machines are available for the duplication of the tubular cylinder type of keys, such as the Chicago Ace and similar keys. Whatever the apparent differences in construction of duplicating machines, the three important elements remain the same.

Vises

Key machine vises may be considered as two vises coupled by an extremely heavy bridge of material, or as a single vise having two jaws. In either case, the movement of the jaws is coordinated by the coupling bridge which moves them in unison toward, away from, and along the cutter. To accomplish the in-and-out movement of the vises, the bridge may be pivoted in a sliding pivot arrangement, thereby permitting the entire vise assembly to slide lengthwise in such a way as to traverse them across the positions of the cutter and guide, while the pivot allows them to move in and out. The pivot point is so positioned as to bring the key blank squarely into the cutter when the cutter is in a position approximately midway between the shallowest and the deepest cut made on the average key which the machine is expected to duplicate.

Referring to Figure 165, we see that not all depths of cut can be exactly square across the key blank, as

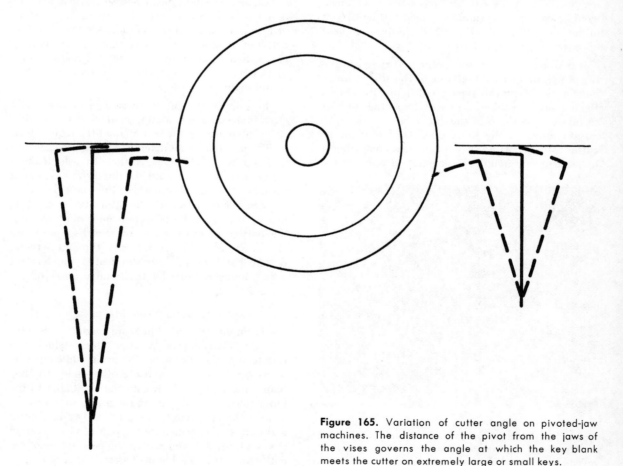

Figure 165. Variation of cutter angle on pivoted-jaw machines. The distance of the pivot from the jaws of the vises governs the angle at which the key blank meets the cutter on extremely large or small keys.

Figure 166. Cutter angle is uniform through all depths of cut. The key duplicating machine shown utilizes sliding vises to produce a straight line motion of the key into the cutter, keeping the cutter angle uniform through all depths of cut.

this pivoted motion tends to cause the key blank to meet the cutter at different angles which vary with the depth of cut. For most ordinary key duplicating, the tipping of the cut across the edge of the duplicate key is an insignificant factor in a well-designed machine. It may become significant if the key blank has an unusually wide or narrow blade and the lock is made to extremely close tolerances.

Any deviation from the norm in this type of machine is ordinarily less significant than micrometer measurements would make it appear. This is so because most key blanks have the center of the blade as carefully centered with the tumbler as possible; and a pin tumbler lock has only the center of the pin seated on the center of the key blade, where the center portion of the cut would be substantially at norm. Those instances which may create problems with this type of machine would involve those keys having a key blade somewhat eccentric to the location of the pin tumblers and either considerably larger or considerably smaller than ordinary. Figure 165 shows, in an exaggerated form, this type of displacement of cut, and how the extent of the angular displacement is dependent upon the distance of the pivot point for the vises from the vises.

The sliding type of vise used in the key duplicating machine shown in Figure 166 utilizes a straight line motion as opposed to the radial motion of pivoted vises, virtually eliminating the factor of angular displacement of cut. The sliding motion is achieved by the use of two sets of slides. One allows the key to slide along the cutter, just as in the

pivoted type. The other set, working at right angles to the first, allows the keys to be fed into and retracted away from the cutter in a straight line motion directly toward the center of the cutter. There is very little, if any, angular displacement of the cut across the key.

Usually the set of slides which permits feeding of the key into the cutter is arranged so that it can be tipped back for inserting the original key and the key blank into their respective vises. The pattern key and the key blank are aligned in the vises with the aid of a *key gauge* which positions the shoulders of the pattern (original) key and the key blank in their respective vises to an interval which is exactly the same as the distance of the key guide from the cutter. The purpose is to position the cuts in the duplicate key so that they will come exactly under the tumbler in a location identical to the positioning of the same cuts on the original key.

The jaws of key machine vises are made in a number of different styles in order to grip the key securely enough for duplicating, while maintaining both the original key and the key blank in correct alignment with the key guide and the cutter. Some vises are designed to grip only the blade portion of the key, and others grip part of the head or shank of the key also.

There are two governing criteria in gripping a key for duplication: 1. The key must be gripped securely enough to not shift during duplication. 2. The key *must not* be tipped in the vise, either before or during duplication. Both the pattern key and the key blank must meet the gauge and the cutter squarely in order to duplicate properly. Those key vise jaws which grip the head of the key contribute to this security against tipping, provided they are kept in proper adjustment.

There are, however, exceptions to this condition, principally where the head of the key is thicker than the blade. Where such a situation occurs, the careful adjustment of the vises can usually compensate for this condition. The jaws of most key duplicating machine vises have a backstop which is part of the lower jaw of the vise. The backstop serves as a solid backing for both the pattern key and the key blank in order to maintain a uniform depth while duplicating the key. This backstop is set back from the true edge of the vise a standard distance corresponding to a value just less than that for the remaining material in the key blank after the deepest possible cut has been made on the smallest size key to be cut on that machine.

Keys which are very narrow of blade, such as those for the No. 7 Master padlock, may not protrude far enough from the jaws of the vise to allow the deepest cuts to be made. In this situation, straight wires may be used behind both the pattern key and the key blank as shims to bring them out far enough for the cutter to be effective, and for the key guide to be able to follow the contours of these deep cuts.

It is important to note that these wires must be of an identical diameter and perfectly straight. In addition, they must not interfere with a shoulder on the back of the key blank. Some keyways have grooves of a dimension and configuration which makes it difficult or impossible to clamp them securely in the vises of a key machine without tipping. When this situation is encountered, there are two means by which it may be overcome. The problem sometimes may be solved by the use of a key machine having a vise which is different in type or depth of jaw.

It may also be solved by shimming the upper groove of the key which is contributing to the problem, by using a short piece of wire of appropriate diameter to bring the groove up to where it may be gripped by the jaws of the vises. Here again, an identical wire must be used for both keys. In this instance, the wires are actually laid into the grooves of the key. It is obvious that the diameter of the wires used must be carefully selected if they are not to become permanently wedged into the grooves of the key, or distorted when the vises are tightened.

Another point of caution is to avoid tightening the jaws of the machine too firmly on a key which has been shimmed in this fashion, or one or both keys

are likely to be distorted, regardless of how carefully the shim has been chosen for diameter. It is always well to leave such a shim sticking slightly out of the open end of the keyway groove, since the shim should fill the groove well; however, even slight sticking can be an annoyance if there is no protruding end to grasp in removing it.

Some key duplicating machines achieve the proper gripping of a maximum number of keyway sections by means of adaptors, commonly called *blocks*. Blocks are used to accomplish secure gripping of keys by either changing the depth to which the key enters the vise, or by gripping the key in one of the grooves. A number of these blocks which are intended for use on one key machine are shown in Figure 167.

Cutters

Cutters for key duplicating machines fall into two major classes: file type cutters and milling type

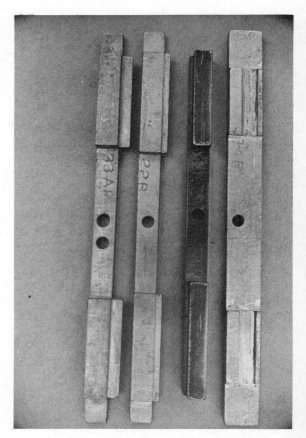

Figure 167. Adaptor "blocks" for gripping different keys in the vises of the key duplicating machine shown in Figure 166.

Figure 168. File and milling cutters in the configuration most commonly used for cylinder key duplication on machines which have the vises traversed by hand.

cutters. The most commonly used cutters for duplicating cylinder keys are those in Figure 168, which are file and milling type cutters, respectively. These two cutters have one feature in common, and that is, one side produces a nearly right-angle cut in the key. The flat side is next to the shoulder of the key, and is made with this configuration so that cuts may be made in the key quite close to the shoulder.

Unless a milling type cutter of this sort has a very small flat on the points of the teeth of the cutter, or the traversing of the cutter is *very* slow across the key, the milling type cutter will leave some striations on the cut surface of the key. These are more pronounced than those left by the file type cutter. However, the milling cutter does cut more quickly and achieves faster penetration in depth than does the file type cutter; hence the prevalent use of milling type cutters in automatic duplicating machines, where the feed of the key into the cutter and that of the key across the cutter is mechanically controlled.

The file type of cutter produces more striations, but they are considerably smaller than those made by a sharply pointed milling cutter. Either of these cutters may be used to make almost any type of cylinder key, since they will faithfully follow a flat bottomed cut, a V bottomed cut, or a round bottomed cut in the key.

Cutters of the type shown in Figure 169 are used exclusively for the making of lever cuts or warding cuts in keys for those types of locks. Key cutters of the side milling type, which are not shown in the illustration, are a recent development and may be used to produce lever cuts, warding cuts, or cylinder

key cuts of reasonable accuracy for any type of cylinder key cut except the sharp V bottom. The cutter is very similar in configuration to that shown in Figure 169, but also has milling teeth on the sides of the otherwise flat cutter, which permit it to be used in the sidewise motion required for cylinder keys, as well as the in-and-out motion required for lever and warded lock keys.

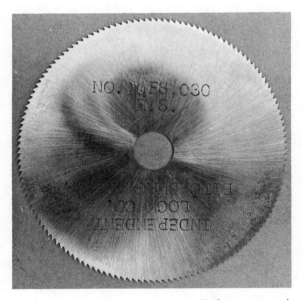

Figure 169. Slotter. The "slotting type" of cutter is used to produce square-sided cuts in keys for lever and warded locks. A similar cutter has teeth on the sides of the cutter as well, and may be traversed to widen a slot or to produce a cylinder key.

The diameter of the cutter used has a bearing on the quality of the keys produced in that the smaller diameter cutters produce a slightly concave cut on a thick key. Cutters of larger diameters produce a less concave cut, giving a more nearly complete bearing surface for tumblers, particularly those of the disc or lever types, where the tumbler extends completely across the bit of the key. This, of course, is less important for pin tumbler keys, since the tumblers contact the blade of the key only in a small portion of that surface. The more concavity in the cut surface of a given key, the more rapidly that key will wear, and for that reason will prove slightly less durable than keys with lesser degrees of concavity.

The speed at which the cutter of the key machine is operated may vary considerably, and is important to the life of the cutter and the proper duplication of keys. Optimum speed for a cutter on a key machine depends upon several factors. The file type does a better job at low speed than does the milling type, as a general rule.

If the cutter is made of carbon steel, the cutter speed will have to be substantially reduced from that which would be permissible for a similar cutter made from high speed steel. The type of material being cut imposes further restrictions on the speed of the cutter; for example, a flat steel key may not be cut at the same cutter speed as would be permissible for a brass or nickel silver key, because the cutting (machining) speed for steel is considerably lower than that for brass.

The speed of the cutter must not exceed that of the most restrictive requirements, yet the speed must not be too low or the key blank will not feed into the cutter smoothly, causing an erratic and heavily striated cut. In the duplication of cylinder keys, the cutter speed should be maintained at the highest speed which these limitations will permit, thus minimizing the effect of the striations which are produced when a key blank is moved across a cutter more rapidly than the following teeth of the cutter can wipe out those striations.

The feed of a key machine, whether manual or automatic, is very highly interdependent with the speed of the cutter's rotation in order to avoid excessive striation in the duplicate key. Inordinate striation would abrade the tumblers as the key is inserted and withdrawn, thus causing an unusual amount of wear in the lock.

The Key Guide

The key guide which operates against the cut edge of the pattern key to move the vises, and consequently the key blank, into and out of the cutter, must have a tip which is identical in cross sectional shape with the cut which will be produced by the cutter when it is moved into the key blank in a straight line. Stated otherwise, the key guide and key machine cutter must be faithful reproductions of each other in cross section. Guides for key machines having sliding jaw vises usually have a straight tip on the key guide, since the motion of the key blank into the cutter is always in a straight line, rather than having a radial motion.

Guides for key machines with pivoted vises usually have a curved tip key guide. The curvature of the tip corresponds to that of a circle, the diameter of which is the same as that of the cutter. Thus, the tip of a guide of this type very much resembles a segment of arc taken from the perimeter of the cutter (but having no cutting teeth, of course). The curved tip on key guides for key machines of the pivoted vise type is highly desirable in that it assists in minimizing the effect created by the segment of arc movement characteristic of the motion of the jaws of vises on this type of machine. Otherwise, cuts might be produced which were not true to depth, because of the pattern key striking the key guide at a thicker portion of the key in those cuts where the key is slightly tipped by the pivoting of the vises.

Automatic Key Duplicating Machines

Automatic key duplicating machines are intended for the duplication of cylinder keys only; that is, keys for pin tumbler or disc tumbler locks. They customarily have the pivoted type of key vise which, during the time the machine is in operation, is arranged with a spring tension to feed the key blank into the cutter automatically, and keep it there while the key is being cut. The limit of travel of the key into the cutter is established by the key guide, as it is with manually operated machines, but, instead of pressure being maintained by hand, it is applied by spring tension.

Another added feature of automatic key duplicating machines is a feed arrangement, (usually contained in the base of the machine) consisting of a worm gear operating to turn another gear which rotates a shaft having a construction very much like the level wind of a fishing reel. The level wind arrangement is used to traverse the key blank across the cutter.

The traversing of the key blank starts the blank into the cutter at the tip of the blank and works toward the bow of the key; then the level wind arrangement returns it back to the original position, where a switching arrangement, sometimes lever operated, cuts off the power to the machine automatically when the cutting of the key is completed. Those with the lever type of switching usually utilize this mechanism to provide for a release of the spring tension holding the key blank into the cutter at the same time that power to the machine is cut off.

There are a number of advantages and disadvantages to this type of construction. Where it can be used, it saves considerable time by allowing the operator to do something else while the key is being cut. Also, the spring tension is more likely to be uniform than is a hand pressure. This is an important contribution to accuracy in the depth of cut of the key.

On the other side of the coin are a few disadvantages, one of which is that some cylinder keys, notably those for the Briggs & Stratton side bar locks, frequently have a square shoulder on the cut which is nearest the bow of the key, in order to minimize the risk of broken keys. An automatic duplicating machine must have a taper on both sides of the guide and cutter in order to feed properly in both directions. The tapered shoulder on the side of the guide next to the shoulder of the key means that the guide, and consequently the cutter, cannot reach to the bottom of a cut having a square side next to the shoulder. This cut must be either cut by hand or done on another type of key machine.

An automatic key duplicating machine frequently has a high rate of travel in relation to the cutter speed so that the cut produced exhibits a considerable amount of striation in the surface of the cut (cutter skips occur). Stated another way, cutter skips occur if the traversing of the key across the cutter is not commensurate with the cutter speed. This is one reason for wanting to see a sample of a particular type of automatic key duplicating machine in operation before purchasing the machine.

Notwithstanding these limitations, a good automatic key duplicating machine is an excellent investment because the production rate of these machines is high; and a good machine produces keys of high quality, while allowing the operator to either prepare for the next duplication, or to do other work while the machine is running.

ADJUSTMENT OF KEY DUPLICATING MACHINES

Adjusting of the key duplicating machine is of critical importance, and should be approached with all of the care of which one is capable. The process is accomplished by first placing two identical key blanks in the vises of the duplicating machine, exactly and as carefully as if one intended to duplicate a key blank. Then the key blank which is in the pattern key position should be brought up against the key guide, and maintained in that position while the cutter is rotated slowly by hand.

The guide is adjusted until the cutter just barely touches the blank which is in the duplicate key position. This provides a rough first alignment for the machine; indeed, it would cut keys which would usually be within .002 in. to .005 in. of being the same as the pattern key. This is the point at which the real alignment of the machine begins.

The key blanks are removed from the machine at this point, and a duplicate key is cut. One of the depths on the duplicate key is measured against the corresponding depth on the pattern key, using a key micrometer or micrometer caliper. The key guide is then further adjusted in or out of its holding post until a key can be duplicated with an accuracy which measures within the limits of plus .0003 (three ten-thousandths) or minus nothing, in the amount of its difference from the pattern key.

This degree of accuracy is quite adequate for all ordinary key duplicating purposes, and results in a key which will fit well and operate properly in any lock in which the pattern key was satisfactory. In most cases, the resulting duplicate key may itself be duplicated. This process is not accomplished quickly, nor is it usually an easy one; it is, however, time well spent. The author maintains his key duplicating machines consistently at plus .0002 in., and finds this setting to be the most desirable for nearly all situations.

USING KEY DUPLICATING MACHINES

Certain precautions are in order before duplication of a key is begun. The first of these precautions is being sure that the key blank chosen matches perfectly with that on which the original key is cut. Second, a key machine which is capable of duplicating the original key must be chosen. This machine must have the proper type of cutter for cutting the key to be duplicated. If the key is a cylinder key, the angle of the cutter must not be greater than the angle of the sides of the cut in the original key.

The vises must be of the proper shape, and the jaws of the proper depth, to accommodate both the original key and the key blank, and to hold both firmly without tipping while duplication is in process. Also necessary is full penetration of both guide and cutter to the deepest depth to which the keys are cut, without their coming into contact with the jaws of the vise. The cutter must also be capable of cutting the material from which the key blank is made, without damage to itself.

This brings us to the actual operation of the machine. The first thing in operating a key machine of any kind is to be sure that it is clean and free from chips and cuttings, particularly in the jaws of the vises and in the area where the vises slide in the process of moving across the cutter. It is always good practice to use a small paint brush to clean the jaws of the vises just before the keys are placed in them. Place the keys in their respective vises, using the key gauge to ensure proper spacing so that the cutter will meet the blank key in a position corresponding to the position of the cuts on the original key.

At the same time the vise is being tightened on one of the keys, a pressure is maintained against the key's blade to ensure that it is firmly seated against the back of the vise, and that any back shoulder on the key does not ride up onto the back of the vise (or the block, as the case may be). Having secured both keys in their respective vises, the key gauge is moved aside and out of the way of guide and cutter.

The machine is switched on, and the original key is traversed slowly and smoothly across the guide, beginning at its tip and ending just short of its shoulder. The direction of travel is then reversed, and the original key is moved slowly and smoothly back across the guide toward its tip; all the while, an even, uniform pressure on the vises is maintained to keep the pattern key in firm but gentle contact with the key guide.

At the completion of this cycle, the duplicate key should be fully and smoothly cut. If the cuts on the duplicate are not smooth, the cycle should be repeated *until they are*. When the duplicate key is smoothly cut, it may be removed from the duplicating machine for finishing.

This procedure covers keys of the cylinder type which are cut on one side only. If the key to be duplicated is of the double bit type (cut on both sides), a special block is used to hold it in the vise, since it has no straight edge to bottom against the back side of the vise, and keep the key straight. A block with a ridge along its outside edge is used; this ridge engages in one of the grooves of the key blank, and serves to position the keys uniformly in the vises. It is essential to use the key gauge to position the keys longitudinally in the vises.

If the key is of the flat steel or bit key type, no sidewise motion across the cutter is advisable. In many cases, it is not possible, due to the square notches, and the absence of cutting teeth on the sides of most slotting cutters. Besides the risk of breaking the cutter, and the consequent risk of personal injury to the operator, moving the slotting cutter sidewise across the key will quickly wear out cutters, unless they are of the type having cutting teeth on the sides. Even then, any sidewise motion should be very slowly and smoothly made.

A straight in-and-out motion is the preferred method of making this type of cut on a key, traversing the key only when it is not in contact with the cutter. Customarily, cutting is begun at the tip of flat steel keys, or with the lever cuts of a bit key. The warding cuts on a bit key are the last cuts made. The procedure for making the warding cuts on bit keys will vary considerably, since key machines which are intended for this purpose may vary widely, both in design and operation. It is recommended that any operating or maintenance instructions for any particular key machine be followed scrupulously, unless the locksmith has made a thorough study of the instructions, understands the reasoning behind them, and can come up with a procedure which he *knows* is better.

The foregoing operating procedure will take the operator through the essential elements of successfully producing a key on a key duplicating machine; but it does not cover one of the most essential and all too often neglected phases of key duplicating, namely, that of duplicating keys safely. There are several safety hazards which are commonly associated with the process of key duplicating, or with using any power driven machine to cut keys.

First, the cutters of a key machine are intended to cut metal, but they are not too particular about that; they are also quite ready, willing, and able to cut, maim, or amputate fingers of the unwary. A key duplicating machine should be turned off when the metal of the key blank is not being cut.

While the key is being cut, and one is maintaining pressure to traverse the vises of the machine and to keep the original key against the guide, this pressure should be maintained in such a manner that, should one's fingers slip, they will not come into contact with the cutter. A finger or hand can be severely lacerated on the cutter of a key machine. Even when the machine is off and one is positioning keys in the vises or making adjustments, the possibility of acquiring minor yet painful lacerations from the cutter or the guide is always present, and should be guarded against.

The second hazard is that caused by the chips produced by the cutting operation. In any cutting operation of this type, chips are produced and will be thrown. The greatest hazard is, of course, to the eyes. Since most of the chips are of non-ferrous metal, they cannot be removed from the eye by magnetic means, if they should become imbedded there. Under these circumstances, eye surgery would very likely be necessary.

As a minimum precaution, glasses should always be worn while operating a key cutting machine. Goggles would be even better, but irrespective of the type of machine or cutter, some type of eye protection is essential. Some machines are provided with a plastic eye shield to protect the operator from chips flying into the eyes. Avoid placing your entire reliance on a shield, since a chip can ricochet from another moving part, such as a belt or pulley, and be thrown into the eye. A clean, serviceable shield should always be available.

A milling type cutter producing cylinder type keys poses another hazard in that the cuttings are usually in the form of long, needle-sharp slivers. Naturally, these are more dangerous to the eye than are the chunky cuttings made by a file type cutter, and they also create a special hazard for fingers. The operator should take care to keep his fingers out of such cuttings, whether they are on the key machine, or scattered about on the bench or work area. To leave cuttings on a chair is likely to pose problems of a profoundly personal nature.

Another serious hazard in the operation of key machines may be the presence of open pulleys, belts, or gears in which fingers may become entangled and injured. Many machines provide safety guards against this situation; where they exist, they should be left in place and kept in good condition. Whether or not there are guards on the machine, the rule is to keep fingers clear of all power-driven parts.

The traversing vises or feed mechanism of an automatic key duplicating machine can cut off fingers or severely pinch, even break bones if one is so careless as to allow his fingers to become entangled in the parts. In a machine where the key is traversed manually across the cutter, painful pinching of the fingers is likely if one puts them in the path of travel of the vises.

FINISHING THE KEY

After the completely cut duplicate key is taken from the machine, there is a certain amount of finishing to do in order to make it functional. To avoid possible damage to the lock by cuttings or burrs which may be left on the key, deburring is all that is usually required.

In a cutting operation of this type, and regardless of the type of cutter used, a certain amount of the metal is pushed aside and bent over at the edge of the key instead of being cut cleanly away. The folded-over metal is the burr. There are several means of removing this distortion to the key blank. One method is to run a file lightly across the distorted portion of the key. It is necessary to remove the distortion in such a way as to avoid any further cutting of the tumbler depth positions on the key. This means that all cutting or metal removing is done only on the side of the key where the metal moves out of its normal position in the process of cutting.

The various types and degrees of sharpness of the cutters will produce a corresponding variance in the amount and type of burr on the key. Careful filing will remove burrs; or, a sharp-edged scraper, such as the sharp edge of a knife or a triangle file ground smooth, may be drawn lenthwise along the burrs to remove them. A power-driven wire brush may be used, but the wires on the brush must be extremely fine so that the depth of the cut made by the cutter is not increased by any measurable degree. This requires not only the use of an extremely fine wire on the power brush, but also, that it be kept in good condition. A short brush, even one with fine wire, will have a cutting action in the edge of the key, resulting in depth changes.

When deburring of a key has been completed, it is always a good idea to check for any excessive roughness in the cut which the machine has made. If the cut is rough (heavily striated), the key should be returned to the duplicating machine and run through the entire duplicating process again. Square shoulders must not be left in any cut or any position, because a square edge will jam against one of the disc or pin tumblers, and impede the proper insertion and withdrawal of the key.

Some customers require that a code number or other identifying mark, or both, be made on the finished key. One should always be prepared to mark these keys in accordance with any reasonable request of the customer. Usually, a set of steel letter and number stencils will suffice for this purpose. It reflects credit on the locksmith if such markings are kept as neat and symmetrical as possible.

Another marking which is always highly desirable should be made. All keys made by the locksmith should be stamped with the name and location of the shop. Steel stencils with which they may be stamped neatly and uniformly, are available from various supply houses. Unitized stencils are extremely worthwhile for two reasons: Everyone who carries a key with this marking has the locksmith's advertising in his pocket. Many firms expend large sums in hopes of accomplishing this aim, but it's free to the locksmith. This situation should be taken advantage of. The key not only advertises, but also provides a positive product association.

An additional benefit from stamping is that it helps avoid any possibility of the locksmith's having to make good on someone else's sloppy key duplicating. If the locksmith's keys are marked with his shop name, they are completely identified as being made by him. No other keys are of his make, if *all* his keys are marked. Customers will often bring a poorly fitting or misfit key to a locksmith who had nothing to do with making it. If *all* keys are marked, it can be proved to the customer that he is mistaken as to the source of the bad key. If this is explained in a diplomatic way, it usually builds good will (and business) for the locksmith.

Code Key Cutting

A key code is a symbol (usually composed of letters, or numbers, or a combination of letters and numbers) which represents a specific plan for cutting a particular key. A code series is a planned arrangement of a specific number of these key codes, all of which are cut on the same key blank while maintaining sufficient difference in the depth of cut from one key to another, so that the possibility of accidental interchange of any two of the keys is minimized.

The code series in which a particular key code is found gives certain information; it tells us what key blank to use, the position in which each cut may be found, and the measurements of the various depths which may be used in cutting a key in that code series. The key blank does not change from one key to another, nor does the position of the cuts.

How to Use Key Codes

The individual key code tells the locksmith which standard depths to use in each cut position along the key blade. The depths are usually designated by a single digit, so that each digit in a code number signifies the depth of a particular cut. The standard depth symbols usually begin with the smallest digit representing the shallowest cut; that is, the *smallest number* represents the *greatest actual measurement* taken from the back of the blade to the bottom of the cut. The digits have a definite order in their appearance and represent the order in which each successive depth is cut along the key. Usually, the order of cutting is from bow to tip with the first digit representing the cut nearest the bow.

A few code series reverse the process to start at the tip and work toward the bow; however, this necessitates taking nothing for granted about the cutting sequence. A code with which the locksmith is not completely familiar should be specifically checked for cutting order.

There are two common classes of code series: One of these is the *direct digit series*, which frequently consists of a letter or two as a prefix, followed by digits which refer directly to the depths of the individual cuts, in sequence, progressively along the key blank. Common usage has this progression from bow to tip, although, again, the reverse may be true in isolated instances. An example of the direct digit type of code series would be the code number LC 52363, referring to a Sargent key, cut on the Sargent LC key blank (or keyway section), with the LC of the code number standing for the keyway designation.

Since there are five digits following these letters, the lock for which the key is to be cut is a five tumbler lock. The cut nearest the bow will be cut in the standard position for Sargent keys, and to the No. 5 depth which is standard for Sargent locks. The second cut on the key will be cut in the No. 2 position for the No. 2 standard depth; and the process will continue in sequence until all five standard cuts are completed near the tip of the key.

The second class of code series would be referred to as *indirect codes;* that is, the code number on the lock or on the key does not stand directly for the depths which are to be cut on the key, nor does it necessarily stand directly for the key blank to be used. For this type of coding, we must refer to a *code book* which contains the code series required. A large variety of code books is available through locksmith supply houses.

Let us suppose, as an example, that we have a key code number of B 73. In checking our code book, we find a key series, Briggs & Stratton B 50-249, in which this key code will fit. We also find a code series Corbin B 1-200 in which this same B 73 code will fit. Here we have a by no means uncommon situation in which a single code number will fit into more than one code series. How, then, is one to determine which code series is to be used? Obviously, we need more information.

The code book also tells us that the Briggs & Stratton code series utilizes Briggs & Stratton key blank 72584 and is used principally for automotive equipment, and to a lesser degree, for steel storage cabinets. The Corbin code series is used on cabinets, drawer locks, and desks, and is cut on Corbin key blank 5864JVR. The name of the lock manufacturer would, of course, tell us immediately which code series to use. The name of the manufacturer is not, however, always, or even often, apparent from the outside of the lock. If the lock is readily available, we can try the two different key blanks and see which one goes into the lock correctly.

Suppose your customer has brought you only the number. There is still a way to tell which code series to use. Asking the customer what the key is to fit will often provide a clue as to the code series to be used. For example, suppose that the lock is on a desk. The Briggs & Stratton B series is not used for desk locks, but the Corbin B 1-200 series is, therefore, the latter would be the series for this particular example. For the Corbin B 1-200 series key cutting information, we find that the key blank to be used is Corbin 5865JVR; that it uses the Corbin Wafer Gauges or depth keys; and that the key is cut from tip to bow, rather than using the customary bow to tip cutting sequence. Tracing the particular number in the series which is required, B 73, we find that it is cut 22021; this, then, will be the key called for.

In order that it may be made perfectly clear what these cuts stand for, let us consider the means by which they are established. For a given manufacture and type of key, the cut nearest the shoulder is always a given, specific distance from the shoulder to the center of the cut. This distance never varies within a code series, and seldom within keys of the same type by the same manufacturer. The next cut to this one is at a given distance from it; this

distance is called the *interval* and it is likewise invariable. The interval, or distance from one cut to the next, is the same for all cuts on the key. The depth of the cut is in terms of the distance from the bottom of the cut to the back of the blade of the key; this is expressed as the depth of the initial cut, and a *depth interval*.

In the present example, the initial cut's depth is .250 in., and the depth interval is .030. This means that the depth sequence for the cuts used in making keys in this series would be: 0 = .250, 1 = .220, 2 = .190; therefore, the coding, *tip to bow*, would translate from 22021 into measurements like this:

2   2   0   2   1
.190 .190 .250 .190 .220 and this is the key shown in Figure 170. Sometimes the key codes show these measurements, but frequently they do not. More complete information is often available from the catalogs of key blank manufacturers, or manufacturers of key cutting equipment; the latter give the specifications in terms of distance of the shoulder cut from the shoulder, or *lead space, spacing interval, initial depth,* and *depth interval*. These measurements vary from one manufacturer to another, even from one type of lock to another within the same manufacture. Thus, in cutting keys by measurement, one must be careful to establish that he is working from the right set of measurements.

By the use of a rule which is graduated in hundredths of an inch, the locations of the cuts can be marked along the length of the key blade. By careful filing with a round Swiss pattern file, and careful checking of the depths with a micrometer, it is possible to make a key from code by hand filing.

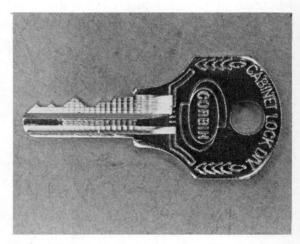

Figure 170. Code cut key. Key number B 73 is cut 22021, from tip to bow.

Considerable care is required, both in keeping the depth correct, and in keeping the positioning of the cuts from straying away from their proper location.

GUIDE KEYS

A quicker and easier method of producing keys from code is by the use of guide keys. Guide keys are simply keys which have been cut to the standard spacing and depth specification for a particular lock, with all depths on any one key cut to the same standard depth. For example, a typical guide key would be cut 33333; each cut would be in the proper position on the key, and would be identical in measurement of depth. This means that a set of guide keys must contain one key for each different depth of cut used for that particular type of lock. A lock having 5 different depths would require a guide key set consisting of 5 different keys; or, a lock using 10 different depths would require 10 different keys in its guide key set. A set of guide keys is shown in Figure 171 for a lock using 10 different depths.

To use the guide keys, one merely clamps the key blank in a key duplicating machine which is *not* of the automatic type, and selects the guide key whose number corresponds to that of the shallowest depth to be used in the finished key. For example, in a key whose code is determined to be 53476, the No. 3 guide key would be selected, and this key would be duplicated in its entirety on the key blank. The next shallowest depth of cut is determined, and the corresponding guide key placed in the vise ordinarily used for the original key (when duplicating). In this instance, the No. 4 guide key would be the one selected. The No. 4 cut would then be made in the proper tumbler position only. The next shallowest depth being a No. 5 in our example, that guide key is substituted in the pattern key vise of the machine, and it is used to produce the required cut in the key blank.

This process is continued until all cuts have been made. One must be careful to keep the cuts from lapping over into the adjoining cuts and to avoid leaving a square shoulder at any point along the blade of the key. This is not always an easy task, particularly when deep cuts adjoin shallow ones, but it must be done. When the cuts are arranged thusly, it may be desirable to leave the square shoulder until the code key is removed from the machine, then remove the square shoulder by careful hand filing. This is particularly true in the case of duplicating machines which do not use a full V type of cutter, but tend to leave the side of the cut nearest the shoulder of the key either flat, or nearly

Figure 171. Guide keys for Sargent locks. This is one of two different sets of guide keys used for different models of Sargent locks.

so, and at near right angles to the length of the key bit. The result is that the key would not be capable of insertion into the lock unless the square shoulder is removed.

This type of shoulder in the blade of the key is not ordinarily removed in the key machine. The key with all depths cut is removed from the machine and the square cut areas are filed to a taper as shown in Figure 172. It is, of course, essential to do this in such a manner that the depth of none of the cuts is changed, and that one cut does not run into another.

CODE MACHINES

Another means of producing keys from code is by the use of a *code machine*. These machines customarily use a milling type cutter of the truncated V type and do not produce a square edge cut in any position on the key. Variable stops to establish the different positions of cut are provided, either intrinsic in the machine, or as an accessory disc or gauge, to enable the machine to be stopped precisely at the location for the cut which is being made. Similar positioning means are used to allow the machine to cut into the key blank for a specified depth, and to stop at precisely that point. This produces a finished cut on the key with no further hand filing usually required, other than deburring when all of the cuts have been made.

There is, however, a problem area in cutting some

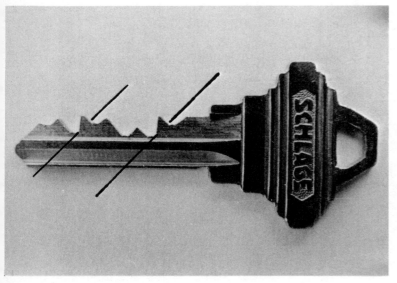

Figure 172. Finishing code keys cut from guide keys. Keys produced from guide keys often have square sides on the cut, which must be filed away carefully as shown.

keys on this type of machine, in that occasionally the variations in depth from one cut to the next exceed the depth of the cutting teeth on the cutter. When this occurs, a square shoulder is left which will either keep the key from going into or being withdrawn from the lock once it has been inserted. When such a square shoulder is left, it must be

Figure 173. A code key machine. The small dial at the far right controls the position of the cut and the large dial at the far left controls the depth. The dials are calibrated in terms of standard spacing and depth increments respectively, rather than in terms of actual measurements. Each dial may be replaced quickly and easily to conform to variations in depth or spacing requirements of various locks.

removed by filing. The filing process requires considerable diligence in order to keep one cut from running into another on the key, and to avoid altering the depth of the position being filed.

Another means of producing certain types of keys from code is the punch type of key cutter. It utilizes a shaped punch and die construction which removes a wedge of material from the key blank, all in one piece, and corresponding in depth and shape to that of the required cut. The key machine shown in Figure 174 is one of the most popular of its type. It utilizes a series of carriages which are shaped to accept different types of key blanks. The carriage traverses the key across the punch and die set (or punch and anvil, as it is called), with detents provided in various positions corresponding to location of the cuts on the blade of the key.

The depths are set by moving the lever at the rear of the machine to engage in various detent notches on the large dial of the machine. When both the tumbler position and the depth position have been set, the handles are squeezed, actuating the punch mechanism to remove the unwanted metal from the key. These machines are light, compact, and easy to use. Their degree of accuracy is quite within acceptable limits if the machine is properly maintained.

There are certain limitations to the usage of this type of machine which are unique in that one must have the proper blocks for each type of key cut; although, in some instances, one block may be used for more than one specific keyway. Primarily, the machine is intended for keys of the automotive or desk lock type. Keys for Master padlocks may also be cut using this machine. The principal problem area in making keys on this type of machine is that in some keyways the key blank is somewhat distorted by the pressure of the punch. This is particularly true in those key blanks in which the side of the key blade next to the anvil curves away from the anvil at the edge where the punch engages it.

The situation may be corrected after the key is cut and removed from the machine by filing away the distorted portion of the key blade in such a way as to allow the key to enter the keyhole without binding. Other than this, the cut produced by this type of machine is smooth and comparable to that made by the milling cutter type of machine.

It should be understood that all of the material presented here on code key cutting is primarily directed at the cutting of code keys for disc or pin tumbler locks. The cutting of keys from code for lever tumbler locks requires, of course, that the cuts

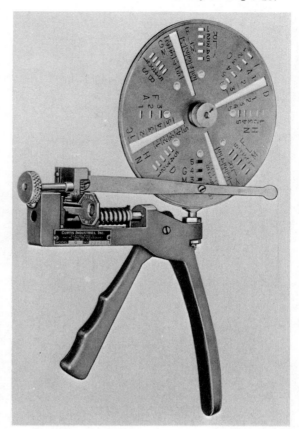

Figure 174. A "punch" type of code machine. This type of machine is intended primarily for producing automobile keys, but it is quite useful for making a number of other keys as well. Illustration courtesy of Curtis Industries, Inc.

be square on the sides rather than sloped. Other than this, the principles of the various means of code key cutting all apply to lever locks, as well as to disc and pin tumbler locks, remembering that because of the square sides on the cuts, a different cutter is required, and the machine cannot be traversed along the key blade without disengaging the cutter from the key blank.

VII

Lock Coding and Masterkeying

LESSON 25

Lock Coding

Lock coding is the systematic means by which a group of locks is keyed so that any one key will operate only the lock within the group it was intended for. Other locks within the same group not to be operated by the key will not be operated by it. The plan by which this is accomplished for a group of locks is called a *code series*.

Depth, The Variable

For locks having individually replaceable tumblers, that is, disc, pin, or lever tumbler locks, a given code series utilizes a single keyway, and the spacing from one tumbler cut to the next does not vary from one lock to the next. (An exception is the Sargent "Keso" pin tumbler lock, which utilizes a computerized coding system of monumental complexity.) In a straight coding system which involves no masterkeying, all of the codes for a given system of locks are built around one keyway.

The spacing of tumblers for similar types of locks of the same manufacture is usually uniform for that type of lock through the manufacturer's line. There are in this, as in so many rules, exceptions. In this case, they are frequently associated with an obsolete line of locks of that manufacturer, or with certain design variations which may require different spacing to accommodate a particular design. Within a single code series, however, the keyway, tumbler spacing, and the depth standards to which tumbler cuts are made, do not vary.

The variable factor, then, in a code series which serves to make one lock differ from another is in the arrangement of the standard depths by which the use of standard tumbler dimensions will permit the keying of the lock to vary. Thus, the introduction of a variable depth into a code series is the sole variable factor in achieving that differentiation required to avoid unplanned interchanges within the same series. Even the differences between the depth of the cut required by one tumbler and another are standardized as the depth interval.

The use of a depth interval, rather than making intrinsic changes, is caused by two factors: first, the economics of standardization are such that it is more economical and feasible to limit these variations in depth to a certain number of sizes or configurations of tumblers to conform with mass production principles; second, and even more essentially, these depth intervals are limited by their relationship to the tolerances to which the lock is built.

Depth Interval in Relation to Tolerances

A lock, like any other mechanical device, requires a certain amount of clearance, or slack, between two parts, one or both of which is movable. Clearance is required in order to achieve the proper degree of freedom and smoothness of motion which is mandatory to any smoothly operating mechanical device. For example, in locks having a rotating plug encased in a cylinder (as in a pin or disc tumbler lock), the diameter of the rotating plug must be smaller than the diameter of the hole of the cylinder in which it rotates, if the plug is not to be difficult or impossible to turn. Such a plug would never turn smoothly if the clearance is insufficient.

241

Usually the clearance in a lock with a rotating plug is sufficient to allow a piece of shim stock to be inserted on one side of the plug which may, for very closely built locks, be as small as .001 or .0015 in. thick, and for other locks may be as high as .007 or even .010 in. If tumblers are going to be aligned at the surface of the plug, they will be operable in positions from slightly below the surface until they are extended far enough above the surface to become engaged immovably with the upper tumbler chamber, preventing the plug from turning. For this reason, then, these depths of cut on the key, effected by the tumbler position, have a certain minimum interval inherently imposed upon them by the design tolerances of the lock.

Lever locks have similar depth intervals imposed upon them by other requirements of tolerances which are just as effective in imposing these depth intervals as is the clearance of the plug in a disc or pin tumbler lock.

It is universally true of all locks that the imposed depth interval becomes smaller as the precision with which the lock is built becomes greater. It is likewise true that the keyway tolerances which allow a key to move slightly within the keyway also contribute to the imposed depth interval, and this increment to the imposed interval becomes progressively greater as the keyway becomes worn.

The depth intervals most commonly used for various types of locks are as follow: for lever locks, .015 − .025; for disc tumbler locks, .015 − .030; for pin tumbler locks, .0125 − .020.

Coding Systems

As we have already seen, the substitution of a single digit to represent a considerably more complex measurement gives rise to a coding system which can be stated in sufficiently simple terms to be meaningful. Conversely, a code series expressed in terms of actual measurements would be unwieldy. A number of types of coding systems are used in the process of encoding a lock series. Obviously, if interchanges are to be avoided, a code series must be systematic. The simplest of these systems is a straight progressive system.

The Straight Progressive System

A straight progressive system is one in which the depth of cut is changed in a true linear progression, beginning with the last digit of the combination, and progressing until all of the available depths are used on that tumbler. The next to the last tumbler is then changed by one depth (unless a two depth

interval is used), and the changes previously made on the last tumbler are repeated. This process is carried out until all of the available depths are used on all of the tumblers. The partial code series that follows is an illustration of a straight progressive system:

1−3413**4**	progression to the
2−3413**5**	next tumbler at
3−3413**1**	this point)
4−3413**2**	7−3414**5**
5−3413**3**	8−3414**1**
6−3414**4** (Note	9−3414**2**

The Modified Progressive System

The next stage of increasing complexity in code systems is the modified progressive code system, in which the key changes are initiated on the last tumbler, as in the straight progressive system. All available combinations are used on that tumbler, then the next last tumbler is changed. These changes are not, however, always to the next succeeding depth, but may be to a much deeper cut, or to a shallower one. Instead of being in a straight progression, the depths on any one tumbler follow a certain *change order*; for example, the change order for one tumbler may be 5, 2, 3, 1, 4. Such a code series may be set up in outline form:

Tumbler	1	2	3	4	5
Initial Combination	3	4	1	3	5
	4	2	4	1	2
	2	5	5	4	3
	1	1	3	2	1
	5	3	2	5	4

The following partial code series is derived from this outline.

1−3413**5**	6−3411**5**
2−3413**2**	7−3411**2**
3−3413**3**	8−3411**3**
4−3413**1**	9−3411**1**
5−3413**4**	

The Mixed Progressive System

Increasing the degree of complexity of a code series yet another step, we have the mixed progressive coding system. This is created by the selection of any tumbler position for the first tumbler change, progressing through all of the possible changes on that tumbler, making the changes in a random order as in the modified progressive system. A second tumbler to undergo the next succeeding changes is selected *at random,* and all possible changes on it are used before progressing to another tumbler. Progression continues in a random, yet systematic manner until all of the changes have been used. The outline for such a system would look like this:

Tumbler Preference Order;	5	2	4	1	3
Tumbler Number;	1	2	3	4	5
Initial Combination;	3	4	1	3	5
	4	2	4	1	2
	2	5	5	4	3
	1	1	3	2	1
	5	3	2	5	4

A partial code series derived from this outline would take this form:

1—34135	6—32135
2—34115	7—32115
3—34145	8—32145
4—34125	9—32125
5—34155	

This type of progression would be followed until all of the key changes have been used on all tumblers.

Scrambled Code Series

This leaves the most complex of all of the coding systems, the scrambled code series, as the ultimate in complexity for a standard lock. This system is the most difficult of all the code systems to decipher, or *decode,* by having available only a part of the locks or keys in that code series. It can be created by building a complete code series for a group of locks in a straight progressive system, then thoroughly mixing all of the combinations.

It is very much like, and indeed, it can be arrived at by writing each combination on a separate piece of paper, then drawing one out of a box to determine which combination is first in the series. This piece of paper is then thrown away, and a second combination is drawn for the second combination; the process continuing in this fashion until all combinations have been used. Obviously, in such a coding system, there is no systematic progression which can be worked out from the mere possession of part of the locks or keys in the code series. Access to all of the locks or keys used would be required in order to completely reconstruct such a code series. The partial code series following is an example of a scrambled code series:

1—35514	6—51322
2—21332	7—13343
3—43225	8—41243
4—44421	9—52453
5—22314	

Master Code Series and Derived Series

Certain code series are known as *master series,* not because there is masterkeying involved, but because the same combinations are used in the same change order for several different code series. The differences may be limited to changing the prefix by which the series is known; or it may be that the change in the prefix calls for the key to be cut on a different key blank. For example, key A 238 may be identical with key B 238, or it may be that they are identical *except* that key B 238 is cut on a different key blank.

Another use of the master series is in obtaining a *derived series.* This situation is not at all uncommon with key codes. For example, in a master series, A 238 may stand for a key cut on key blank A to a combination which is 31452, while key B 487, which is cut on key blank B, is also cut to a combination 31452, exactly as was A 238. Other code numbers in the B series can be determined by subtracting 249 from the B series code number, and looking up the resulting number under the A series. In this instance, series A would be a master series, while series B would be a *derived series.* Partial codes showing the parallel qualities of such series are shown below:

Series A 1 — 250	Series B 251 — 500
A 238 — 31452	B 487 — 31452
A 239 — 35515	B 488 — 35515
A 240 — 21332	B 489 — 21332
A 241 — 43225	B 490 — 43225
A 242 — 44421	B 491 — 44421
A 243 — 22314	B 492 — 22314

Keying Locks Alike-Convenience Keying

Convenience keying may be accomplished within the framework of any code series by keying more than one lock to the same combination, or using *keyed alike* locks. This is done by simply using tumblers of the same dimensions in all tumbler positions of identical locks, doing this for as many locks as are to be keyed alike. This permits one key operation for all locks which are so keyed. Often the general public will refer to this key as a "master key," but such a key is not a master key. A key which operates more than one lock is a master key only if other keys exist which will operate only a portion (or one) of the locks which are operated by the master key.

Frequently a customer will ask for his locks to be masterkeyed when all he actually wants is for one key to work them all. Under these circumstances, his needs would probably be better served by merely having them keyed alike. This is particularly true in most residential keying, as there is small reason to have individual keys for a number of locks if the master key is the only one ever used.

Forbidden and Undesirable Combinations

No discussion of lock coding is complete without some reference to forbidden and undesirable combinations. A *forbidden combination* is one which cannot be used because the variation in depth between adjoining tumblers is too great to allow the key to be cut and to operate the lock properly. Usually this is encountered because the angle of the cutter which makes the cut in the key for the deep tumbler is such that an extension of that angle would cut away the next adjoining cut altogether, if it is a shallow one such as shown in Figure 175.

This could, of course, be rectified in some situations by the use of a more sharply angled cutter, but a steeper cutter angle could make the key difficult to insert and withdraw. In the design stages of a lock, the situation could be taken care of by extending the spacing between tumblers; but a lock having some six or seven tumblers with extended spacing would require an exceedingly long blade on the key. Usually, these forbidden combinations are simply avoided in keying the lock.

The forbidden combinations are expressed in terms of the maximum number of depths by which adjoining cuts on a key are permitted to vary. For example, in a lock having the capability of varying through 10 different depths of cut, the forbidden combinations may be those which incorporate a variance of more than seven depths of cut in immediately adjoining positions. In such a lock, the combination 41829 would be an acceptable combination; whereas combination 42819 would be a forbidden combination because of the 1-9 relationship between the last two cuts. Since 9 minus 1 equals 8, and the permissible variation is only 7, the key could not be cut, hence it is forbidden.

Undesirable combinations are those which will operate the lock, but can give rise to undesirable characteristics in that operation, such as being able to withdraw the key without the lock being firmly locked in position. As an example, the 44444 would

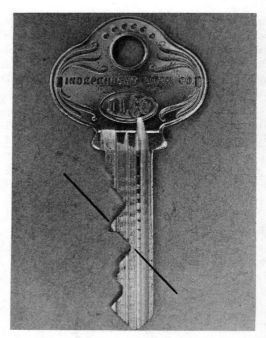

Figure 175. A forbidden combination. Extension of the side of the cut in this key would wipe out the adjoining cut, creating a forbidden combination type of situation.

be such a combination; 13446 would be another. Since the depths are all the same in one case, this key can, when slightly worn, be withdrawn from the lock, even though the lock is in a partially turned position. When the key is new and freshly cut, there may be enough raised portion between the bottoms of the various cuts to prevent this; however, the normal use of a key will soon wear these raised portions down to the point where withdrawal of the key in any position is possible.

The same situation prevails with the second combination of 13446, because each of the succeeding depths toward the tip of the key is deeper than the previous depths, resulting in the same type of situation, regardless of the changed appearance of the key. If the combination were 13443, the situation would be altered because the No. 3 cut at the tip of the key would not clear the No. 4 tumbler while the lock is being rotated and the No. 4 tumbler is held fast in position.

Several manufacturers of pin tumbler locks and a few of disc tumbler locks have depths and spacings which give rise to forbidden combinations, and of course, all pin and disc tumbler locks have some undesirable combinations. It is well for the locksmith to be aware of these situations and avoid them.

Master Key Systems

THE purpose of a master key system is to permit a group of locks to be keyed differently for use and operation by various persons, and yet allow one key operation of all of the locks by those authorized to have a master key.

Such a system of lock keying finds but small application for home purposes. It is easier, and usually better, to key locks alike for home use, rather than to invite the problems and deterioration of security which might result from a more complicated system. For these reasons, master key systems are usually favored more for commercial, industrial, and institutional use than for residential.

Potential customers for master key systems include public buildings, office buildings, hotels, motels, the larger commercial and industrial firms and institutions which may find housing for various departments in more than one building or room. Many firms and clubs having individual employee or member lockers also find need for master key systems.

Even discounting the fact that the demand for residential masterkeying is almost nonexistent, it can be seen that there is, nonetheless, a large and sustained demand for master key systems for other applications and purposes.

The methods and techniques for masterkeying vary from one type of lock to another. Lever locks and disc tumbler locks may be masterkeyed to a limited extent by the locksmith in his shop, but more extensive masterkeying of either of these types of locks requires special provisions by the factory. A supply of special tumblers which are standard only to that particular type and manufacture of lock is required for extensive masterkeying. A rather large and diverse inventory is a necessity. The superior versatility of pin tumbler master key systems make it the preferred medium for masterkeying; however, the three types will be discussed in this section.

LEVER LOCK MASTERKEYING

Masterkeying of lever locks in the shop where no provision for this work has been made by the manufacturer is usually done by the wide and double gate system. As the term implies, and as we have seen in the discussion of lever locks, it simply means widening the gate through which the fence enters the tumbler. Wide gating of both the notch and T type of tumblers is shown in Figure 176.

Double gating of tumblers is even more restrictive as to available combinations for use, as is shown in Figures 176 and 177. These depict wide gating and the difficulty of double gating the T-notched gate in that type of tumbler.

Even with the restricted availability of key changes for individual keys in this type of system, extreme care must be taken to prevent unintentional cross-operation of two or more individual keys. For wide gated tumblers, a master key combination must be selected which is not more than one standard depth from the change key combination. If the change key combination is two depths away from the master key combination in either direction, then two different change keys will operate the same lock.

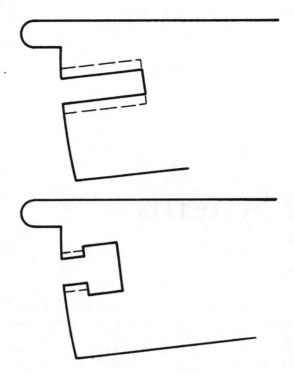

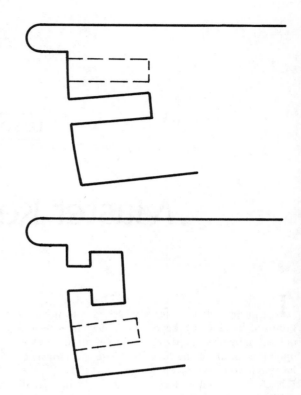

Figure 176. Wide gating of notch and T gated lever tumblers. The gate may be widened on only one side, not both.

Double gating may be done only when the variation between the master key depth and the change key depth is extreme. No masterkeying can be done unless both the keying and the internal construction of the locks are all the same in type and design, whether or not alike in specific keying. For purposes of illustration, let us assume that we have such a group of locks, all of the same manufacture and model.

Wide and Double Gated Systems

The first step in the masterkeying of these locks is to determine the number of different tumblers on hand in the different available depths in the entire group of locks. This may be done by reading the depths from the keys, if keys are available for them; or by disassembly of the lock and removal of the tumblers for comparison purposes and sorting. For this particular masterkeying operation, the depths used on the masterkey will depend entirely upon the tumbler depths available. Figure 178 illustrates the means by which the depths available for use on a master key are determined, based on the use of available tumblers.

Beginning with the shallow cuts, for the sake of

Figure 177. Double gating of both notch and T gated lever tumblers is even more restrictive than wide gating. Notch type tumblers must have enough space between the double and original gating for some of the tumbler material, or it degenerates into an exceedingly wide range single gate which includes all tumbler depths between those which are intended to be operable. The material left between the gates must have adequate strength to resist the turning of an improper key which is right everywhere except in that particular depth position. The same problem is even more acute with T notch tumblers.

progression, we attempt to utilize these tumblers. Using the principle that only one depth of cut differential in the gating of the tumbler may be used on either side of the master key, we first use 10 of the No. 1 depths and 10 of the No. 2 depths. The No. 1 depths are cut out in the gate to accept a No. 2 cut on a key, so that they will accept keys cut either No. 1 or No. 2 in depth. The No. 2 depth is cut out to accept a No. 3 depth in similar fashion. This leaves a No. 1 and a No. 3 for the change key; the No. 2 is reserved for the master key depth, since it will operate all 20 tumblers.

Since we have enough left for another group of 20 on the No. 1 depth as well as on the No. 2 depth, we select 10 of the No. 1s for cutting to the No. 2 depth for 1 and 2 operation. This leaves 3 of the No. 1

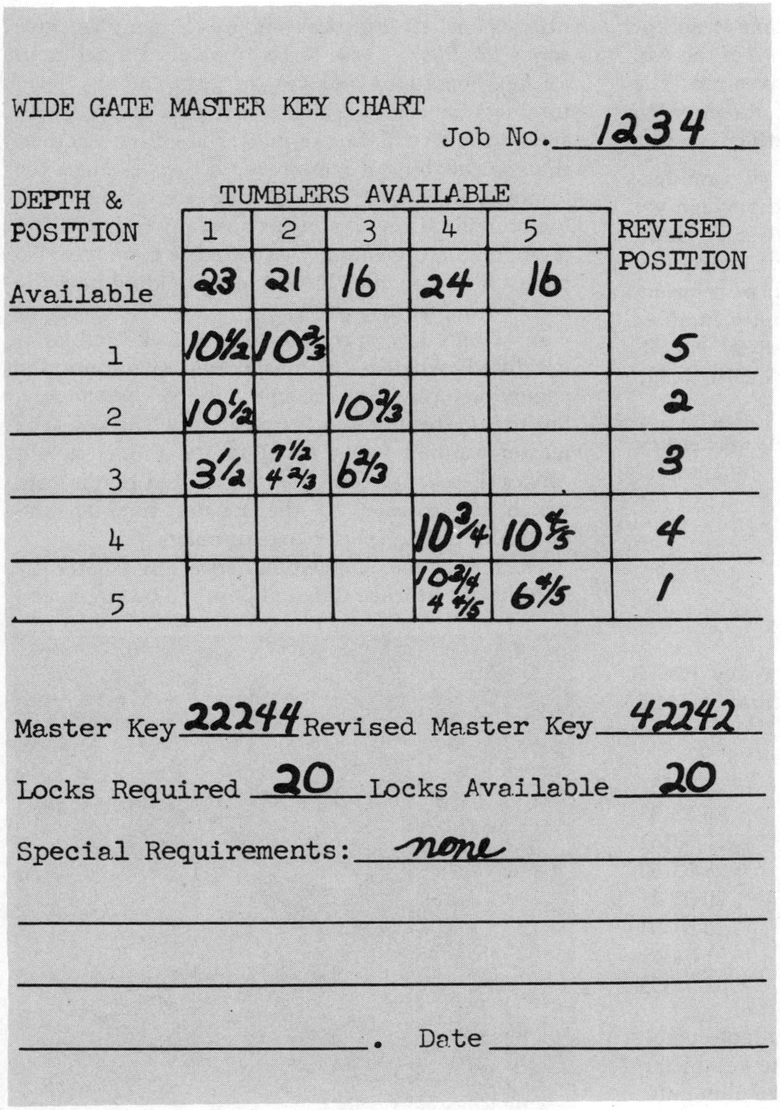

WIDE GATE MASTER KEY CHART

Job No. 1234

DEPTH & POSITION	TUMBLERS AVAILABLE					REVISED POSITION
	1	2	3	4	5	
Available	23	21	16	24	16	
1	$10\frac{1}{2}$	$10\frac{2}{3}$				5
2	$10\frac{1}{2}$		$10\frac{2}{3}$			2
3	$3\frac{1}{2}$	$7\frac{1}{2}$ $4\frac{2}{3}$	$6\frac{2}{3}$			3
4				$10\frac{3}{4}$	$10\frac{4}{3}$	4
5				$10\frac{3}{4}$ $4\frac{4}{5}$	$6\frac{4}{5}$	1

Master Key **22244** Revised Master Key **42242**

Locks Required **20** Locks Available **20**

Special Requirements: *none*

_____

_____

_____

_____. Date _____

Figure 178. Chart of a wide gate master key system. This chart shows a system of twenty locks with a common keyway and five tumblers per lock. The total number of tumblers of a given depth is listed to determine availability of various depths used in establishing a master key which will operate with these tumblers when wide gated. Blank forms of this type are available from your locksmith supply house.

tumblers unused. The No. 2 tumbler can be cut to accept either a shallower or a deeper cut on the key. Therefore, let us write the number "3" in the third column of the No. 1 depth row, since they, too, will have to be used.

Progressing to No. 3 depth, we find that we have more than 10, namely 16 tumblers already cut to that depth. These No. 3s may be cut out wider to accept a No. 2 cut on the key also. Let us use 10 of these No. 3s in the second tumbler position, completing the needed tumblers for that position. This leaves us with 6 No. 3 tumblers left over. The No. 3 tumblers may also be used for the change key when cut out for a No. 2 master key depth, so let us write

the remaining 6 tumblers in the third tumbler positon row. This gives us a total of 3 tumblers shallower and 6 tumblers deeper than the master key depth.

We note that we have 11 tumblers left over in the No. 2 depth. This is exactly sufficient to complete a group of 20 tumblers, and since they can be used for either the No. 1 depth or the No. 3 depth on the change key, let us write that total in the No. 3 position column. This leaves us with 24 No. 4 and 16 No. 5 depth tumblers. Distributing 10 of each of these in the fourth tumbler position column leaves a balance of 14 for the No. 5 tumbler position in the No. 4 depth and 6 in the No. 5 depth. Since the No.

5 depth can be cut only in the direction to accept the No. 4 depth also, this is done; then 4 of the No. 4s are cut to include the No. 5 depth position. The remaining 10 are cut in the opposite direction to include the No. 3 depth of cut in the gating.

Having completed our availability of tumblers study, we go back and check the depths which we have selected for master key use. We find those depths are 22244. This is not a fortuitous master key combination; therefore, let us juggle these columns around a bit. For example, we can trade tumbler position No. 5 with the tumbler position No. 2, giving us 24242; or we could trade No. 5 with No. 1, and have 42242.

Let us assume that we have chosen the master key combination to be 42242. This allows us tumbler changes as shown:

Master key	4	2	2	4	2
Changes	3	1	1	3	1
	5	3	3	5	3

Taking the first change row, change key No. 1 becomes 13313, and progressing as shown in the code series developed from this:

1-31131	9-33131	17-51131	25-53131
2-31133	10-33133	18-51133	26-53133
3-31151	11-33151	19-51151	27-53151
4-31153	12-33153	20-51153	28-53153
5-31331	13-33331	21-51331	29-53331
6-31333	14-33333	22-51333	30-53333
7-31351	15-33351	23-51351	31-53351
8-31353	16-33353	24-51353	32-53353

Since these key codes provide 32 available combinations which may be used with the tumblers, and we require only 20, we can afford to eliminate the more undesirable combinations. First, let us eliminate No. 1, because all of the cuts on that change key are shallower than the master key depths; therefore, the change key could be cut down to make a master key. Then, let us eliminate Nos. 6, 10, 13, 14, and 16 because in these particular keys, the cuts are too nearly all the same to make a good, presentable key. Any of the remaining cuts may be used and still show a reasonably presentable key, although it should be stressed that the protection afforded by the lock is considerably diminished.

The Master Tumbler System

The second method for masterkeying lever tumbler locks is known as the *master tumbler system*. It utilizes one, two, or occasionally as many as three *series tumblers*. These series tumblers are active in all key operations and are the same as any lever tumblers for a lock which is not masterkeyed, and serve to differentiate one master key from another, thus establishing a master key series. Change key tumblers in a master tumbler system operate with the individual key, exactly as does any conventional lever tumbler; however, these change tumblers also have a hole through which a peg extending from the *master tumbler* may pass.

In change key operation, the master tumbler is not lifted; whereas, in master key operation, the change key tumblers are not operated by the key, but the master tumbler *is* operated by the key. The master tumbler being lifted to the proper height causes the change tumblers to be lifted by the peg, which is attached to the master tumbler and protrudes through the change tumblers.

The hole in the change tumblers are at exactly the right height for each different depth to be aligned by

Figure 179. A lock using the master tumbler system. The master tumbler illustrated is at the top of the tumbler pile and acts to support the tumbler face, rather than at the bottom of the pile and extending up through a hole in the tumblers as described in the text. See Figure 96 for a lock made to utilize the method described in the text.

the lifting action of the master tumbler's peg. In this type of situation, the change tumblers are all lifted as a unit by the action of the master tumbler.

Change keying within a master series of this type is limited only by the same keying standards that apply to locks of a similar type which are not masterkeyed. In the change tumblers, any ordinary combination which is acceptable in any regular code series of any type is equally acceptable in this sytem of masterkeying.

It is important to note, however, that the series tumblers remain the same throughout an entire master key series. For all change keys in a master tumbler system, the change key cut at the position of the master tumbler must always be deeper than that required to move the master tumbler toward the operating position. Preferably, this means that the cut under the master tumbler in such a system should be cut to maximum depth in each and every change key.

DISC TUMBLER MASTERKEYING

As is the case with lever lock masterkeying, disc tumbler masterkeying is commonly done by two different methods. One of these methods may be followed without special provision by the factory for that particular lock to be masterkeyed. The other method, just as with the lever locks, requires a factory provision at the time of manufacture in order for this type of masterkeying to be done.

Short Tumbler System

The first method for masterkeying of disc tumbler locks is the *short tumbler method*. This is very similar, in many respects, to the wide gating method used in lever locks. Immediately adjacent tumbler depths are used, and the overall length of the disc tumbler is short enough for either of the two depths of cut on the key to operate the lock. This dual operation is accomplished by selecting a tumbler which is correct for the deeper of the two depths to be used.

The selected tumbler is placed in the lock. The key having the shallower cut in that position is inserted into the lock plug, causing that tumbler to be lifted slightly above the position where it aligns with the surface of the plug. The excess of the tumbler is then filed off until it is smooth at the shear line; next, the tumbler is deburred, at which time both the shallow cut and the deeper cut keys will be found to operate the lock. By referring to Figure 180, we will be able to see why the deeper depth of tumbler is required.

Figure 180. Preparation of a disc tumbler plug for short tumbler master keying. A shallow cut on a key will lift a deeper tumbler out of the lock plug and support it while filing is done, whereas a tumbler shallower than the cut of the key will protrude from the other side of the plug with only the tumbler spring for support.

The more shallow cut key operating on the tumbler for the deeper cut pushes the tumbler part way out of the plug and the tumbler continues to rest on the key all through the filing process; whereas, if the shallower tumbler were used, only the spring tension which operates the tumbler would be pushing it out of the lock plug (with the deeper cut key in position as a filing guide), making it very difficult to file the tumbler correctly. Additionally, in some instances where the depth of cut is quite deep to begin with, filing off an additional depth from the *top* of a tumbler, already thin by reason of being for a deep cut, might well cut the tumbler in two at that point, or come so nearly to doing so as to cause the tumbler to lose substantially all of its strength at that point. Either of these situations would render a tumbler useless.

Assuming a depth availability of five different depths of cut, the No. 1 and No. 5 cuts are not usable for master key purposes in this type of keying, although they may be used for change key combinations. Using either of these cuts for a master key combination in this type of masterkeying would allow only the master key cut and one change key cut in that particular tumbler position, allowing no differing whatever on such a tumbler position. Merely one change key cut on a tumbler position would realize no actual changes in the change keys;

it would be better not to master the tumbler at all.

As an example of the method by which master-keying of this type is accomplished, let us suppose that we have selected a No. 2 depth to be the master key depth for that particular tumbler position. If the change key depth is to be a No.1 in that position, then the tumbler selected should be a No. 2 depth. This will allow the tumbler to protrude from the lock plug when the key with a No. 1 depth is inserted into that tumbler position, allowing it to be filed off properly. If the change key cut in the same position is to be a No. 3, and the master key is to continue to be a No. 2, then the No. 3 tumbler is selected for that position and is filed off with the master key in place in the lock plug.

In coding for the change keys for a master key system of this type, the methods and limitations are precisely the same as were shown for the wide gate method of lever lock masterkeying. Indeed, the sample code series shown in that section is as applicable to disc tumbler locks as it is to lever locks. It should, however, be noted that there is no possiblity of anything equivalent to double-gating being done with a disc tumbler lock of the standard type. It should be remembered that when applying the short tumbler masterkeying to a disc tumbler lock the tolerances of the lock may not be sufficiently close to permit this type of adjoining depth arrangement. Should it be determined that this type of situation prevails, it would be advisable to abstain from masterkeying such a group of locks.

The Complementary Keyway Method

The second method of masterkeying disc tumbler locks is known as the complementary keyway method. The lock plug for locks intended for this type of masterkeying is made at the factory in such a way as to accept two different but complementary key sections. Complementary keyway sections are shown in Figure 181. It should be noted from this illustration that the keyway sections are identical up to the point where the cut portion of the key blade departs from the root portion of that key section. In other words, the paracentric portion of the keyway section is the same for both key sections.

One of these complementary sections has the cut portion of the key blade on the right side of the keyway, and the other has it on the left. The purpose of this arrangement is to permit the blade of the key to lift on a different portion of the cut out in the center of the disc tumbler. For this system of masterkeying, one of these two keyway sections is

Figure 181. Complementary keyway sections used in disc tumbler masterkeying. Illustration courtesy of Eaton Yale & Towne, Inc.

used as a master key; the other keyway section is used as a change key. This type of masterkeying utilizes a special *stepped tumbler*, as shown in Figure 182, in order that different depths may be used under the change key position, serving to allow the differing of change keys from one another, as well as from the master key.

Unlike other systems of masterkeying, the position of the master key depth has absolutely no bearing on that of the change key cut. The change

Figure 182. Stepped tumblers. This is the configuration of tumblers used in complementary keyway masterkeying. Note that these tumblers will place the deep cut on the same side of the keyway if used in adjoining positions.

key depth is utterly independent of the depth of the master key in any and all positions. If desired, a change key could be cut to the master key depth in all positions, without compromise to the master key itself, since the change key does not touch the tumbler at the master key position. This allows the use of a full code series, just as if the locks were not masterkeyed. Coding of this type of lock may proceed without regard to the master key combination, insofar as the change keys are concerned.

The mechanics of keying such a lock plug must, however, take cognizance of the master key combination, and is somewhat complicated as a result of the fact that disc tumblers are not all inserted into the lock plug in the same direction. That is, the spring is alternately on opposite sides of the keyway.

As an example of how this complicates the keying and masterkeying of these locks, let us assume that we desire a master key depth of No. 4 in a particular tumbler position, and a change key depth of No. 2 in the same position. Referring to Figure 182, the tumbler which we wish has the No. 2 depth on the same side of the tumbler slot as the arm of the tumbler, with the No. 4 depth established on the opposite side of the slot.

If the same combination is desired in the next adjoining position, the positions of both of these depths of cut would have to be reversed because the spring is on the opposite side of the keyhole. If the same combinations on both change key and master key were used (master = 4, change = 2), the tumbler used in one position would be 4-2, but in the next tumbler position, the required tumbler would become 2-4, because of the reversing of the tumbler from one tumbler position to the next. These step tumblers are available for locks of various manufacture through their distributors. The tumblers should be stocked if sufficient masterkeying of this type is contemplated, since deliveries usually require some time.

Assuming that we have a master key combination of 41232, and a change key combination in the same lock of 13521, let us write them out in this fashion:

```
    *     *     *
    4  1  2  3  2
    1  3  5  2  1
```

By experimentation with the keys, we find that the desired first tumbler is marked 41. Remembering that each tumbler is reversed from the one on either side of it, as shown by the alternating asterisks on the set of figures above, the tumbler list for the lock becomes: 41,31,25,23,21.

Pin Tumbler Masterkeying

In pin tumbler masterkeying, a single tumbler position is made to operate on two different depths by the insertion between the bottom pin (or key pin) and the top pin (or driver) of a short pin segment, which is variously known as upper pin segment, master pin, master tumbler, master key chips, or, in

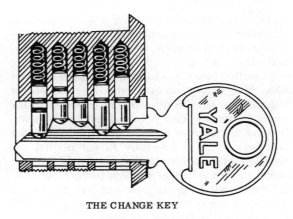

THE CHANGE KEY

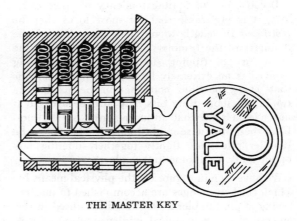

THE MASTER KEY

Figure 183. Mechanics of pin tumbler masterkeying. The use of differ bits permits two different keys to align the tumblers at the shear line. Illustration courtesy of Eaton Yale & Towne, Inc.

British terminology, as *differ bits*. Since the British term is quite descriptive, let us consider them to be differ bits for the purpose of this discussion, even though the same tumblers which are differ bits may also serve as drivers, depending solely upon their position in the lock. The length of the differ bits is in multiples of the depth interval of the lock which is being masterkeyed. Figure 183 shows us the mechanical means by which masterkeying is achieved in the pin tumbler lock.

Certain rules and principles must be observed in the implementation of any pin tumbler master key system, regardless of the coding or type of system in which it is used. The first is that each key must be different enough from the next closest key to it to avoid the possibility of an unintentional interchange of keys which are supposed to be different and not mutually operable. For most locks, this means that each key must differ from the one next closest on at least one tumbler by a minimum of two depths of cut. In practice, this amounts to anywhere from .025 to .040 in.

Secondly, there is the hazard, in using the shorter differ bits, that these extremely short sections of. tumbler may tend to tip and turn crosswise in the tumbler chamber. This is caused by the clearance, between the tumbler and the walls of the tumbler chamber, which is necessary to ensure that the tumblers will move up and down in the tumbler chamber with sufficient freedom to be functional, not forgetting that in the process of using the lock, a certain amount of wear will be incurred by both the tumblers and the tumbler chambers.

Because of the tipping tendency of short differ bits, it is preferable that for most locks they be restricted in length to not less than 40% of the diameter of the tumblers. This will minimize the possibility of tipping, unless the system being evolved is an extremely large one which demands that every possible combination be used. The tipping of short differ bits may also be avoided if one uses a somewhat stronger spring on the tumblers than otherwise might be used. This holds the tumbler pile more tightly together, offering less opportunity for tipping to take place.

We have seen in Figure 183 the physical means by which the key changes are accomplished in masterkeying of pin tumbler locks. The differs used in the process of the mechanical implementation of this type of masterkeying are achieved solely on the basis of mathematical sequencing.

The different methods of mathematical sequencing account for the different types of masterkeying in locks. Each method of sequencing has its use, its purpose, and its limitations. For the purposes of our discussion, we shall assume that each master key system requires the maximum potential of the lock to be used in order to provide us with sufficient master key changes. This is done not because it is a good policy in a small master key system, but because it may become necessary to utilize the complete potential of a system for a particular job. The author feels that this full potential should be shown.

Certainly, if the full potential of masterkeying is understood, it can be modified to achieve somewhat more limited objectives. The different systems of pin tumbler masterkeying which are shown in the succeeding pages provide an indication of the type of situation for which they may most effectively be used, and the more important problems and limita-

tions or advantages with which they may be cursed or blessed, as the case may be. The mathematical mechanics by which these systems are evolved will also be discussed.

Straight Progressive Masterkeying

Of all the master key systems, the straight progressive system is by far the simplest. It finds its principal use in small master key systems (although it is capable of gross expansion), such as homes, rooming houses, and small businesses. This somewhat restricted usage is imposed upon it because of the mathematical means by which the system progresses from one change key to the next. It is quite easy to decipher (or decode) the entire system by knowing the combination to which a few keys are cut, as will become obvious in studying the mathematics on which it is based.

Progressive masterkeying, as the term implies, is simply an orderly progression of the change key combinations from one combination to the next. By systematizing the progression of the series, unplanned interchanges are quite easily avoided, and by an orderly arrangement, planned interchanges can be made to occur where and as needed. Certain other features are required in order to keep all keys both usable and presentable; for example, the masterkeying outline shown below, using a two-depth-of-space progression, alternates the tumblers between odd and even numbers. This has two purposes; one keeps the usage of key pin tumblers approximately even, and the other keeps repetition of the same depths of cut on the key from being in immediately adjoining positions. The effect of these two conditions is to keep the key more presentable in appearance and to increase the resistance of the lock to picking. The high-low-high configuration of a key cut in this fashion helps keep the key in position in the lock after it has become worn, and holds true for very nearly every possible combination in any series. An outline of a progressive master key series follows:

Pin preference:	5	4	3	2	1	
Master key:	0	5	8	3	6	Master: 05836
Change order:	2	7	0	5	8	1- 27058
	4	9	2	7	0	2- 0
	6	1	4	9	2	3- 2
	8	3	6	1	4	4- 4
						5- 78
						6- 0
Interval: 2 depths of cut						7- 2
Forbidden combinations: none						8- 4
Depth progression: 0 − 9						9- 27098

This code series may be carried out until all available key changes in the series have been used. How many is that? For coding series which have no forbidden combinations, the formula for figuring the key changes is:

$$C = D_c^T$$

Where C = the number of change key combinations available.

D_c = the number of change key depths usable on each tumbler.

T = the number of tumblers which are masterkeyed.

In our own sample code series, by substituting values in the equation: $C = 4^5 = 4 \times 4 \times 4 \times 4 \times 4 = 1024$. Notice that the master key combination depths are not included in these figures, nor do we use the master key depth in any change key combination. This is very important, and we shall return to it presently. It should be noted in the outline for our sample master key system that one end or the other of our progression through the various change key depths comes out just one depth away from the end of those available, like this:

odd used	even used
	0 +
− 1	
	2 +
− 3	
	4 +
− 5	
	6 +
− 7	
	8 +
− 9	

Note that tumbler positions with the odd progression *could* use a 0 cut just as well as the No. 1 when it is called for in this progression and would gain the advantage of having a No. 3 differ bit instead of a No. 2. Precisely the same situation is evident in the even progression at the other end of the progression where the last number used is an 8, but there is a 9 beyond it which could be used without affecting the number of key changes available. In situations where the master key is the one which uses one of these No. 1 or No. 8 cuts, no such substitution could be made unless it were made through the entire series because the master key combination must not vary.

Let us now consider another aspect of the master key system which we have used as our example. For this purpose, we shall repeat that same outline and study the relationship of the master key to the change key combinations in some detail.

Pin preference:	5	4	3	2	1	
Master key:	0	5	8	3	6	Master: 05836
Change order:	2	7	0	5	8	1- 27058
	4	9	2	7	0	
	6	1	4	9	2	
	8	3	6	1	4	

Now, suppose we were to cut a key to the change key combination No. 1, but instead of following that combination all the way through, we cut the master key depth on the last tumbler position (the one which is the first one changed in the change key progression), giving us a key cut 27056. Since the first four depths of this key are in the change key progression, it can operate only those locks using the first four depths of cut which are 1705_, but the cut to the master key depth in the last position means that this one cut will operate that tumbler position for any and all of the change key depths in that position. Thus, we have a key which will operate all locks beginning with the cuts 2705, whether the final digit be 8, or 0, or 2, or 4; when the progression of the series enforces a change in the No. 4 tumbler position, however, that depth becomes No. 7 instead of No. 5 and our key is no longer operable. The key 27056 which we have cut is, then, partially a master key, and is called a *sub-master key.*

Carrying the process just a bit further, we can cut a key which hits the master key depth on pins with both the first and second order of pin preference: 27036. This key will operate all locks in the series the first three depths of which are 270__. The result is that whereas our first sub-master key operated 4 locks, this one will operate 4×4, or 16 locks. This process may be carried as far through the key series as may be required to fit the situation.

Now let us suppose that we cut a key just a bit differently. Instead of hitting the master key depth on the first preference order, we hit it only on the second, cutting our key 27038. This key will operate all locks in our series having depths of 270_8, or one in every four until the progression of the series forces a change to be made in the tumbler position which is the third preference order. This particular key will operate combinations number 1, 5, 9, and 13 in our series, but not the intervening combinations because they do not end with a No. 8 depth.

By moving the single tumbler position over to the third preference order for our single master key depth, we have a key numbered 27858. This key will operate one lock in every 16 in our series until the progression of the series forces a change to be made in the tumbler position which is in the fourth preference order. This means that this particular key will operate combinations number 1, 17, 33, and 49 in our sample series. The technique described could be called skip-masterkeying, but is usually referred to as cross-keying.

A great variety of operating conditions may be obtained by carrying this technique through, picking up master key depths on various change key combinations, for example, by using master key depths on pin preference positions 1 and 3, or positions 1 and 4. The possible variations may be further increased if 6 or 7 tumbler locks are used.

Sub-masterkeying and cross-keying such as we have demonstrated here find considerable use in apartment and hotel masterkeying, in providing master keys for the maids for the various floors with cross-keying for the supply closets on all floors. Commercial and industrial institutions also find use for similar keying arrangements to suit operating requirements which may be unique to their own particular operation.

Retrograde Keying (or Maison Keying)

Yet another keying technique which is of considerable worth in certain situations involves the creation of a single lock which is operated by all of the individual keys for a certain group of locks. This technique is known as retrograde keying, and is usually found in conjunction with a master key system. Its principal use is in the masterkeying of apartment houses where such a lock is used for the main entry door to the building, although its usefulness and value may apply equally well to other places and situations.

Common practice in creating such a lock is often to simply leave all tumblers out of the lock except in one or two positions, but this is not a desirable solution to this particular keying problem, particularly for those situations where the number of keys which are used to operate that lock is not too great. A better solution is the use of a three-depth-of-cut interval, adding whatever differ bits are required in each tumbler position to achieve the desired result. Differ bits of two depths of cut are often troublesome where two or more are required in immediate-

ly adjoining locations in the tumbler pile, but differ bits of three-depth-of-cut length are considerably more reliable.

Even though the building may be large enough to require more combinations than may be possible if a three-depth-of-cut interval is observed completely throughout all of the tumbler positions, it may be possible to limit the two-depth interval to one or two tumbler positions, especially if six pin locks are used. Such a master key system should have the tumblers left out of the retrograde lock in the positions where two-depth differ bits are used. This will leave the tumblers intact in all other positions, forcing at least some effort to be made in picking the lock. A master key system of this type is shown in outline form below:

Interval: 3 depths of cut
Forbidden combinations: none
Depth progression: 0 — 9

Pin preference:	6	5	4	3	2	1	
Master key:	3	0	6	3	9	6	Master: 306396
Change order:	6	3	9	6	0	9	1- 639609
	9	6	0	9	3	0	2- 0
	0	9	3	0	6	3	3- 3
							4- 39
							5- 0
							6- 3
							7- 69
							8- 0
							9- 3
							10- 639909

It should be noted that the depths chosen for use in these locks are 0, 3, 6, and 9 in all tumbler positions. One of these has been used for the master key depth in each tumbler position, leaving three available for change key use. Had other depths been chosen for use, only two would be available for change key use if the three depth of cut interval is to be maintained.

Wherever possible in such a system, tumblers not required for masterkeying should be left without differ bits in order to make provision for changing combinations in the event of lost keys, including master keys. An additional reason for not using all of the tumbler positions with differ bits lies in the construction of the retrograde cylinder. Should all of the tumbler positions be masterkeyed, then the retrograde cylinder would contain four different positions where any straight tool could be used to line up the tumblers to pick the lock. They would, in this instance, line up at 000000, 333333, 666666, and 999999. Two tumbler positions which were not masterkeyed would avoid this situation.

Implementation of Master Key Specifications

We have discussed the mechanical method of pin tumbler masterkeying in a general way, and shown the method of arriving at definitive specifications for the keys in such a system. At this point, we should consider the specific requirements of the lock plug for the creation of a pin tumbler master key system. Let us consider the requirements of the lock in which this masterkeying is to be done. There is a requirement for a key pin having a tip suitable for contact with the bottom of the cut in the key. The key pin must be of a length which is suitable for either the change key depth or the master key depth, whichever is shorter. (Remember that pin length may be added to by differ bits, but never diminished.) A differ bit must be added to take up the space remaining between the change key depth and the master key depth, or vice versa.

A driver must be added, and it must be of such length as to keep the overall length of the tumbler

TUMBLER SELECTION FORM

	1	2	3	4	5	6	7
Driver	16	11	10	13	12		
MASTER	0 2	5 2	8 8	3 2	6 2		
KEY	2 0	7 5	0 0	5 3	4 4		

PILE LENGTH, TOTAL: *18*	For Plug Sleeve: Add 10

NO. REQUIRED	MARKING:		
	TYPE	NOTES:	*4 keys required*
2	MORTISE		
	RIM		
	CYLINDRICAL		
1	PADLOCK		
	OTHER		

DESTROY BY FIRE WHEN WORK IS COMPLETE

JOB # *1234*   DATE: *Jan 11*

pile uniform. The uniform length must be long enough to penetrate well into the upper pin chamber, even with the key removed, and to compress the tumbler spring sufficiently to cause it to exert adequate tension on the tumbler pile, even with the key removed, to prevent short differ bits from tipping. At the same time, it must not be so long as to damage the spring when a key blank is inserted into the keyhole.

The author has found it convenient to use a special form for masterkeying of locks which is similar to the charts shown, which have been filled out in accordance with our first sample master key system. By using pencils of different colors, it is possible to keep the tumbler requirements separate from the key cuts on the same form. Forms of this type are available from your locksmith supply house. (See page 255).

It should be noted that in tumbler positions 1, 2, and 4, the master key depth is shallower than that of the change key, and that the key pin length is consequently that required by the master key depth.

The reverse of this is true on tumblers 3 and 5 where the key pin length is governed by the change key depth. A good master key should contain one (at least) tumbler position where the master key depth is shallower than any of the change key depths used in that position. In our sample series the master key cut is in the No. 1 tumbler position. Such keying prevents any of the change keys being filed down to make a master key.

Grand Masterkeying

There are times when, due to the intricacies of a master key system, or due to the great number of change key combinations required, the key changes available from a particular lock are simply not enough to meet the requirements. When this occurs, it becomes necessary to look outside of that system for other available combinations. For this reason, a number of manufacturers provide locks with *compatible keyways,* such as those shown in Figure 184.

These keyways are different from one another, yet a key blank is available which will enter all of them. By cutting the master key on the key blank which will enter all of the keyways used, we have a key which will open all of the locks. Combinations previously used in locks having one keyway may be repeated in one or more additional keyways with no fear of interchange, because only a key cut on the grand master section will enter the keyhole for more than one keyway. Keying of individual key changes may be exactly the same from one keyway section to another, or it may vary widely. The sole requirement is that the master key depths be identical in all keyways used in such a grand master key system.

Forbidden Combinations

Several times, forbidden combinations have been mentioned in the course of discussing various pin tumbler locks and keys. The problems which are presented by these combinations are particularly magnified by the presence of masterkeying in a lock. One reason for this is that in a pin tumbler lock utilizing 10 depth changes, it may be that there are no forbidden combinations, or it may be that an interval of no more than 6, or 7, or 8, or 9 depths is permissible. The situation is not irremediable; while these forbidden combinations can seldom be completely eliminated, even from a simple master key system, they can be minimized by the judicious selection of a master key combination. For example:

Interval: 2 depths of cut.
Forbidden combinations: 8 or more intervals.
Depth progression: 0 — 9.

	—1—	—2—	—3—	—4—
Master key	0 1 0 1 0	8 9 8 9 8	0 5 8 3 6	0 1 8 9 8
Change key	2 3 2 3 2	0 1 0 1 0	2 7 0 5 8	2 3 0 1 0
Progression	4 5 4 5 4	2 3 2 3 2	4 9 2 7 0	4 5 2 3 2
	6 7 6 7 6	4 5 4 5 4	6 1 4 9 2	6 7 4 5 4
	8 9 8 9 8	6 7 6 7 6	8 3 6 1 4	8 9 6 7 6

In the master key outlines above, the first two master keys produce key changes which are without forbidden combinations; but if sub-master keying is to be done, it will be necessary to choose a sub-master key very carefully if forbidden combinations are to be avoided. These are not the best of the possible master key combinations from the standpoint of being difficult to pick; however, something of this sort is occasionally used. Master key No. 3 in our example above is the one used in our sample series for progressive masterkeying. Notice that it has four tumbler depths in its change key progression which could generate forbidden combinations if not carefully watched, and they are scattered through practically the entire key.

A master key similar to that used in the fourth example is a compromise which is often adopted by manufacturers for factory master keys since it has only two tumbler positions containing one depth of cut in each position where it is necessary to be careful of forbidden combinations. Such a system limits and localizes forbidden combinations rather

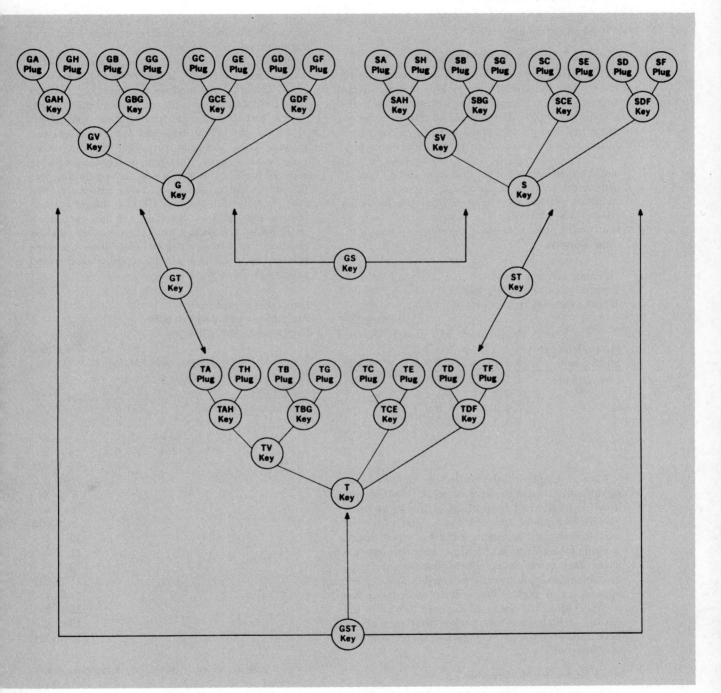

Figure 184. Chart of keyways used in a grand master key system. Illustration courtesy of Eaton Yale & Towne, Inc.

than eliminates them, and produces a better master key combination as a result.

Modified Progressive Masterkeying

The modified progressive system of masterkeying is very like the straight progressive system, except that a random sequencing is introduced into the change key depths. The element of random sequencing introduces an additional variable factor which makes reconstruction of an entire code series from the mere possession of a few of the keys in that series a considerably more uncertain thing. Typical differences between the progressive system and the modified system may be seen from the following comparative outlines. The system is the same as that used to illustrate the progressive system in all other respects.

Interval: 2 depths of cut.
Forbidden combinations: none.
Depth progression: 0 — 9.

	Progressive	Modified	Modified Series
Pin preference:	5 4 3 2 1	5 4 3 2 1	
Master key:	0 5 8 3 6	0 5 8 3 6	05836
Change order:	2 7 0 5 8	6 7 0 9 4	1- 67094
	4 9 2 7 0	4 1 4 5 2	2- 2
	6 1 4 9 2	8 9 2 7 0	3- 0
	8 3 6 1 4	2 3 6 1 8	4- 8
			5- 54
			6- 2

With an outline similar to the one shown above, no difficulty should be experienced in building such a series to the full limitations of the locks used. Forbidden combinations could prove just a bit more troublesome until one gains sufficient experience in working from this type of outline, but with practice it becomes much easier. The difference insofar as forbidden combinations is concerned is merely their appearance in slightly less orderly sequence in the outline. Once their potential has been located in the outline, a little care taken when that part of the development of the series is reached should result in very few problems.

Mixed Progressive Masterkeying

The mixed progressive system of masterkeying is only a little more complex to create and work with than is the modified progressive, but is vastly more difficult to decode if one has access to only a few of the locks or keys of a series. A great many locks are usually required, and even they must be selected from just the right parts of the series. The mixed progressive system merely adds one more variable to the modified progressive system.

The variable which makes this difference is the order of *pin preference*. It will no doubt have been noted that the pin tumbler masterkeying outlines have heretofore given a pin preference order of 5, 4, 3, 2, 1, and that this has been uniform throughout all of the previously discussed systems of masterkeying. This, too, can vary, and in the mixed progressive system, the pin preference order is selected more or less at random. At first glance, this introduces the appearance of being extremely difficult, but the difficulty is illusory, once one has formed the habit of checking the pin preference order before proceeding to change another tumbler position. The keying outlines below show the same master key, done in both the modified and mixed progressive fashions:

Interval: 2 depths of cut.
Forbidden combinations: none.
Depth progression: 0 — 9

	Progressive	Modified	Modified Series
Pin preference:	5 4 3 2 1	5 3 1 4 2	
Master key:	0 5 8 3 6	0 5 8 3 6	05836
Change order:	6 7 0 9 4	6 7 0 9 4	1- 67094
	4 1 4 5 2	4 1 4 5 2	2- 4
	8 9 2 7 0	8 9 2 7 0	3- 2
	2 3 6 1 8	2 3 6 1 8	4- 6
			5- 67092
			6- 4
			7- 2
			8- 6
			9- 67090
			10- 4
			11- 2
			12- 6
			13- 67098
			14- 4
			15- 2
			16- 6
			17- 61094

Here again, should there be forbidden combinations involved in the key changes for a lock selected for use, they can be minimized by care in the selection of the master key combination. At the time the key changes are outlined, each depth of cut which carries a potential for becoming involved in a forbidden combination should be carefully sought and marked *in the outline* before the pin preference order is chosen. The pin preference order may then be chosen to position those forbidden combinations at the point in the code series where they will present the fewest problems to both the change key combinations and any sub-master keys which may be required. Often, it is better to place them at the position where sub-master keying is done, rather than to limit the change key combinations too drastically.

Scrambled Master Key Systems

The scrambled master key system is planned and executed for key changes, just as any straight progressive system would be. Then all of the required sub-master keys are chosen, preferably allowing for a few whose need is not apparent at the time the system is created. Each of the sub-master keys is next assigned a letter designation: A, B, C, D, etc., and the letter is entered in the key codes after each combination which that particular sub-master key will operate.

After all of this has been done, the key codes are cut into individual strips, each of which bears the code of a particular lock. These strips are placed in a box, and the codes drawn one at a time from the box and entered on a new code sheet, together with the letter designating which sub-master will operate them. In selecting locks from among these codes, it is relatively simple to go down the list, scanning it for the letter designation which will give a combination in the desired sub-master key group.

At the time of encoding, the choice of sub-master series is relatively simple, but after the codes have once been scrambled, it is an exceedingly tedious task. At times, particularly in a rather large system, it would be more practical to determine new sub-master keys and the locks to use with them by feeding the entire code series into a computer, together with a notation of each lock which has already been used.

If sufficient planning is put into a scrambled system at the time of its inception, no particular trouble should be experienced, but the modification or addition to such a system after it has been encoded and scrambled is an almost insurmountable task.

Planning A Master Key System

THE planning of a master key system is usually directed into one of two patterns, depending upon the type of business or firm for whom the work is being done. Masterkeying may adhere closely to the physical structure of the building(s) if done for a hotel or motel. Some industrial establishments also prefer this pattern. Ordinarily, however, industry prefers the master key structure to assume a pattern in rather close approximation to their organizational structure. At times, industrial masterkeying tends to become a combination of both the physical and organizational structures.

It should be made very clear at this time that the function of a locksmith in creating a master key system is to please the customer and meet his requirements to the best of his ability and the availability of materials to meet those requirements. It is *not* the function of the locksmith to tell the customer what kind of a master key system he requires, but rather to try to ascertain his customer's needs and try to meet them, or make alternative suggestions which he believes would be helpful. This is not to say that the locksmith should try to teach masterkeying to his customer; however, situations such as those where the customer may want a sub-master key for each employee can usually be handled just as well by keying those particular locks alike, within the framework of the master key system, without the need for sub-master keying.

In brief, the function of the locksmith in establishing a master key system for his customer should be one of helpful advice as to what is possible, and what is not possible or practical. To this end, there are a number of questions which should be asked of the customer:

"Do you want a straight master key system with only one master key operating all locks, or do you want the system subdivided so as to provide as nearly as possible a one-key operation on all levels of responsibility?" If interest is expressed in a one-key operation, the possibilities for sub-masterkeying based on physical or organizational structure should be explained, showing how such a system can save time and money. A small price differential could be charged the customer for sub-masterkeying on the order of a few cents a cylinder over the regular masterkeying charge. It is well worth it, since the time saved a customer is worth far more than the modest fee for sub-masterkeying, and little or no extra work is required of the locksmith.

"What is the structure of your organization or plant?" is the next question, if sub-masterkeying is contemplated. The locksmith who is holding a consultation with a prospective customer for a master key system should be prepared to ask this question and have adequate forms especially prepared for determining this type of information and recording it in a manner which can be understood. At the same time, he should have a format to serve as the basis for a keying outline. If a sub-master key system is desired, it is suggested that one of these forms be reserved exclusively for showing the structure of the sub-mastering required. The change key chart may be used for individual locks under each of the sub-masters, if the customer wishes to go over the layout in detail.

MASTER KEY SYSTEM

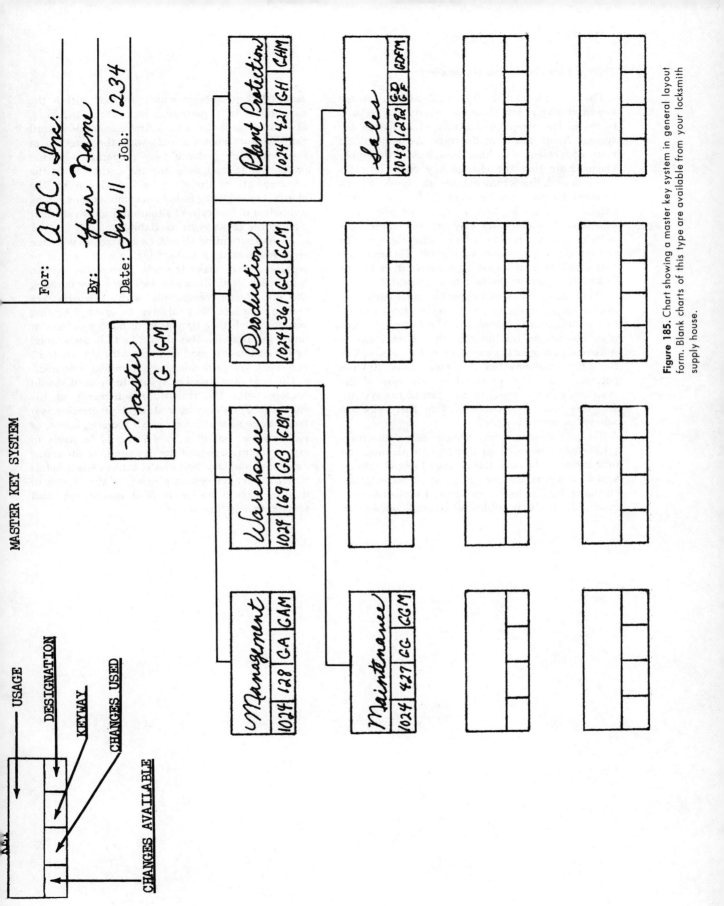

Figure 185. Chart showing a master key system in general layout form. Blank charts of this type are available from your locksmith supply house.

The master key chart, Figure 185, and the change key chart, Figure 186, show how charts may be used to reveal the complete structure of a master key system. Blank charts of this type are available in more convenient size from your locksmith supply house. Note that the change key chart provides space on each line for the number and types of locks required for each combination. This is extremely helpful when actual keying begins, or when cost estimates are made which may require certain types of locks to be replaced in order to bring them within the keyway requirements for the job. This information should be based upon a personal survey by the locksmith to determine the matter accurately.

"How many locks will there be under each sub-master key (or in the system)?" As has been shown, this information is essential if the locksmith is to plan the system intelligently and select locks of sufficient keying or keyway potential to accomplish the job in the way the customer wants. Reserve potential should be provided to take care of the customer's future needs in any part of the system, and this must be planned before the locks are selected or keying is done.

If a particular lock has already been tentatively selected, or if re-use of existing locks is planned, the limitations of the potential of those locks should be kept in mind during the planning stages with the customer, both as to change key and sub-master key potential. It is possible to commit oneself to establish key systems which are not within the capabilities of a particular lock. It is, therefore, always advisable to avoid a firm commitment until the locksmith knows exactly what the potential of the system is, preferably by preparing a keying outline for it, and adjusting the outline to fit the requirements of the situation, as shown by the charts. Locks having forbidden combinations can be a problem in this type of situation, unless sufficient compatible keyways are available.

An understanding should be arrived at with the customer regarding locks which require repair or replacement to make the new master key system fully operational. Even new locks in a new building are not always compatible with a master key system, nor are they always in good operating condition, whether by reason of factory defects or faulty installation. This should be fully understood by the customer if good relations with him are to be preserved, and further business from him is desired.

The type of master key system to be used should conform with the realistic requirements of the customer. For example, a straight progressive system may be adequate for a small rooming house, or for farm use, but in a hotel it would be likely to expose the management and its guests to an undue amount of burglary. Some hotel thieves select hotels with a master key system which is relatively easy to decode, make a master (or floor master) key, and proceed to help themselves.

MASTER KEY SYSTEM

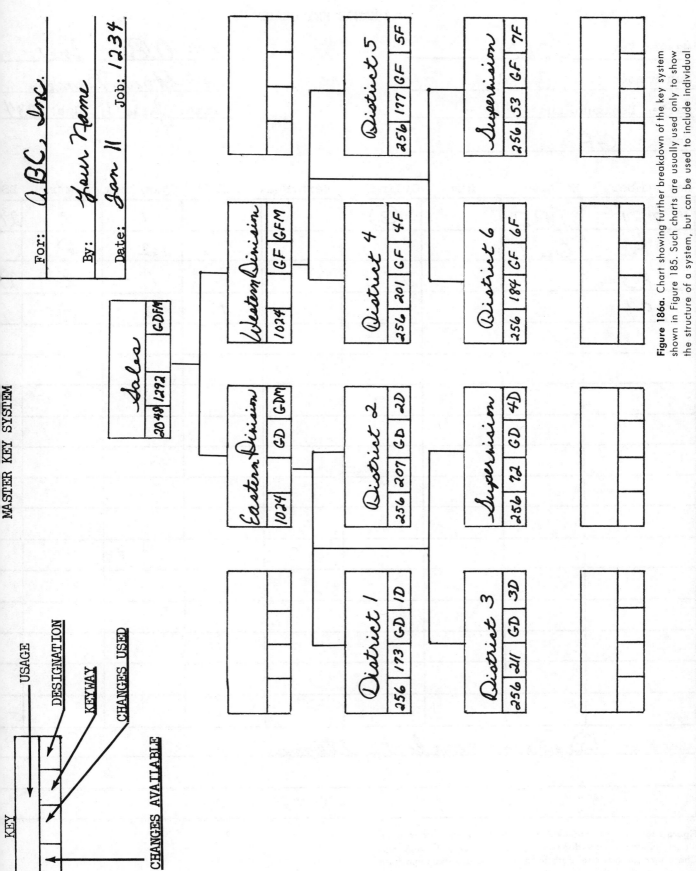

Figure 186a. Chart showing further breakdown of the key system shown in Figure 185. Such charts are usually used only to show the structure of a system, but can be used to include individual locks.

CHANGE KEY CHART

Master Key: _GDFM_   For: _A.B.C. Inc._

Sub-Master Key: _GFM_   By: _Your Name_

Series Designation: _7F_   Date: _Jan 11_   Job: _1234_

Keyway: _GF_

Key Number	Mortise	Rim	Cylind'l.	Vert.Rod.	P.Exit	Cyls/lock	Mfr.	BS
7F1	DB2		③			1	Z	2¾
7F2	DL1					2	P	
7F3			DL1			1	Z	2¾
7F4					1	2	B	

Remarks: _Replace circled items_

Figure 186b. A further breakdown for the locksmith's use in planning a system is usually done on a more informative Change Key Chart such as this one. Both of these forms are available from your locksmith supply house.

LESSON 28

Design Variations Affecting Masterkeying

THE keying of some locks is controlled by a design variation affecting the structure of the lock cylinder itself, such as variations in the spacing of the tumblers; as, for example, the Sargent "Keso" locks. Masterkeying of these locks is done at the factory at the time of manufacture. It involves the drilling of the tumbler holes in a location specific to the keying of the lock series being keyed; since special tumblers are used, these locks do not lend themselves to re-keying by the locksmith.

Master Ring

Certain other locks use a "master ring" around the lock plug. This is a floating ring between the lock plug and the cylinder hull through which holes are drilled for the tumblers to pass. The master key operates at the shear line between the floating ring and the cylinder hull, while the change key operates at the shear line between the lock plug and the floating ring. Use of a floating ring means that the master key depth is usable on change key combinations as well as on master keys, with the provision that not all of the master key depths can be used at once.

It is good practice in this type of a master key system to leave at least two tumblers of the master key combination unused on any change key at all times, not necessarily the same two tumblers, but two in any one key. This is the system primarily used in Corbin locks and requires a somewhat larger diameter lock cylinder than is customary. Utilization of a single master key depth on a change key combination does not create a sub-master key for this sort of lock. Additional differ bits must be used in one or the other shear line positions to create sub-master keying.

Bi-centric Cylinder

Another means of utilizing the maximum number of combinations on the change keys of a lock code series is the use of a bi-centric cylinder, such as the one shown in Figure 187. These cylinders are made by Yale, and have two lock plugs in the same cylinder, one of which is keyed for operation by the master key and is operable only by the master key. The other is the change key plug, and is operated solely by the change key. A double cam arrangement as shown in Figure 187 is used on these locks, one cam being operated by each plug. One of these cams is operated remotely by the master key's plug by means of a geared drive system.

One of the distinct advantages of this type of system is that the security of the lock, insofar as picking is concerned, is not deteriorated by the presence of differ bits, causing each tumbler to have two positions at which the cylinder may be picked. A precaution to observe in the use of these cylinders is that the overall length of the tumbler pile and springs is critical, due to the lessened depth of the upper tumbler chamber in the cylinder hull, by virtue of its being set into the cylinder hull at an angle, rather than squarely across it as in a conventional cylinder. The two cams require more space inside the lock case than does a single cam cylinder, so this factor, too, should be taken into consideration in selecting these cylinders.

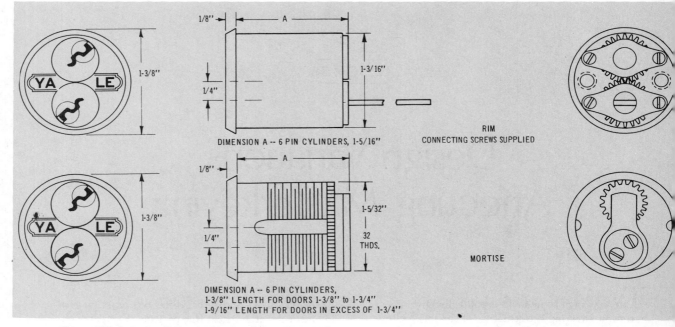

1/8'' A 1-3/16'' 1/4''

DIMENSION A -- 6 PIN CYLINDERS, 1-5/16''

RIM
CONNECTING SCREWS SUPPLIED

1/8'' A 1-5/32'' 1/4'' 32 THDS.

MORTISE

DIMENSION A -- 6 PIN CYLINDERS,
1-3/8'' LENGTH FOR DOORS 1-3/8'' to 1-3/4''
1-9/16'' LENGTH FOR DOORS IN EXCESS OF 1-3/4''

Figure 187. Bi-centric lock cylinder. In addition to two plugs, this cylinder used two sets of upper tumbler chambers placed in two parallel rows, rather than in the usual intersection with the center line of the cylinder. Illustration courtesy of Eaton Yale & Towne, Inc.

Schlage Wafer Locks

The following discussion on the keying and masterkeying of Schlage wafer locks, although it includes material other than masterkeying information, comprises a unit of information which the author thinks advisable to retain intact. The following discussion on the keying of these locks (pages 268-285) is presented through the courtesy of the Schlage Lock Company.

Removable Core Locks

Locks of the removable core type are keyed and masterkeyed just as are conventional pin tumbler locks, but the *control key* operation (removing the core from its cylinder) varies from one manufacturer to another. One of the more common types of the removable core is typified by those manufactured by Best Universal Lock Co. and Falcon Lock Co.

These locks utilize a sleeve around the lock plug, shaped as in Figure 188, to retain the core in the cylinder. The sleeve is rotated about 15° clockwise to retract its corner to a position within the cross-sectional configuration of the core to allow the core to be removed.

The rotation of the sleeve is accomplished by the addition of an extra set of differ bits operating at the shear line which is formed by the junction of the

sleeve with the body of the removable core. Care should be taken that no change key or master key may duplicate the depths of the control key in all

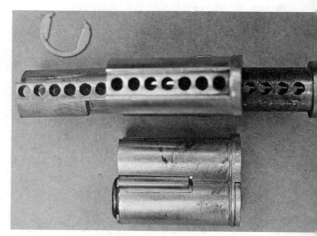

Figure 188. A removable pin tumbler core. Removable cores of this type may be used with the many different cylinders and locks made to accept them. The protruding corner of the control sleeve engages behind a matching protrusion on the cylinder or lock in which it is to be used. This holds the core securely in place until the control key is used to retract the corner of the control sleeve. At the bottom is an assembled core; above it is one partially disassembled to show construction.

positions. The change key and master key both operate at the shear line formed by the plug and the sleeve, hence may be keyed exactly as in a conventional master key system.

To key the control key combination on one of these locks, "10" is added to the total length of the tumbler depths to permit the control key to line up at the control shear line. This "10" is not added to the pile after selecting a key pin and differ bit for the master key, but is added to the depth shown on the cut of the control key; for example, suppose that we have a change key depth of 3, a master key depth of 7, and a control key depth of 5. The key pin would be a No. 3 (for the change key); the differ bit would be a No. 4 (for the master key); another differ bit would be a No. 8 (for the No. 15 depth of the control key); the driver would be another No. 8 differ bit, bringing the total tumbler pile length of 23, which is standard for these locks. Springs are added and the core is capped. If many of these cores are to be keyed, special tools should be obtained for the purpose from your locksmith supply house. As pin diameters are special, one should not plan on using standard size pins in these cores. Depth interval is .0125 and must be closely maintained if key operation is to be satisfactory.

DISASSEMBLING AND REMOVING KEYWA

The Schlage wafer keyway unit, after years of research and development, was first marketed in 1927. Eleven hundred and twenty stock keys can be combinated in this keyway and four hundred and eighty different changes are possible with a single master key. It can also be grand master keyed.

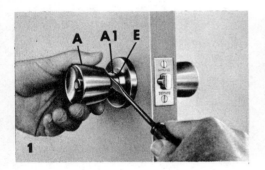

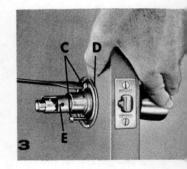

Step 1 Depress catch (A1) through the hole in the shank of inside knob (A) with a screwdriver to release inside knob from spindle (E).

Step 2 With knob removed, position screwdriver into small notch (B1) usually located on the bottom edge of inside rosette (B) and, with prying motion, snap off the inside rosette.

Step 3 Remove the two machine screws (C) and inside mounting plate (D) will slip off over the inside spindle (E). The lock will now slip out of hole.

Step 4 The lock housing (F) is attached to the lock by small cotter pin (G) or by twisted lugs. Remove cotter pin (G) or straighten out the lugs. Lift housing above lugs and, with slight turn, rotate housing (F) ¼ inch and remove.

Step 5 With housing removed, the lock frame is now exposed. Hold lock in both hands, positioning fingers as shown in illustration. (When performing this operation, hold palms of hands carefully around lock to prevent springs from escaping the retractor slide.) To remove thrust plate (I), press forward with thumbs against frame tabs (H1), push upward with index fingers against thrust plate. This will disengage plate.

Step 6 In order to free the plunger unit (L) as you remove this assembly, it is necessary to push the slide (J) all the way to the rear against the compression of the two slide springs (K) and hold down on the slide with the thumb.

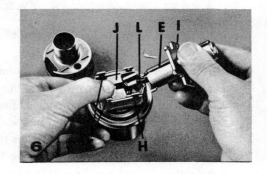

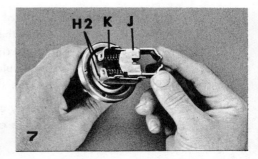

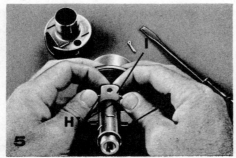

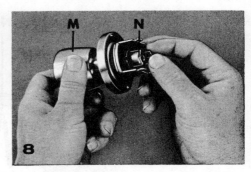

Step 7 After the inside spindle (E), thrust plate (I), and plunger assembly (L) have been removed, let the slide (J) and two slide springs (K) ease forward gradually and remove them from the lock frame (H).

Step 8 To remove wafer keyway unit (N) from lock, push in on face of the wafer keyway unit from the outside knob (M). Unit will then slide inward, where it can be removed from the knob assembly.

Step 9 To facilitate reassembly, remove outside knob (M) from lock frame by rotating knob ¾ of a turn while pulling out.

THE WAFER KEYWAY UNIT
Description and Terminology

In working with the wafer keyway unit, hold it in the left hand with the "V" grooved dividing strip (N1) facing you. In this position, the spring comb is on the underneath side and cannot be seen. The protrusion at the extreme left is the keyway cam (N2). The most common form of this cam is illustrated below. To the right is the plunger spring (N3). Next to this is the keyway frame (N4) which includes the entire steel area from the plunger spring to the finished cap of the keyway unit. In this steel framework are located the wafers which are activated by the insertion of the key.

The first column (N5) to the right of the plunger spring, containing two slots is the location of the master wafer. Notice that the slots in this column have a different proportion than do the rest of the columns in the keyway frame. In the remaining columns, the top slot is shorter than the bottom slot in this particular unit. In some units, the relationship of these slots is reversed, the long on top and the short on bottom.

On the reverse side of the keyway unit, notice that there is a metal spring rack (N6) illustrated below, which looks like a comb and upon which are seated the wafer springs (N7).

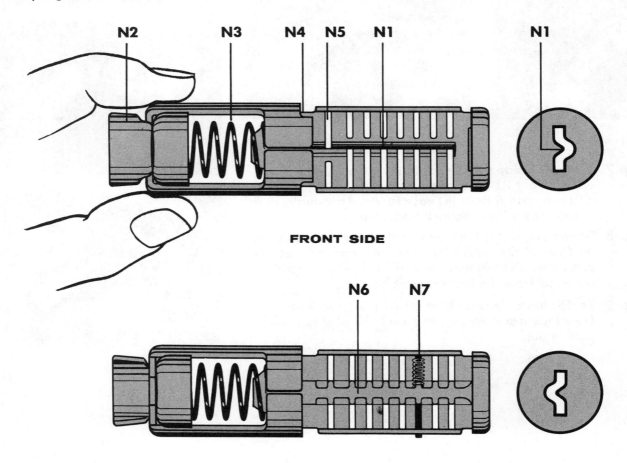

FRONT SIDE

REVERSE SIDE

WAFERS

To more easily recognize the three types of wafers, always arrange them so that the small protrusion is upward and the opening is to the right. Notice that each of the three types has a definite silhouette. The master wafer (N5) has a notch cut out at the base of the protrusion just inside of the spring seat. The combination wafer (N8) has a protrusion on the rounded shoulder opposite the spring seat location. The series wafer (N9) has the protrusion at the top of the wafer, close to the spring seat but does not have the small notch, as does the master wafer (N5). Each of these wafers performs in a different manner and it is most important to recognize each type before it is inserted in the keyway unit.

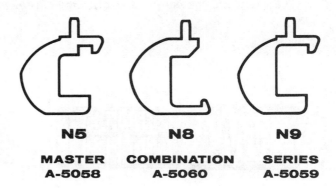

N5	N8	N9
MASTER **A-5058**	**COMBINATION** **A-5060**	**SERIES** **A-5059**

OPERATION OF THE KEYWAY UNIT

Here are two wafer keyway units set to the same combination. Figure 1 is in the relaxed position or with the key out of the keyway. Figure 2 has the proper key inserted. Notice in Figure 1, there are four protrusions of the wafers, one at the bottom and three at the top. The first protrusion to the right of the plunger spring (N3) is the master wafer (N5). This remains out except when it is retracted by the uncut portion of the tip of the key, which is designated in ▨. The cut portion of the tip is necessary to allow full insertion of the key into the keyway. The three protrusions (N9) at the top of the keyway are the series wafers. When the key is inserted, the uncut portion opposite the protrusion (indicated in ▨) acts upon the series wafers to pull them into the keyway.

The remaining four wafers (N8) in the keyway are combination wafers. Both in the relaxed position (Figure 1) and with the key inserted (Figure 2), these wafers lie within the confines of the keyway unit. Therefore, cuts on the key adjacent to their protrusion are required to prevent them from being pushed out into the locking position. An improper key will fail to draw back all the protrusions of the master and series wafers and will extend some or all of the protrusions on the combination wafers.

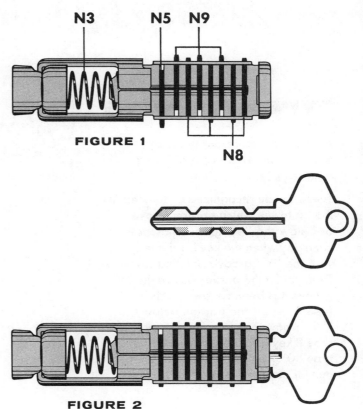

FIGURE 1

FIGURE 2

TYPES OF KEYWAY UNITS

Schlage wafer keyway units are made in two distinct types, type 1 and type 2. To distinguish between these two types, look first at the master wafer column. In type 1 the elongated slot will be at the top. Type 2 keyway has the elongated slot of the master wafer column below the "V" groove.

TYPE 1

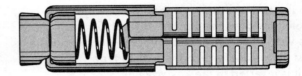

TYPE 2

TYPES OF KEYS

As there are different types of keyways, there are also different types of keys.

These can be recognized by looking at the tip to see which portion above or below the "V" groove has been cut away when the key is oriented with the "V" groove pointing away from you. If the portion above the "V" groove has been cut away, this is a type 1 key. If the portion below the "V" groove has been cut away it is a type 2 key. If the tip is uncut it is a type "0" key, usually used as a master or grand master key.

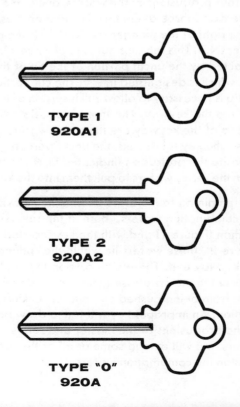

TYPE 1
920A1

TYPE 2
920A2

TYPE "0"
920A

KEYWAY CODING

As explained before, the first column to the right of the plunger spring is the master wafer column. Its proportions give a clue as to the type of keyway with which we are dealing. In illustration "A" at right, you notice the longer slot is below the "V" grooved dividing strip, indicating a type #2 keyway.

The 14 slots to the right of the master wafer column are assigned code numbers corresponding to the placement of the combination wafers (N8). The first slot to the right of the master wafer column and located above the dividing strip is given the designation code #1.

The slot directly below this is given the designation code #2. The code numbers alternate between odd and even, continuing to the right of the keyway. All odd numbers are on top—1, 3, 5, 7, 9 and the last two numbers are 1' and 3' (read as 1 prime and 3 prime). All the even numbers are located below the dividing strip—2, 4, 6, 8, 0—2' 4'.

All factory cut keys have a combination number stamped on the bow which indicates the notching on the key.

In illustration "B" at the right, the key carried the number 203823. The first digit indicates a type #2 key with its tip cut away, as explained on p. 272. The second digit, in this case 0, indicates a stock key not related to any masterkeyed system. The last four digits indicate the location of the notches cut in the key. These same four digits (3823 in the illustration) also indicate the position of the combination wafers in the keyway since these wafers must rest within the cut away portions of the key.

If a key is not stamped with a factory combination number, take an empty wafer keyway and insert the questionable key to determine its combination number. In illustration "C", at the right, the shank of the key may be seen through the slots in the keyway except at those code locations where the key has been notched.

Looking first at the master wafer column, the slot below the dividing strip is unobstructed indicating the tip of the key has been cut away at the bottom. This key is a type #2 (see p. 272). Therefore, the first digit of the combination number would be 2. The second digit of the code number is not related to any cuts on the key and is always 0 for stock (non-masterkeyed) keys. Other numbers are used as the second digit to designate masterkeyed systems, as explained on p. 279.

Next look at the 14 code slots in the remaining seven columns to locate the four cut away portions of the key. In illustration "C" at the right, these cuts occur at code positions 3, 8, 2', 3'. The complete combination number which would be stamped on this key should be 203823.

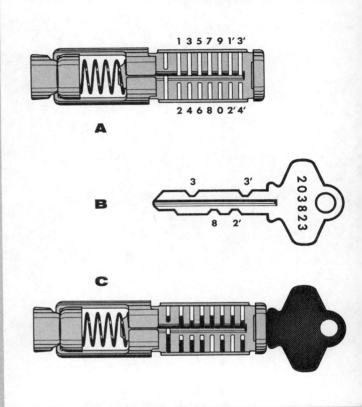

A

B

C

SETTING UP STOCK WAFER KEYWA

Using the combination number 101450, the first dig designates a type #1 keyway. Select such a keywa and into this, insert a master wafer in the first or maste column with the protrusion pointing up. The secon digit being 0, indicates this will be a stock keyway uni and should be set up in accordance with the following procedure.

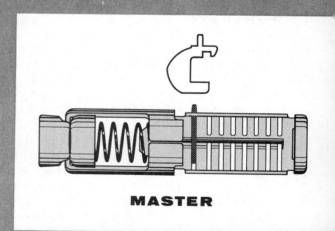

MASTER

Taking four combination wafers, insert them in the po-sitions indicated by the last four digits of the combina-tion. The combination wafer is unique in that it may be inserted with the protrusion pointing either up or down in the seven combination columns. The code number designates in which of the 14 slots the protrusion should be inserted. For example, #1 would be inserted point-

NITS BY COMBINATION NUMBERS

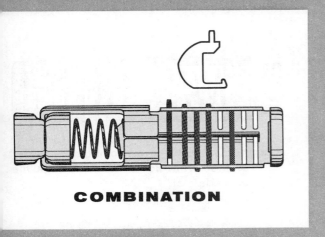

COMBINATION

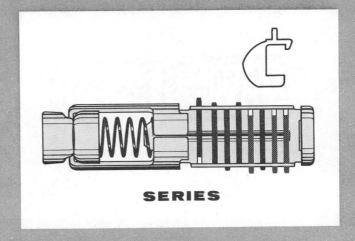

SERIES

g upward in the first column after the master wafer
lumn. #4 would point downward in the second col-
nn. #5 would be pointing upward in the third combi-
tion column and the 0 would be pointing downward
the fifth combination column. After the four combi-
tion wafers have been positioned, the remaining
ree empty columns should be filled with the series

wafers. Note that the protrusion of the series wafers
can be inserted only in the longer slot of the empty columns. The protrusion of all the series wafers, therefore,
will point in the same direction within any one keyway.
When springs have been properly attached to all
wafers (see p. 278), the keyway unit is then ready to
be operated by a key cut to combination 101450.

KEYING ALIKE WAFER KEYWAY UNITS

Frequently it becomes necessary to alter the combinations of one or more stock keyway units to exactly match that of another. The procedure used to accomplish this is termed "keying alike" and should not be confused with "masterkeying" which is discussed on page 279.

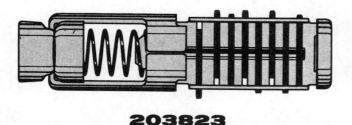

203823

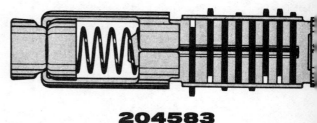

204583

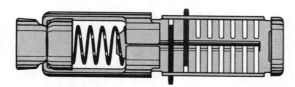

First Column—no change is necessary.

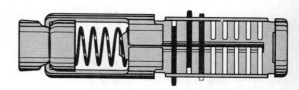

Second Column—invert combination wafer so that protrusion extends through the #3 code slot.

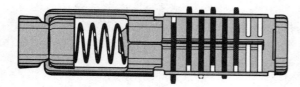

Fifth Column—no change is necessary.

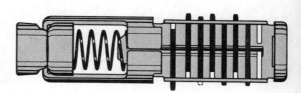

Sixth Column—replace series wafer with combination wafer — protrusion to extend through #2' code slot.

One of the simplest methods of keying alike a group of stock wafer keyway units of the same type, either type #1 or type #2, is to put one aside as a control, empty the series and combination wafers from the others, then "set up" these units to the code combination of the control keyway, using the procedure explained on pp. 274, 275. An alternate method involving fewer operations consists of rearranging only those series and combination wafers in the random keyways which differ in position from those located in the control keyway. Illustrating this method, we first select two keyway units of the same type—type #2 in this example. The keyway coded 203823 will be used as the control. Next, examine the seven combination columns to determine what rearrangement of the wafers is necessary to match the random keyway unit to the control keyway.

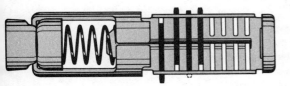

Third Column—replace combination wafer with the series wafer.

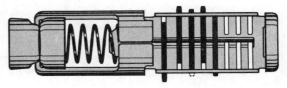

Fourth Column—no change is necessary.

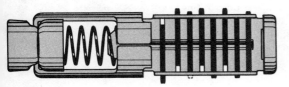

Seventh Column—no change is necessary.

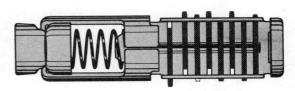

In this typical example, illustrating the alternate method of keying alike two keyway units, only three rearrangements of the wafers were necessary.

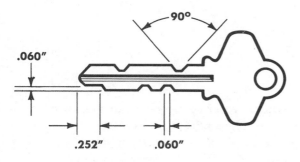

Keys accurately cut from genuine Schlage #920A key blanks insure sm_ operation if the dimensions of the notching, as shown on the accompanying _ trations, are closely observed.

The first operation normally performed on a 920A key blank is to cut aw_ portion of the tip to correspond to the type of keyway unit with which it is _ used. (Note: key blanks may be purchased with this notch already cut by s_ fying 920A1 blanks for type #1 keyway units, or 920A2 blanks for typ_ keyway units.) All other cuts on the key are made to the same depth of ._ and have the same width .060″ at the bottom of the notch. All of the ang_ the cuts should have a minimum of 90 degrees.

It is best to use either a factory cut key as the basis for the duplicate, or fu_ pattern keys available from the factory. With full cut pattern keys it is nece_ to select only those notches corresponding to the specific combination nur_ to be cut. After the keys are cut, dress them lightly with a file to remove s_ edges and check the keys in the keyway unit to make sure they operate _ erly. All the protrusions on the wafers should be flush with the keyway whe_ proper key is inserted.

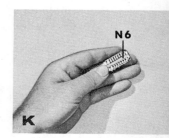

As in illustration "H", always hold the keyway unit in the left hand with the keyway cam (N2) to the left and the "V" grooved dividing strip facing you. Then as you insert the wafers (see "I") you will find the series wafers will only go in the elongated slot of the column. They should not be forced as they will not fit into the incorrect position. All wafers should be inserted in the slots with the protrusion first. This protrusion projects between the two side sections of the steel keyway frame.

When all of the wafers have been inserted into their proper location, hold the finished cap of the keyway unit with the right hand (see "J") and transfer the left hand so that the first two fingers cover the wafers and hold them in position as you rotate the keyway unit (see "K") to expose the spring rack (N6). If these wafers are not held in position as you rotate the keyway, they may drop out and cause you to rework the setup.

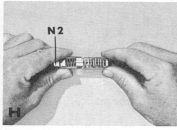

After the keyway unit has been rotated, hold it tightly against the fingers with pressure by the thumb and then exert force downward against the wafers with the index finger to open the distance between the spring rack in the center of the keyway and the spring seat on the wafers, to provide space for the insertion of the wafer springs.

The easiest way to insert a wafer spring is to use a pair of fine needle-nose tweezers (see "L") and grasp the spring at the second coil back from one end. Holding it in this position, you can guide the free end of the spring over the spring rack seat and use the needle-nose tweezers to guide the other end over the spring seat on the wafer.

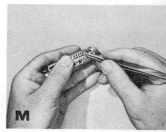

When all springs have been inserted in the bottom of the keyway, exert pressure upward against the wafers with the middle finger and open the distance between the spring rack and the spring seat on the wafers. Again install the springs, as explained above (see "M"). Take the key (see "N") and run it in and out of the keyway several times as it is important that all springs are fully seated before the keyway unit is reassembled into the lock.

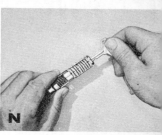

MASTERKEYING

Frequently it becomes necessary to provide a group of locks with keyway units, each having a different and non-interchangeable key, but all having in common one masterkey capable of operating each of the locks in the group.

Combinating locks in this manner is termed "masterkeying" and is quite easily accomplished with Schlage wafer keyway units. To illustrate the principle of masterkeying, consider first, two stock keyway units having combination numbers 203823 and 204823. These units, in common with all other stock keyway units, contain one master wafer, four combination wafers and three series wafers. Their code numbers indicate that these units are identical except for the position of the combination wafer in the second column. This difference is sufficient to prevent interchangeability of their keys but a third key could be cut which would operate both of these keyway units by providing it with notches for both the #3 and #4 combination tumbler positions. In other words, their masterkey would be cut with five notches corresponding to code positions 3, 4, 8, 2 and 3.

This principle of masterkeying can be expanded to include many more combinations. However, as the number increases, it will become necessary to eliminate one or two series wafers from the keyway units. Series wafers cannot be used in any column for which the key has been notched since it is the unnotched portion of the key which retracts the series wafer. At no time should a keyway unit be set up without any series wafers, as this decreases the security of the masterkey system.

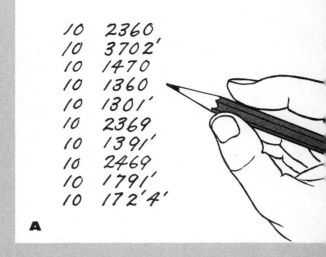

A

One method of laying out a masterkeyed system involves five basic steps:

Step 1 Select the required number of keyway units, all of which should be either type #1 or type #2. (If more than 240 keyway units are required, refer to p. 284 under "Extended Masterkey Systems".)

Step 2 Of the seven columns available, arbitrarily select either 3, 2 or at least 1 column which will be reserved in all keyway units for the placement of series wafers. 16 masterkeyed units are available with 3 columns reserved for series wafers. 80 masterkeyed units are available when 2 columns are reserved for series wafers. 240 masterkeyed units are available when 1 column is reserved for a series wafer.

Step 3 List the code numbers of all the slots in the remaining columns. These slots are available for the placement of the four combination wafers. These code numbers also represent the cuts in the masterkey which will operate all the keyway units having combinations determined by the procedure given in step #4.

Step 4 Utilizing the digits obtained in step #3, tabulate a list of four digit combination numbers. Although each column contains two code slots, it can accommodate only one wafer. Therefore, be careful to select not more than one digit from any available column.

Step 5 Using the code numbers determined above, follow the procedure given on pp. 274, 275 and set up the masterkeyed keyway units, placing series wafers only in those columns reserved in step #2.

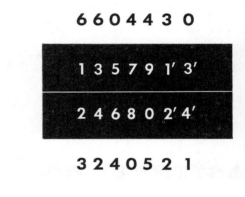

B

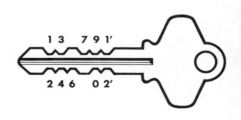

C

There is an alternate method of masterkeying which is particularly useful when there is at hand a number of assembled stock keyway units, or a supply of factory numbered stock keys.

The following seven basic steps illustrate this method:

Step 1 Select approximately one-third more keys or keyway units (all of one type) than your anticipated needs and list their combination code numbers as illustrated to the left. (Illustration A).

Step 2 In this method, only one series wafer is used, so in order to determine which column should be reserved for its placement to best utilize the maximum number of listed keys or keyway units, represent the keyway unit by drawing a rectangle and indicate the 14 slots by writing in their appropriate code numbers (see illustration B).

Step 3 Above and below the rectangle and opposite each slot code number, write the figure which indicates the frequency of occurrence of the particular code number in the list of combinations (see illustration B).

Step 4 Mentally add the two figures in each column outside the rectangle obtained in step #3 and determine the column having the lowest sum. This column will be the best location for the placement of a series wafer in all keyways selected to be masterkeyed. In the illustration B, the 7th column with a sum of one is the most desirable location for the series wafer.

Step 5 Review the list of combination code numbers and strike out as impractical all combinations which contain digits corresponding to code slots which are located in the column selected in step #4. In the illustration, a combination 101724 would be eliminated since it contains the digit 4', the code slot for which is in the 7th column.

Step 6 The required quantity of keyway units to be masterkeyed should now be selected from the remaining group of combination code numbers. Remove from each of these selected units, the two series wafers which are not located in the column reserved in step #4. Each of these keyway units may still be operated by its own key, but now, in addition, they all can be operated by a masterkey.

Step 7 The masterkey which will operate these units is made by notching the key only at each location where a combination code number occurs. The remainder of the key must be left unnotched. In the illustration C, the masterkey would be notched at combination code locations 1, 2, 3, 4, 6, 7, 9, 0, 1', 2'.

MASTERKEY STAMPING

Some specific designation, such as the letter "M", should be stamped on the bow of the masterkey. Factory cut masterkeys are stamped with a registry number prefixed with the letter "R". This number is an index to the factory records which contain detailed information, such as the job name, location and the combinations controlled by this masterkey. Factory cut change keys which operate the individual keyway units in a masterkeyed system are stamped with a combination number. This number has the same meaning as that stamped on stock keys (see page #7).

The 2nd digit, always "0" on stock keys, now assumes the number of the column in the keyway which contains the series wafer. Stock keys whose keyway units have been converted into masterkeyed units should reflect this change by overstamping the 2nd digit with the appropriate number 1 through 7.

Occasionally it becomes necessary to provide two or more groups of masterkeyed keyway units with each group having its own masterkey and, in addition, a key called a grand masterkey which would operate the keyway units in all the groups. One method of laying out a grand masterkey system involves seven basic steps.

Step 1 Select either type #1 or type #2 keyway with which to build the system.

Step 2 Of the seven columns available, arbitrarily select one column which will be reserved in all keyway units for a series wafer.

Step 3 If only two masterkeyed groups are desired under the grand masterkey, arbitrarily select another column which will provide two slots, each of which will be assigned exclusively to a masterkeyed group and is the means for separating the two groups and preventing interchangeability of their keys.

Step 4 List the code numbers of the slots in the remaining five columns.

Step 5 Tabulate a list of four digit combination numbers, each of which must contain one of the unique slot code numbers selected in step #3. The remaining three digits may be drawn from the available code numbers obtained in step #4, but remembering to select not more than one digit from any one available column. This list of combination numbers represents one masterkeyed group.

Step 6 Repeat step #5. However, substitute the other unique slot code number for the one previously used. This list of combination numbers represents the second masterkeyed group. While these two groups may appear similar, their masterkeys are not interchangeable since one will not have the notch which is necessary to operate that combination wafer which is used exclusively in the other group.

Step 7 The grand masterkey has all of the notches found on both masterkeys and will, therefore, operate all of the keyway units in both groups. This illustration represents but one of a large variety of possible grand masterkeying arrangements. More complex systems or those involving a greater number of masterkeys should be referred to the factory.

STEP 1

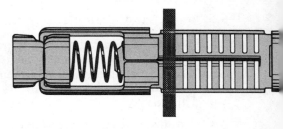

TYPE 1 KEYWAY

GRAI

STEP 7

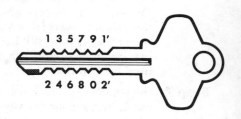

135791'
246802'

STEP 2

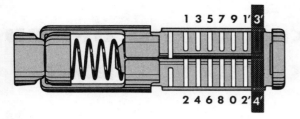

1 3 5 7 9 1' 3'

2 4 6 8 0 2' 4'

**SEVENTH COLUMN RESERVED
FOR SERIES WAFERS**

STEP 3

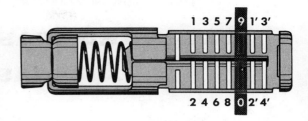

1 3 5 7 9 1' 3'

2 4 6 8 0 2' 4'

**ONE SLOT IN THIS COLUMN
RESERVED FOR EACH
MASTERKEYED GROUP**

ASTERKEYING

STEP 4

1 2 3 4 5 6 7 8 1' 2'

STEP 6

SECOND MASTERKEYED GROUP

1350	3570
1360	3580
1370	3501'
1380	3502'
1301'	4570
1302'	4580
2350	4501'
2360	4502'
2370	4701'
2380	4702'
2301'	5701'
2302'	5702'

Etc. Continuing To a
Maximum of 80 Combinations

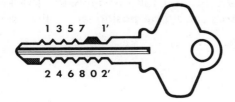

1 3 5 7 1'

2 4 6 8 0 2'

STEP 5

FIRST MASTERKEYED GROUP

1359	3579
1369	3589
1379	3591'
1389	3592'
1391'	4579
1392'	4589
2359	4591'
2369	4592'
2379	4791'
2389	4792'
2391'	5791'
2392'	5792'

Etc. Continuing To a
Maximum of 80 Combinations

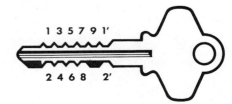

1 3 5 7 9 1'

2 4 6 8 2'

EXTENDED MASTERKEYED AND GRAND MASTERKEYED SYSTEMS

Masterkeyed or grand masterkeyed systems may be doubled simply by utilizing both type keyways (type #1 and type #2) and providing a key capable of operating both. In such an extended system all keyways must be provided with a series wafer in the master column. This does not, in any way, impair the operation of any existing keys. However, it does make possible the use of an additional key (type "0"—see page #6) capable of acting as a master, grandmaster or great grandmaster key. When a type #1 keyway unit is set up with a series wafer replacing the master wafer, both the keyway unit and its key are redesignated as type #3. Similarly, a type #2 keyway unit and its key are redesignated as #4.

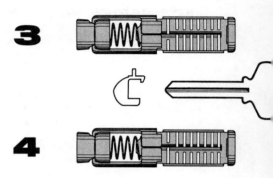

REASSEMBLING LOCK MECHANISM

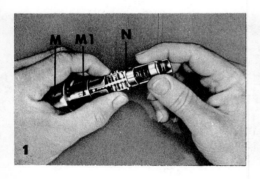

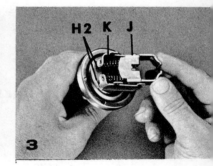

Step 1 Grasp wafer keyway unit (N) in right hand. In other hand, hold outside knob (M) with slot (M1) facing you. Insert wafer keyway unit (N) into outside knob spindle, aligning keyway cam as illustrated in figure #1. Depress wafer protrusions between thumb and forefinger while pushing in the keyway unit until it is bottomed.

Step 2 Insert knob through outside rosette and frame, pushing and rotating simultaneously until fully assembled. The keyway cam and spindle ears should be positioned at the open end of the lock frame.

Step 3 Place springs (K) into the slide (J) and assemble both into the frame locating the free ends of the springs on the ears (H2) of the lock frame. Be sure the bridge of the slide is facing you, as illustrated.

Step 4 Replace thrust plate (I) with its curved side over the springs (K) by inserting the bottom side between the bottom tabs of the lock frame and by pressing down snap into position between top tabs (H1).

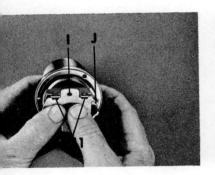  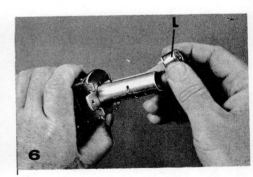

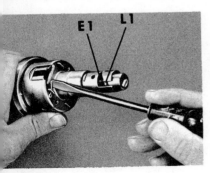

Step 5 With the thumb depress the slide all the way back against the spring tension. Insert empty spindle (E) as illustrated. Release slide and spindle will lock into position.

Step 6 Slip plunger (L) inside empty spindle, making sure that the tab (L1) on turn button aligns with locking slot (E1). Again depress the slide and work plunger in all the way. Release slide and the plunger will lock into position.

Step 7 Replace housing over lock assembly with its opening in line with the jaws of the slide and replace cotter pin. Lock is now ready for installation in the door.

VIII Opening Doors, Equipment, and Automobiles

LESSON 29

Doors

THERE are times when it is either necessary or desirable to open a locked door, or another locked item, without the necessity of picking the lock. Under those conditions, it is necessary to bypass the locking function in some way, or otherwise negate it. If one has a pin tumbler lock on a door which has a stuck pin that cannot be freed up, all picking attempts are utterly useless, as are attempts at impressioning. Under those circumstances, one must look for another way to unlock the door and get at the source of the trouble.

This section will deal with opening locks without having to pick them; it is not intended as a substitute or a replacement for the more professional methods of opening these things, but is intended for use when those methods fail, or when a mechanical breakdown of some type is involved, or simply when the consideration of time makes it economically unfeasible to use those conventional methods.

Several techniques are available for opening doors without picking the lock. Although none of these techniques will work in all situations, one or more will succeed in a great many situations. There are a number of vulnerable points for locks, and these are the points attacked.

LEVER HANDLES

One of these vulnerabilities is concerned with the resurgence, after a long dormant period, of lever handles for mortise lock sets. These lever handles are attached to the spindle, just as are conventional knobs. In the mortise deadlatch function, the lever handles employ a stopworks operating off the knob cam. In many cases, the lock itself is simply an ordinary mortise lock of a type standard to a manufacturer, and employing identical stopworks. The case is generally of cast iron, but the addition of lever handles has altered the situation to the extent that the same stopworks and cast iron case which furnished adequate strength for a knob (which would be more difficult to grip and which has much less leverage than a lever type handle) becomes too weak to cope with severe pressure on the lever handle.

Often, a heavy turning pressure on a lever handle will be sufficient to snap off the post which is extruded from the cast iron lock case to hold the stopworks in position in the lock. When the applied pressure is sufficient to break off this post (or stump), the handle is free to turn and the lock may be opened. This results in breaking the lock case; however, in the majority of instances, the break is only at the point where the stump joins the lock case, and may be readily repaired by welding if nothing else is wrong. On occasion, however, this method will break a lock case so that it is not repairable. It is recommended only as a last resort.

MORTISE DEADLATCH AND SPRINGLATCH SPINDLES

Mortise deadlatches and springlatches may often be entered by removing the outside knob, if it is of the type which has a set screw, or if the escutcheon may be sprung far enough to remove a retaining pin from the knob bearing with a small punch and hammer.

With the outside knob removed, it frequently will be found that the installation of the lock has been made incorrectly, with the stop on the knob shaft on the inside of the door. If this is the situation, the knob spindle may be pushed through to the inside of the door, completely out of the lock. One may then reach through to the inside half of the knob cam with a T-shaped pick, or a square knob spindle which has been ground down round except for a square portion at the end, and operate the inside cam, just as with the knob.

There are a number of situations where the knob spindle has been properly installed with the stops on the outside of the spindle to keep it from being pushed through to the inside. Remove the escutcheon screws and move the escutcheon to one side far enough so that a very slim chisel can go through the knob hole. Cut off the pins or lugs which keep the knob spindle from being pushed through. The knob may then be removed using the same technique as previously outlined.

Cylindrical Locksets

Cylindrical locksets are frequently designed with a knob which is removable from the outside when the key is turned partially. One common construction of these knobs passes the knob retaining disc over the cylinder connecting bar of the lock cylinder, using an eccentric shaped hole in the center of the knob retaining disc. This normally prevents the knob retaining disc from being depressed when the cylinder is in the fully locked position.

It is frequently possible in this type of situation to use a small punch and hammer to force the knob retaining pin to push the cylinder connecting bar sufficiently out of the way to release the knob. Once the knob is released, it may be removed, and the lock may be opened with a screwdriver or other appropriate instrument. A word of caution: not all outside knobs are removable from cylindrical locks without removing the lock from the door; others may use a more solid backstop for the cylinder retaining pin, in which instance such efforts will be fruitless and may damage the lock.

After removal of a knob by this method, the cylinder connecting bar and the knob retaining disc should be inspected for damage, and any damage, however minor, should be corrected.

Escutcheon Vulnerability Mortise Deadbolts

The escutcheon is another commonly vulnerable point for door locks. Mortise deadbolts of the type which have a turn knob (or a provision for it) on the inside which penetrates the lock case rather than being used with a turn knob disc, are particularly vulnerable.

The first step in taking advantage of this situation is the removal of the knob and knob spindle, using the techniques mentioned under mortise latches (deadbolt knobs do not usually have stops on the spindle). It may be necessary to wedge between the knob and the escutcheon bearing to gain access to a pin if the knob is attached to the spindle in that manner. After removal of the knob and spindle, the screws are removed from the escutcheon.

A thin sheet of plastic or pasteboard is inserted under the escutcheon to avoid scarring the door, and the escutcheon is rotated around the lock cylinder for about 90° to expose the area below the cylinder where the turn knob penetrates the lock case. A great many metal doors are already cut out on both sides to accept a turn knob shaft; therefore, it may not even be necessary to drill into the door at the position to expose this point on the lock. When the turn knob location has been exposed, whether by drilling or otherwise, a screwdriver is then turned to retract the bolt. If a turn knob already is in place in this hole and extends deeply enough so that the screwdriver point will not enter to retract the bolt, a piece of flat spring stock, filed to a sharp edge (much like a wood chisel) may be wedged into the hole alongside the turn knob shaft and used in the same manner as a turn knob.

If the knob attachment to the shaft is of a type which is not readily removable, and a replacement escutcheon for that particular type of lock is available, the escutcheon may be drilled at a point which will give access to the turn knob mechanism. This method requires replacement of the entire escutcheon.

Cylindrical locksets are sometimes equipped with an outside escutcheon which can be loosened from the outside. In this situation, one may loosen the escutcheon fully and go into the door at the edge of the escutcheon where the latch tube joins the lock housing, using a flat piece of spring steel to engage the T-bar at the end of the latch bolt to retract it part way. This releases the deadlocking mechanism in the latch tube, allowing the latch bolt to be shimmed back as described in the next few pages.

If the outside escutcheon is fixed and cannot be loosened from the outside, it may be drilled at a point to give access to the T-bar area at the junction of the latch tube and the lock housing. Replacement of the outside escutcheon is necessary if this method is used.

Another method of releasing the deadlocking mechanism of cylindrical locks calls for prying the knob as far away from the edge of the door as possible without damaging the lock. This tends to separate the lock housing from the latch tube. If the installation does not provide firm backing for the lock housing, the separation will be sufficient to release the deadlocking mechanism of the latch tube. This will allow the latch bolt to be shimmed, just as if a deadlocking device were not present.

It should be noted that tension on the T-bar of a cylindrical deadlatch must be maintained while shimming is in process, or the deadlocking mechanism will automatically relock the latch bolt so that it cannot be shimmed.

The locksmith should be cautioned that all of the methods described for cylindrical locks are intended for use only on locks in which the lock housing is joined to the latch tube by interleafing tangs. They will not work on locks where the latch tube penetrates into or through the housing of the lock, nor will they work on locks using the interleaving joining of tube and housing. They are effective often enough to produce consistent if not 100% results.

SHIMMING

Shimming is one of the oldest techniques for opening a lock without the use of a key, and it is still useful in a wide variety of situations. Shimming may be defined as applying pressure to one of the operating parts of a lock in such a way as to cause the lock to function without resorting to a key or picks. Most frequently, this pressure is applied to the latch bolt to push it back into the lock. It can be done by scratching the latch bolt with a knife blade or a screwdriver bit if the door happens to open outward, but a better technique is to use a piece of sheet metal (or stiff plastic) cut roughly in the shape of a linoleum knife, as shown in Figure 189.

The curved point is inserted behind the latch bolt, and the curved back edge is rocked against the door stop, while applying pressure against the sloping face of the latch bolt, and pushing it back into the lock. This method does no damage whatever to the lock or the latch bolt, but if sheet metal is used, the edge in contact with the latch bolt must be carefully smoothened.

DOORS OPENING INWARD

Shimming of doors opening inward requires a different, but hardly more complicated technique. A piece of spring wire or other stiff wire is bent into a shape resembling a right angled Z. The Z is laid flat

Figure 189. Latchbolt shim for doors opening out. Although the example shown is metal, stiff plastic may be used in situations which do not require an appreciable force to be applied to the latch bolt, further reducing the amount of scarring. The curved point of this tool may be used with a rocking motion, or the wedge point may be used between the door stop and the latchbolt, whichever is more convenient.

Figure 190. "Z" wire shimming. This tool is intended for use on doors opening inward, away from the lock cylinder. The wire is rocked with the offset resting on the door stop.

on the door, with the center part of the wire horizontal, immediately above or below the latch bolt. One vertical arm of the Z is slid between the door stop and the door until it comes up against the door frame.

The position of the vertical arm should be just about directly behind the latch bolt at this stage. The wire is then rotated so as to engage and depress the latch bolt. This method will not, of course, work with a deadlatch *in good operating condition,* unless the deadlocking device has previously been "gimmicked" by a method similar to that described. It will, however, be effective against springlatches of most types, and deadlatches which are not in good operating condition.

Keyway Vulnerability

A mortise deadlatch having a lock cylinder with a keyway which is open completely through the cylinder may be released without picking by taking a piece of spring wire small enough to pass through the keyway warding, and bending it into a J shape. The curved end of the J is then fed through the lock cylinder with the curved part in a horizontal position. When the tip of the J clears the open end of the rear of the lock cylinder, it will resume its bent shape, but the tip will come into contact with the opposite side of the lock case. (This requires the point to be very smooth in order to slide on the lock case surface.) Continued pressure inward will cause

Figure 191. Use of J wire. This illustration shows it in position in an open keyway lock cylinder.

the tip to slide over the surface of the lock case, then curve sightly away from it.

From this position, the J wire is rotated to bring it into contact with the latch bolt retracting lever. If the latchbolt spring is not a stiff one, the lock may be opened directly by merely turning the J wire; otherwise, it may be turned merely enough to release the deadlocking mechanism without retracting the latchbolt. The latchbolt is then vulnerable to the shimming techniques previously described for latchbolts.

Certain types of cylindrical locks using push button locking rather than turn knob locking on the inside knob, may be opened by simply inserting a suitable length of very small, flat, spring stock through the keyway and pushing to trip the locking mechanism. When the mechanism is tripped, the outside knob will be unlocked. The method is not applicable to all cylindrical locks, nor even all having this type of locking; but on occasion, this is a handy technique.

Push Button Cylindricals

Panic exit devices are subject to all of the techniques which may be used with similar mortise deadlocking devices, and have, in addition, a few special vulnerabilities of their own. Some of the panic exit devices which are equipped with a thumbpiece on the outside instead of a knob may be opened by inserting a flat piece of spring stock, shaped as shown in Figure 192, alongside the thumbpiece on the side *away* from the edge of the door. This device is used to lift the latch lever lift block which is normally operated by the push bar from the inside of the door.

Even though it may not be possible to lift the latch lever lift block far enough to retract the latch bolt, it is usually possible to lift it sufficiently to release the deadlocking device and allow the latch bolt to be shimmed.

Panic Exit Devices

Entry locks using thumblatches also are often vulnerable to this technique. At times, if the thumblatch is loose or worn, it may be possible to push the thumbpiece sidewise, using the pivot point as a fulcrum, to such an extent that the thumblatch will engage the latch lever lift block which is normally operated by the panic bar. Once again, this may provide good enough contact to fully retract the latch bolt, or it may be merely enough to release the deadlocking mechanism so that the latch bolt can be shimmed.

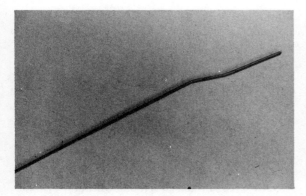

Figure 192. A thumblever shim. It is intended for use alongside the thumblever of a panic device or entry lock and is shaped to reach the unlocked latch lever lift block.

Frequently, this type of thumblatch operation has an outside escutcheon which, due to poor installation, has not been anchored on the inside of the door. When such a situation is found, the escutcheon screws may be removed and the entire escutcheon rotated about the lock cylinder enough to permit the thumblatch to engage the latch lever lift block which is normally operated from the inside of the door. If enough space can be obtained (without rotating the escutcheon) for the insertion of a Z wire between the escutcheon and the door, the latch lever lift block may be operated in this way. The Z wire may require somewhat different dimensioning than that for shimming inward opening doors and is subject to some experimentation for a particular lock.

If locks having thumbpiece operation from the outside are installed on a hollow metal door, it may be possible to remove one of the escutcheon's attaching screws near the thumbpiece, on the side *away* from the edge of the door. A piece of stiff spring wire inserted through the hole vacated by the attaching screw may often be used to trip the unlocked latch lever lift block.

Quite often, rim panic devices are also vulnerable if the escutcheon plate can be removed from the outside, or loosened and turned out of the way sufficiently to expose the hole where the thumbpiece goes through the door. In one type, this is accomplished by lifting a latch lever lift block, very much as for a mortise device. Another type has a stop which can be manipulated by a wire from the outside to move a slide which unlocks the latch lever lift block.

When rim or mortise panic devices are mounted on double doors which are butt-joined (without an overlap), it is usually feasible to insert an L shaped

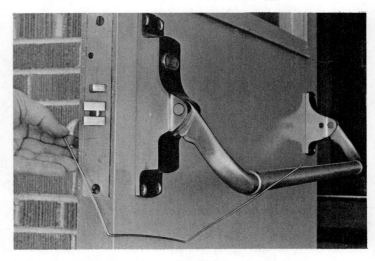

Figure 193. Use of the three dimensional Z wire. This tool is used to manipulate a push bar. Although it may not have enough power to fully depress the push bar and retract the latchbolt, the Z wire is usually capable of depressing it far enough to release the deadlocking mechanism, allowing the latchbolt to be shimmed. Since these wires are handed, a pair of them are needed for right and left hand doors.

rod between the two doors, turning the rod to catch the push bar or its end bracket. A pull on the rod is usually sufficient to depress the bar and release the door.

If the door is a single door, there is often a door stop to contend with, but a stiff wire, bent as shown in Figure 193, may be worked around the stop into position to engage the push bar end, and merely rotated to depress the push bar. It may be necessary to pry on the door a bit with a large screwdriver in order to get enough room to insert the wire at the proper point. However, the technique works unless weather stripping precludes insertion of the wire.

If the door is "fit up" extremely tight, this method may not be possible, unless enough room can be gained to insert the wire. Such a wire does not always have the strength to fully depress the push bar, but will usually move it enough to release the deadlocking mechanism.

CYLINDER REMOVAL

Mortise locks having a brass or bronze cylinder set screw can be circumvented by the use of water pump pliers to remove the lock cylinder. This is done by first turning the lock cylinder in such a way that the top turns *toward the edge of the door*, regardless of whether this is in the tightening or loosening direction. Turning the lock cylinder this way will distribute the force toward the central (stronger) part of the lock case. It will minimize the risk of breaking the lock case.

Ordinarily, what happens when this technique is applied is that the cylinder set screw, being engaged in a notch in the cylinder, is sheared off flush with the threaded part of the lock case when the cylinder is turned with sufficient force. Threads on the lock cylinder will be damaged to a certain extent, but usually not so severely that they cannot be straightened out with a triangle file or a cylinder rethreading tool. The cylinder set screw should receive little damage outside of the point of the screw, and this may be cleaned up with a file sufficiently to allow its re-use. Although this method may occasionally cause a broken lock case, it usually does not; however, if the cylinder is turned in the wrong direction, the case may be broken.

After turning the cylinder in the proper direction just enough to break the cylinder set screw, turning is discontinued, then resumed in the direction required to remove the cylinder from the lock, whether the turning is in the same direction as the initial turning or in the opposite direction.

Several problems are incurred in applying this technique, however. One of these is the scarring of the lock cylinder by the water pump pliers; this can be minimized by lining the jaws of the pliers with leather strips and rubbing the face of the leather with rosin. Another problem concerns lock cylinders which use a sheet metal cap over the face of the cylinder. You will frequently find that this cap turns before the cylinder does, and turning it over the raised surface of the lock plug tends to distort it. Sometimes caps can be repaired satisfactorily, and sometimes not.

Mortise lock cylinders which are held in place in the lock by a cylinder clamp are not likely to yield successfully to forcible removal. One must be prepared to accept a severely broken lock case if the cylinder clamp is the means used to secure the lock cylinder.

In order to remove the cylinder from such a lock, it is usually necessary to deliberately break the lock case by turning the top of the lock cylinder *away* from the edge of the door. This throws severe strain onto the weaker portion of the lock case. Lock cases of cast iron, as in most current manufacture, will break under these conditions and allow the cylinder to be removed. Occasionally, a lock case with a break of this nature can be salvaged, but considerable skill is required in the welding operation to keep cylinder hole threads in alignment, even if the break is clean.

Rim lock cylinders may frequently be removable in a similar manner, by turning the lock cylinder forcibly with water pump pliers. The turning force should always be applied in the same direction as the key is supposed to turn to unlock the lock. If the cylinder is properly mounted, the cylinder clamp screws will be broken by the turning force, or, rather infrequently, the ear on the cylinder through which the cylinder screws are screwed will break off the cylinder. Usually, the broken screws can be drilled out or removed with a broken screw extractor and replaced.

Quite often, a rim lock is found where the cylinder clamp plate is not as tight or as well anchored as it should be, and the lock cylinder will turn in the door. This has the effect of rotating the cylinder connecting bar at the same time. If the cylinder connecting bar is cut sufficiently long to maintain its engagement in the lock, the lock will be unlocked. If the cylinder connecting bar is not long enough to maintain its engagement in the lock when the cylinder is rotated in this manner, it will become disengaged and the cylinder set screws must be broken, either by continuing rotation, or by prying on the cylinder, moving it in a rocking motion with the pliers until the clamp screws break.

Even if the cylinder clamp plate moves and the door is unlocked with no damage to any part, the lock should be removed and the cylinder clamp plate reinstalled properly and firmly on the door. At the same time, the cylinder connecting bar should be inspected for possible damage.

Many rim locks of the jimmy-proof type have a sliding shutter which covers the hole through which the cylinder connecting bar passes to actuate the lock mechanism. The sliding shutter is made in such a way that pressure against the shutter from outside the door will cause it to engage a shallow notch in the lock case, deadlocking the shutter against its sliding back to obtain access to the hole in which the cylinder connecting bar operates.

Ordinary means are ineffectual to slide the shutter back; however, it can be done by using two very sharp ice picks. The first ice pick is driven into the material of the shutter tightly enough so that a tiny amount of pulling pressure on the ice pick will not remove it completely from the metal of the shutter. The ice pick is pulled on *very* gently, at the same time moved crosswise to push the shutter aside.

The shutter is then held in that position by once again applying pushing pressure on the ice pick while the second ice pick is pushed into the shutter and maintained in place by pushing, since the first one will usually have moved the shutter sufficiently to clear the deadlocking notch. The first ice pick is removed and the second is used to move the shutter further aside. The ice picks are used alternately until the conecting bar's operating hole is fully exposed. A screwdriver may then be used to open the lock.

HINGE PIN VULNERABILITIES

When a broken lock prevents normal means of entry through a door which opens outward, the easiest way through the door is usually by removing the hinge pin. Loose pin hinges are generally found in older buildings, although some are still being used, even on outward opening doors. When a loose pin hinge is encountered, it is only necessary to use a punch or screwdriver and a hammer to remove the hinge pin. The door may then be removed from the hinge side, without damage to either the door or the lock.

Fast pin hinges, which are commonly found in today's buildings, have a set screw which is concealed when the door is closed and which enters into a groove in the hinge pin to keep it from rising out of the leaves of the hinge. Frequently, because of sloppy workmanship, these set screws have not been tightened; if this is the situation, the hinge pin may be removed just as readily as that of a loose pin hinge.

If the set screw has been tightened, in most cases it simply means resorting to the old adage "If at first it doesn't work, get a bigger hammer." The set screw for most fast pin hinges does not usually penetrate to any great depth into the hinge pin; consequently, the hinge pin may be scored somewhat in the process of removing it, but usually it will come if sufficient force is applied.

Another type of hinge which is designed to prevent a door being removed is the interlocking leaf type, sometimes used with either fast or loose pins. The hinge pin usually can be removed in either case, but once it is removed, the fingers of the hinge are so shaped as to obstruct removing the door. The gingers of the leaves of these hinges interlock in such a way as to avoid presenting sufficient clearance to allow the door to be slid out of the frame. This condition is easily circumvented by the use of a small crescent wrench to bend the fingers of the leaves which are on the door frame one by one, out of the path which the edge of the door must take as it is removed from the door frame.

Doors which open outward in this fashion are usually found on public, commercial, and industrial buildings, and once the owners realize that vulnerability is present, an excellent sales potential is available for riveted pin hinges. A riveted pin hinge is quite immune to ordinary means of removing the hinge pin. It would amount to grinding off one end of the hinge pin before it could be removed. Such a process would be extremely tedious.

Another special job which can be generated when business men understand this type of vulnerability is the application of a stud to the hinge side of the door, drilling a matching hole in the door frame for the stud to enter. The stud may be screwed into the edge of the door after drilling and tapping the door to receive it. One or more of the studs should be used per door in a location as near the hinge as possible in order to make it difficult to reach with a tool such as a hacksaw blade.

If hardened studs of a suitable type are available, they should be used. Such remedial action is, of course, most effective if both door and door frame are of metal. All such studs must be long enough to keep the door from being removed, even though the hinge pin is removed; a close fitting door would not require as long a stud as would one which fits the door frame more loosely. If the fit of the door is sufficiently loose to permit gripping the stud and unscrewing it after the hinge pins have been removed, another means of anchoring the stud must be used. Welding or cross-pinning of the screw threads are two methods.

Another method is the use of a cap screw, with a piece of tubing under the head of the screw so that if it is gripped, it will merely roll on the cap screw without unscrewing it. The head of such a cap screw must be concealed, even with the hinge pins removed.

OTHER MEANS OF ENTRY

The best means of entering a given building or structure is not always through a locked door. In buildings with windows which can be opened, one may be found to be unlocked, or inadequately locked. If the building contains ordinary vertical rising window sashes with rotary type window latches, it is frequently possible to find sufficient space between the upper and lower sash to insert a stout knife blade, or other type of flat, smooth metal strip. With sufficient strength one can move an incompletely rotated window latch into the unlocked position.

Other means of entering may be found through louvres which are screwed on from the outside, and through coal or utility entrances, open drains, and an unlocked skylight. Wall panels of the building itself may be removable. Some of these methods are practical only when entry by more normal means would result in excessive damage to the building, or if an expensive lock is involved, or for a special reason such as not wanting to mark or mar a lock or door which may figure in an attempted burglary investigation by police authorities.

LESSON 30

Office Equipment

Fɪʟɪɴɢ cabinets have some unique vulnerabilities, one of which is in the lock itself. Some filing cabinet locks have a keyway which is open at the rear and which gives direct access to the locking dog, just as in some padlocks. When this situation exists, as in Figure 194, a straight pick through the keyhole to engage the locking dog and to pull it down will open the lock. It is, however, desirable to maintain a slight inward tension at the time that the locking dog is being pulled down in order to avoid an excessive strain on the straight pick.

A heavy piece of piano wire, bent as shown in Figure 195, and sharpened to a chisel type point at the end of the curve, may also be used to open filing cabinets. In use, it is inserted into the cabinet at the top of the top drawer, the curved end worked

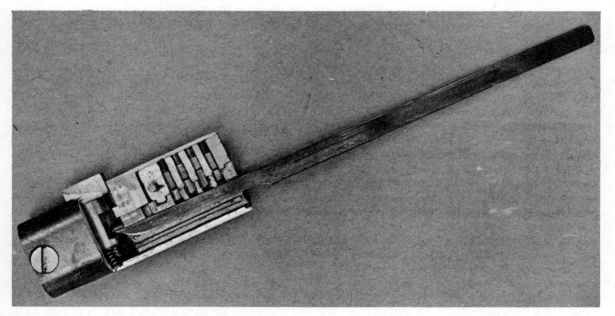

Figure 194. The use of a straight pick through an open keyway. This technique permits the locking dog to be moved to unlock the lock without having to pick the tumblers, as they are by-passed. A number of padlocks are also vulnerable to the same technique in a virtually identical manner.

297

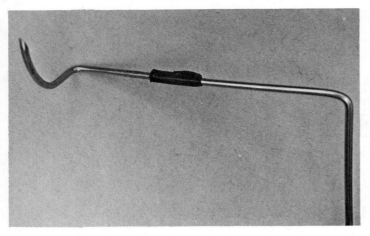

Figure 195. Trip wire for file cabinet locks. The wire is inserted in much the same manner as a Z wire in a door. The tape on the straight shank serves as a depth gauge.

completely into the cabinet; then it is rotated so that the point at the tip of the curve is wedged between the filing cabinet top and the locking dog. Some jiggling may be necessary at this stage, and a steady rotating pressure maintained until the locking dog is forced down into the body of the lock. This will allow the lock to pop out, unlocking all drawers. The tape on the shank of the tool in the illustration serves as a depth gauge and helps to locate the locking dog.

Another means of entering a filing cabinet without picking the lock is to tip the filing cabinet sufficiently to expose the bottom. Occasionally, filing cabinets are made without a bottom, regardless of how solid the cabinet may otherwise appear. If this is the situation, sometimes a person with small hands can tip the cabinet, reach up through the open bottom and trip the mechanism which locks the bottom drawer. After removing this drawer from the cabinet, the procedure may be repeated progressively until the top drawer is taken out and the lock is accessible. If there is not enough hand room, it may be necessary to use a stiff wire, either curved or straight, to trip the drawer locking mechanism on each successive drawer.

A more simple method is to trip the drawer locking mechanism on the top drawer where it engages the side of the drawer, usually some 5 in. or 6 in. back from the front on the right hand side, just above the drawer rails. Tripping of this mechanism may generally be done with a power hacksaw blade from which the teeth have been removed and the end squared. Frequently, a long, slim screwdriver will do equally well, or, for some types of mecha-

nism, even better. It is usually necessary to find the locking device and to pry it slightly away from the drawer.

At other times, however, it may be possible to simply push the saw blade straight in, wedging it between the locking mechanism and the side bracket on the drawer in which it engages, forcing the locking mechanism away from the drawer in a wedging action. When the locking mechanism has been disengaged from the drawer, pressure must be maintained on the locking device until the drawer has been pulled partially open.

Occasionally the locking mechanism is of such a nature that it is necessary to reach behind the vertical bar on which it is mounted and to trip a spring-loaded plunger from the back side. This is usually done best with a very long, slim, sharp-pointed screwdriver blade, by wedging the tip of the blade between the locking dog and the bar which carries it.

Desks

Should a desk lock be too badly damaged to be picked, the drawer can often be opened by prying upward with a wedging action between it and the top of the desk. Pry near the center of the drawer giving the drawer a light, outward impetus. The wedging action of the pry will usually move the top of the desk upward and the drawer downward far enough for the bolt of the lock to clear the desk top and thus allow the drawer to be worked carefully open.

If the operation is meticulously done, it is generally possible to open a locked desk drawer by this means without either excessive marring or damage to the desk or drawer. The operation does require some care to avoid damage, since it is easy to impart a permanent "set" to either the desk top or the drawer. Plastic tape will help minimize tool marks, if it is applied to the prying tool before starting the operation.

Alternatively, the bolt of the lock can sometimes be sawed off with an ordinary hand hacksaw blade. This method is undesirable because it frequently results in severe scarring of the top of the desk drawer. Inserting plastic tape and sticking it along the top edge of the desk drawer helps, but considerable scarring is still likely.

The lock plug may be drilled out, depending upon the type and nature of the damage to the lock. If the damage is to the bolt and is such as to completely prevent its retraction, drilling the lock plug will accomplish little.

LESSON 31

Automobiles

Automobiles, being extremely complex, vary to a considerable degree in locks and locking arrangements. Despite, or perhaps because of, this complexity, there are a number of methods for gaining access to an automobile which are frequently quicker and easier than picking the lock.

One of the most common points of entry is the wing window, and of the various kinds, probably the simplest to open are those with the lever handle type of latch. Tools for opening wing windows from the outside, when locked, are available from a number of locksmith supply houses.

The author recommends that anyone intending to engage in automotive lock work acquire a full set of special tools. A basic set is shown in Figure 196. Both the curve and the dimensioning of the tools are of importance to their effectiveness. It is therefore recommended that these be purchased rather than shop-made.

The tool used for opening the simple lever latch has a compound curve, with one or two notches toward the tip for the purpose of engaging the handle of the latch and turning it upward. In order to accomplish this, the tool is inserted through the weather-stripping which separates the wing window from the door window, and manipulated largely through controlling the depth of penetration along the compound curves. This motion, coupled with a rocking action, suffices to manipulate the lever latch into the unlocked position.

A customary practice among locksmiths is to lubricate tools to be inserted through the weather stripping with glycerine (or a glycerine based hand lotion). This not only makes the tools work easier, but helps avoid scarring the weather stripping.

Another type of lever latching wing windows has a plunger at the pivot of the latch. The plunger deadlocks the latch against rotation unless the plunger is pushed in and held until the initial stage of rotation has been accomplished. This means that another tool must be inserted through the weather stripping between the wing window and the door window to depress the plunger and to hold it in

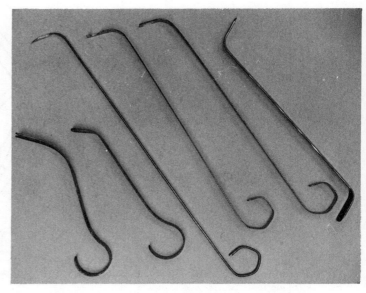

Figure 196. A basic set of car opening tools. The longer tools required to reach across a car are not shown.

299

position (depressed) while the other tool is used to rotate the latch. Commonly, a tool intended for depressing the deadlocking plunger has a compound curve and no notches, since this is its only purpose.

On most automobiles, when the wing window is unlocked and open, one may reach through and open the door from inside by hand. In other cases, it may be necessary to use another tool to reach the inside door handle and to release it. A variation of the wing window tool has a shape that, with expert manipulation, makes it possible to both depress the deadlocking plunger and to rotate the latch. Thus, only one tool is required.

Still another make of wing window is equipped with a crank to close it. In addition, there may be a bolt (which is very much like the barrel bolt used on building doors) to secure the wing window more firmly. The handle for this bolt has a spring loaded sleeve around it which drops into a notch in the bolt housing, making the bolt secure against end pressure.

A tool, similar to the one for opening the lever latch, is used to pull out the spring loaded handle of the bolt, as it pushes it back into its slot in the unlocked position. This tool is then removed. A special stiff wire, with the tip curved in an S shape and of sufficient length to reach the window crank, is inserted. The window crank is rotated by using different parts of the S wire in an alternately pulling and pushing fashion. The procedure is continued until the window is open sufficiently to reach in by hand and complete the entry.

The large door window of an automobile poses problems of a different sort insofar as forcible entry is concerned. If the door window has been left ajar slightly for one reason or another, or if the wing window rubber permits sufficient clearance, a tool is available to gain entry. It can enter through a very narrow crack at the top of the window and reach across the car to the opposite door push button.

The tip of this tool consists of two spring steel yokes tightly held together by a retaining device. When the coupled yokes are pushed astraddle of the push button, the retaining device is automatically released, allowing the yokes to spring apart. The

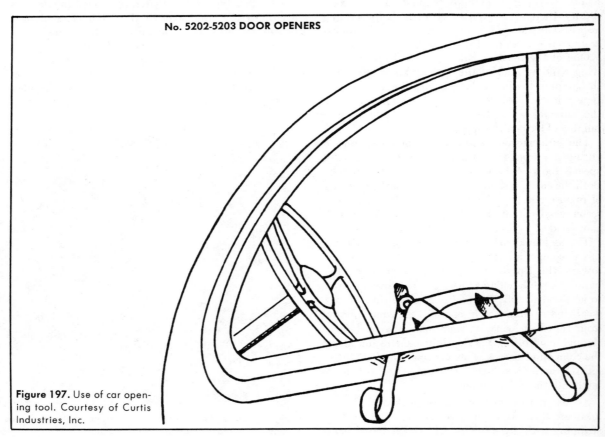

No. 5202-5203 DOOR OPENERS

Figure 197. Use of car opening tool. Courtesy of Curtis Industries, Inc.

No. 5206 DOOR OPENER

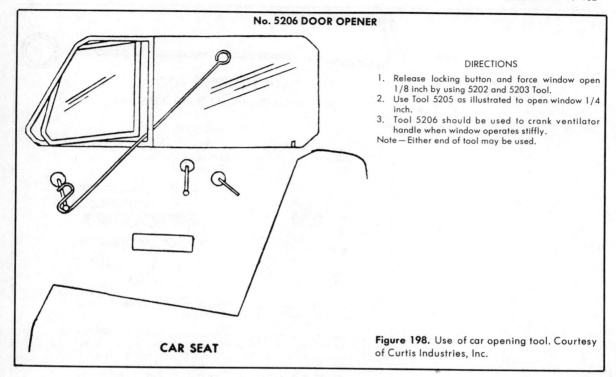

CAR SEAT

DIRECTIONS

1. Release locking button and force window open 1/8 inch by using 5202 and 5203 Tool.
2. Use Tool 5205 as illustrated to open window 1/4 inch.
3. Tool 5206 should be used to crank ventilator handle when window operates stiffly.

Note — Either end of tool may be used.

Figure 198. Use of car opening tool. Courtesy of Curtis Industries, Inc.

No. 5205 DOOR OPENER

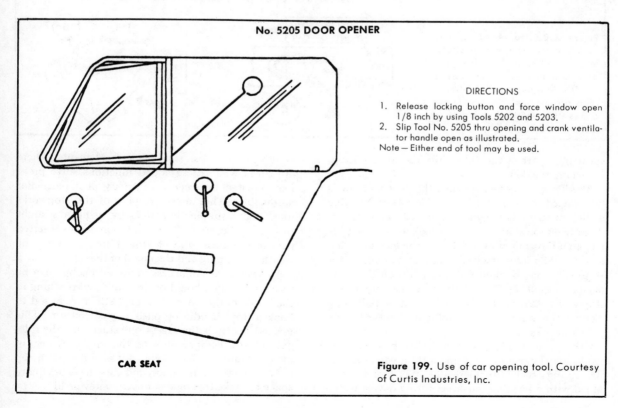

CAR SEAT

DIRECTIONS

1. Release locking button and force window open 1/8 inch by using Tools 5202 and 5203.
2. Slip Tool No. 5205 thru opening and crank ventilator handle open as illustrated.

Note — Either end of tool may be used.

Figure 199. Use of car opening tool. Courtesy of Curtis Industries, Inc.

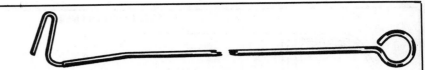

UNLOCKING TOOL

(For Curved Windshield General Motors Cars)

INSTRUCTIONS

From driver's side of car push loop end of tool through rubber seal at top of ventilator wing, as shown at "A". To prevent tearing of rubber, start loop end of hook through first, as shown at "B".

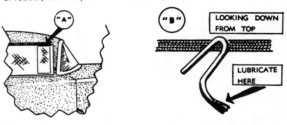

After entering loop, push rod through to reach opposite front door lock button, then hook button as shown at "C".

Turn the tool clockwise to lift button, turning the tool counter-clockwise will cause the rod to unscrew. It is possible with some models of two door cars that the back of front seat may interfere with access to the lock button; in this case place the hook end of tool over seat back as shown at "D" and pull the back of seat forward.

Figure 200. Use of car opening tool. The cross-car tool illustrated is of a different type from that described in the text and serves to show how a number of different tools are available for similar purposes. Illustration courtesy of Curtis Industries, Inc.

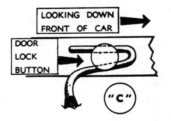

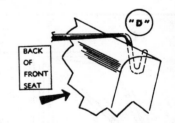

springing apart of the yokes lifts the push button, unlocking the door.

Another means of entry when the door window is fully raised and the door locked consists of inserting a stiff wire bent as shown in Figure 201, and using it directly on the stopworks of the lock by reaching the lock in a position at or just below the lock cylinder, or at the door latch itself. Sometimes tripping of the stopworks may be done by pulling up on the wire, once the tip of the L has been positioned under the stopworks. At other times, the L must be pulled up against the stopworks and rotated slightly to trip the stopworks.

Other automobiles using a rocker type of stopworks may be tripped by using a thin, wide piece of flat spring stock. Commerically, this tool is combined with a hooking, pushing, lifting device which is superior to most shop-made implements. There are times when these added functions solve problems created by some eccentricity of a particular automobile. They make tripping of the stopworks much more simple than it is by using a straight piece of spring stock. The tool is sometimes inserted next to the glass; and at other times, between the weather stripping and the metal of the door.

Another means of access to an automobile can be through the floor board or fire wall, and reaching to the inside of the car with a long, stiff wire or rod to hook a door handle or push a push button. This means of entry is the exception rather than the rule, due to the compactness of the engine space in modern automobiles. Some automobiles with six cylinder engines, and many old cars, have adequate space to make this mode of entry noteworthy.

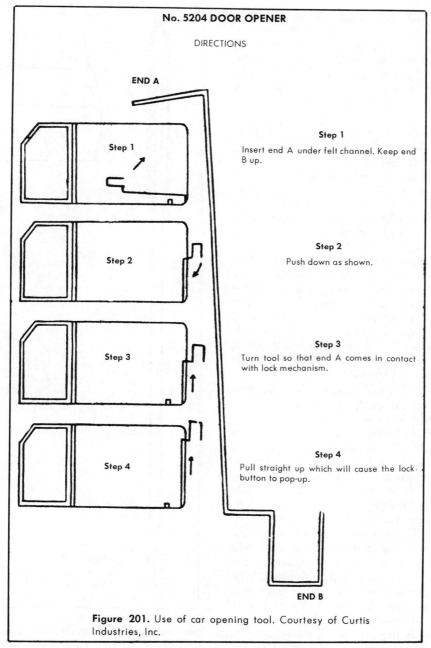

Figure 201. Use of car opening tool. Courtesy of Curtis Industries, Inc.

If ignition keys are required, it is usually easier to remove the door cylinder, and since the doors and ignition are almost universally the same, this is customarily done. Door locks are commonly held into the door by a retainer clip located under the weather stripping and are generally easily removed. At times, the door cylinder may be retained by a set screw; loosening the set screw will release the cylinder, but care should be taken to avoid the set screw coming completely out, as it may drop into the bottom of the door. Unfortunately it is frequently difficult to replace without dropping it into the inside of the door.

If the door lock is one of the type using a cylinder connecting bar, an examination of the tip of the bar will frequently show that there is a small hole in the end. If not, a small, shallow hole can be drilled. A small-diameter piece of piano wire, blunt on one end and filed to a very sharp point on the other, is placed blunt end first into the hole in the cylinder connecting bar. The sharp point is run through the hole in the lock in which the cylinder connecting bar operates and pushed on through the upholstery. The blunt end of the wire serves as a guide to position the cylinder connecting bar and to keep it in alignment so that it will enter the hole.

The cylinder connecting bar will usually turn about a half turn without rotation by the key, so it is possible to get it into the hole in the wrong position. For this reason, the guide wire should not be pulled completely through the upholstery until the key has been tried and the lock found operational.

Other types have the lock set into the door handle. Removal calls for taking off the entire handle; and this, in turn, calls for removal of the inside window frame and loosening or removal of the upholstery. An alternative to the removal of some types of door cylinders on which a code number is stamped is to use a light and mirror arrangement which allows reading of the code number, as shown in Figure 202.

Code numbers are the simplest and quickest means of making automotive keys when they are available. If the code numbers are not available, keys may be made by the regular methods. Sight reading is a quick and easy method for conventional disc tumbler locks. If a hole is drilled (about 1/16 in. diameter) in a position to use a wire for applying pressure on the side bar, the Briggs & Stratton side bar lock may also be sight read.

Sight reading is also a good method for the pin tumbler type of side bar used on some automobiles. Regular automotive pin tumbler locks may be sight read to a certain extent, but if they cannot, it is better to make the keys either from the lock or by code. Many locksmiths use one of the impression systems for this work, but the author prefers to avoid it wherever possible because of the die cast material of which the cylinder is made. If it is absolutely necessary to impression such a lock, the wiggle system tends to create fewer stresses than does the pull system of impressioning and would be preferred.

If the trunk lock is inoperable, about 90% of the time the trouble is dirt. The first thing to try is to wash the trunk lock out thoroughly with a volatile solvent to remove the dirt, running the key in and out several times in between washings. A small

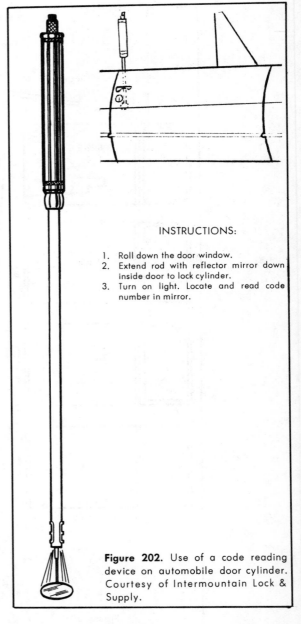

INSTRUCTIONS:

1. Roll down the door window.
2. Extend rod with reflector mirror down inside door to lock cylinder.
3. Turn on light. Locate and read code number in mirror.

Figure 202. Use of a code reading device on automobile door cylinder. Courtesy of Intermountain Lock & Supply.

amount of graphite, worked well into all parts of the lock with the key, both before and after the last washing, is helpful. A worn key is always a possible contributory factor and this should be considered. Dime store keys are also a frequent source of trouble in this type and should be viewed with suspicion.

If these measures prove ineffectual, access to the trunk may be obtained by removing the back seat. First, the cushion is removed; this exposes the fastening of the backrest cushion, usually two bolts immediately below the seat back. Removing the bolts will allow the back of the seat to be lifted and disengaged from a hook at either side to which it is attached. With the seat back removed, another obstacle is exposed: there is usually a heavy paperboard panel between the back of the back seat and the trunk space. This must be removed carefully, by pushing it into the trunk and disengaging the retaining clips one at a time. The clips are strong and incaution could lead to tearing the panel. When the procedure is completed, an agile person can scramble between the cross members at the seat back into the trunk and remove its lock from the inside. If there is any suspicion that something in the trunk has shifted into such a position as to jam the lock, and the cylinder seems to be operating well through all or part of its travel, the car can be driven on the street, accelerated to a moderate speed, and the brakes jammed on. Often the inertia of a sudden stop will jar the jamming object out of the lock. Need it be said that the driver has inertia too, and should be prepared for it? Or that the location for such driving must be carefully chosen?

If keys for a trunk are desired, the glove-box lock of most automobiles uses the same key as does the trunk. The glove box lock is much more accessible, and usually easier to pick. It may have one or two tumblers less than the one on the trunk. Sometimes there is a code number on the glove-box lock which may be used to cut a key for the trunk. Since the glove-box lock is usually a disc tumbler lock, or a pin tumbler lock with fewer pins than the trunk, it is often easier to pick; however, the glove box is sometimes unlocked to begin with.

If the glove-box lock is a disc tumbler lock containing five tumblers, whereas the trunk lock contains six, it is only necessary to cut the remaining tumbler space one depth at a time until it operates the trunk. If the glove box is lacking two tumblers, the missing tumblers may be cut using the order shown below.

By following this process of *elimination cutting*, all possible depths are used for the remaining two tumblers, whatever their position in the lock. The most key blanks that can be required for two tumblers in a five depth lock are five. It should, however, be noted that 21 of those 25 possible combinations are accounted for in the first three

First Cutting First Key	Second Cutting First Key	First Cutting Second Key	Second Cutting Second Key
1-1	2-5	2-1	
1-2	3-5	2-2	3-4
1-3	4-5	2-3	4-4
1-4	5-5	2-4	5-4
1-5			

First Cutting Third Key	Second Cutting Third Key	First Cutting Fourth Key	Second Cutting Fourth Key
3-1	4-3	4-1	5-2
3-2	5-3	4-2	
3-3			

Fifth Key

5-1

keys, giving a better than 80% chance of having to use not more than three key blanks to fit the lock. These keys are also set up in such a way as to make progressive cutting quite easy in any type of code machine.

For tools to use in opening automobiles, and for information on how to use them, contact the salesman of your locksmith supply house. He is usually up-to-date on such matters. Excellent data along these lines is also available in regular monthly locksmiths' publications, such as *The National Locksmith*, 2902 West Arthur Ave., Milwaukee, Wis. 53215. The latest news on developments in locksmithing appears in the magazines, as well as new methods for dealing with old problems, and the author recommends them highly.

Another source of timely information on automotive locks is your local automobile dealer. A cooperative working arrangement may prove highly beneficial to all concerned. Another source of valuable information is the E. D. Reed & Son General Code Set. A "tricks" section of this book is largely devoted to tips on to how to open and remove automotive locks, and make keys for them. This information covers older model automobile locks as well as newer types. The book is available from your locksmith's supply house.

IX

Lock Engineering Standards

Design Modules

Locks are usually designed from a few basic *design modules*. Some of these modules will be shown in the succeeding pages, together with a brief explanation of principles. These modular designs are not intended to be representative of any specific unit, nor are they necessarily complete, in the sense that they do not show the means by which forces are applied to them, but merely how they react to forces applied in a a given direction. Some of these modules will show how certain specific purposes may be accomplished, and should be considered as illustrating an application of a basic principle.

DEADBOLTS

The deadbolt itself, without referring to the mode of actuation, may be made in one of two ways, as shown in Figure 203. The bolt may be solid, or it may be of the saw resistant type. The latter is a deadbolt which is drilled from the back side to accept two or more hardened pins which run more than the full length of the exposed part of the bolt (when a door is locked).

These hardened pins are put into a drilled hole which is slightly larger than the diameter of the pins in order to allow the pins to roll freely under a saw blade, and to prevent attempts to cut the bolt.

Thus, in the intervening space between the pins, there will always be a solid portion of the original, parent metal of the deadbolt to which access to cutting is denied.

SPRINGBOLTS AND SPRINGLATCHES

A springbolt is a very simple and much used device in the design and construction of locks. It

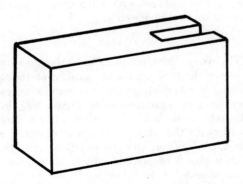
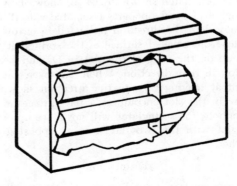

Figure 203. A comparison of solid and saw resistant deadbolts. Solid deadbolt at left; saw resistant deadbolt at right. The hardened rollers in the bolt at right provide excellent saw resistance.

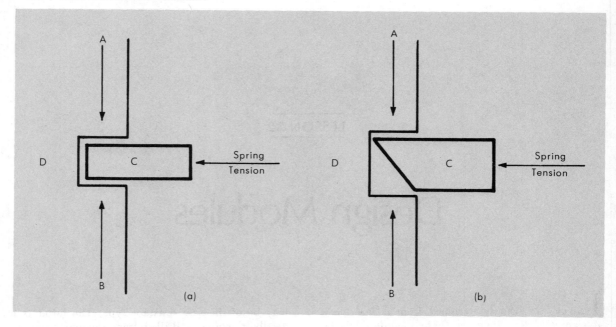

Figure 204. Springbolt and springlatch principle. Springbolt (a); springlatch (b).

consists of a square end bolt, with spring tension applied in such a way as to cause it to enter a predetermined notch, slot, or hole, with the objective of securing two separate bodies to each other to prevent their moving across one another in either direction.

Referring to Figure 204, (a), we see a spring bolt under tension which retains it in the slot and effectively stops motion between the bolt, C, or any parts attached to it, and part D, which is not a part of the same mechanism. It can be readily seen that, because of the square sides on the bolt, and the spring tension, the bolt cannot move out of the slot in D to release the lock — unless an actuating force operates to oppose and overcome the tension.

The springlatch shown in 204 (b) shows how the springbolt, C, and any parts associated with it, may be moved in the direction indicated by arrow, A, because the beveled surface of the springbolt will ride up on the edge of the slot in D, and permit motion in that direction. When, however, the attempt at moving C is in the direction shown by arrow B, the slot encounters a flat surface on the springbolt; the springlatch will not move up out of the slot with the two flat surfaces opposing one another.

DEADLATCH

As we can see from Figure 205, the deadlatch principle is merely an adaptation of the springlatch

device, but because of its importance in the family of locks, it merits consideration as an independent principle. Essentially, the deadlatch principle is composed of two springlatches, one operating to engage a stop or strike exactly as in the foregoing springlatch illustration.

The second springlatch, however, is used to provide a positive stop and antiretraction feature for the main springlatch, A, when it is fully extended.

The springlatch, B, or secondary springlatch, will slide into a position behind a square shoulder of the primary springlatch, A, when it is in the fully extended position, thus causing it to become locked against retracting force until such time as springlatch B has been retracted from its position of opposition to springlatch A.

WEDGE

The wedge principle in locks, although very old, is one of the least commonly used, even though its locking is extremely positive in nature, as can be seen from an examination of Figure 206. In this illustration, the bolt at A is shown in the extended position. In the retracted position, it would either touch, or come very close to an adjoining solid part. With wedge B in place as shown, it is manifestly impossible for A to come any closer to the solid obstruction than it is.

If B is retracted in the direction shown by its

arrow until it is no longer between A and the solid obstruction, then A will be free to move in the direction of its arrow as far as the solid obstruction, or A's own limits of travel will permit. Obviously, in a locking situation, the movement of A would have to be carefully coordinated and timed with the movement of B, just as it must be in a deadlatch construction.

Cam Bolts

Figure 207 shows how an eccentric cam (part C in the illustration) may be rotated around its pivot, A, to cause a straight line motion in bolt B. In actual construction practices, the cam, C, no longer maintains a truly round appearance, because of the necessity for having various provisions for applying a motive power to rotate the cam.

Rotary Bolts

Figure 208 shows a rotary bolt and how it is used to lock a door or drawer in a closed position. The arrows flanking bolt A in the illustration show how the bolt is rotated out of the way to unlock the locked drawer or door while the arrows at the top of the illustration show the direction in which the rotary bolt provides resistance to motion.

If part B is the one which moves and part C is

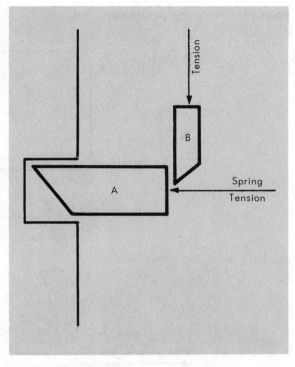

Figure 205. The deadlatch principle.

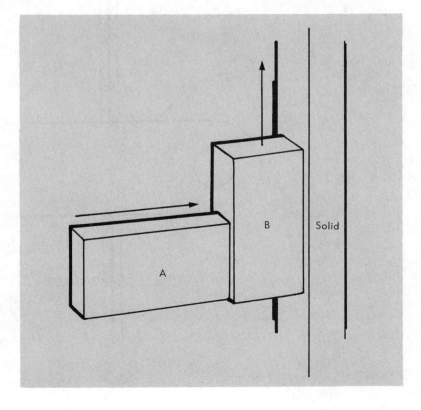

Figure 206. The wedge-lock principle.

Figure 207. The eccentric cam principle.

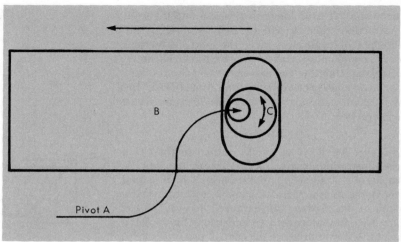

Figure 208. Rotary bolt locking method.

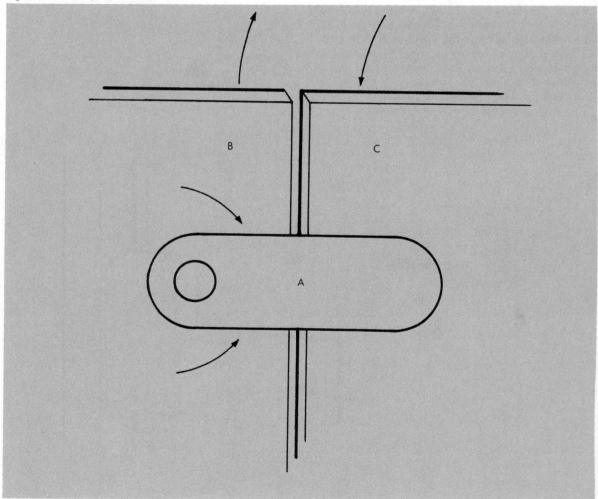

stationary, resistance to the motion of part B is provided in only one direction by the arrangement shown here. If, however, the bolt, instead of extending behind part C, extends into a notch or slot, the resistance to motion is provided in both directions.

LEVER SYSTEMS

Lever systems are extremely important in the design and construction of locks because they provide a means of operating various functions, of transferring motion from one point to another, of synchronizing various motions which are required to the essential operation of the lock, and last, but not least, they provide a means of ensuring the proper distribution of forces at various critical points in the lock.

In order to be of help in your understanding of the principles of leverage and the functions of levers in locks, we present here a variety of types of levers, as well as some idea of how each of these types may find usefulness in lock construction. Please remember that any of these levers may be designed in such a way as to increase the amount of motion provided in a given movement, or to decrease it; or to increase the amount of force applied in direct proportion to the decrease in the amount of motion, or vice versa.

Operating Levers

In the simple lever shown in Figure 209, a force applied at F will produce a resultant force or motion as shown at R. The force produced at R will be smaller than the applied force, but the motion produced at R will be greater than the motion applied at F. This type of lever is frequently used in locks to move a given part further *in the same direction* than the applied force is capable of providing.

Motion Transfer

There are places in locks where it becomes necessary to transfer motion from one point to another which is remote from it in the lock case. This is often done by means of motion transfer levers. Such motion transfer levers are shown in Figure 210 (a) and (b). The illustration at (a) shows a center pivoted lever used to transfer motion with a resultant, R, which is opposite in direction to the applied force, F. Occasionally, it is necessary that the resultant force be in the same direction as the applied force. In this instance, two center-pivoted levers may be used in the arrangement shown at (b).

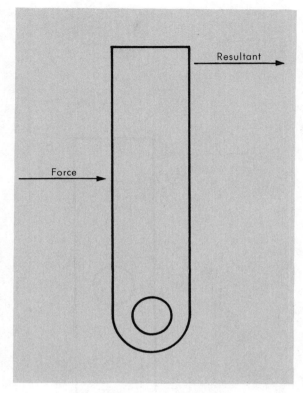

Figure 209. A simple operating lever.

The applied force F rotates lever A about its pivot with an intermediate resultant, IR in the illustration, which becomes the applied force for lever B, moving that end of lever B in the same direction as the intermediate resultant. This produces a rotation of lever B about its pivot, producing a final resultant, R, which is in the same direction as the original applied force, F. The magnitude of the resultant force or motion may be adjusted to the needs and requirements of the situation by changing the location of the pivot points for A and B.

Compensating Levers

The levers shown in Figure 211 may be referred to as a compensating lever module, because they permit the application of equal forces at two different points, F_1 and F_2, to produce an identical resultant force, R, irrespective of which of the two points originates the applied force. In order to accomplish this, the distance of F_1 from A and of F_2 from B must be equal, and the slip-joint pivot, C, must be equidistant between A and B. When this situation prevails, a balance of forces has been achieved, enabling the application of like forces to produce a like resultant.

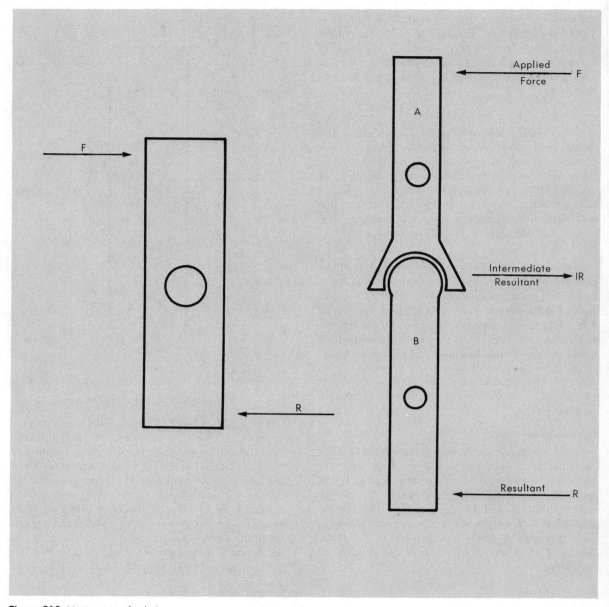

Figure 210. Motion transfer levers.

Floating Piston Levers

The author has chosen this appellation for this type of lever because it is used to produce a motion which is very much like the motion of the piston in an automobile engine. In operation, when a force, F, is applied to rotate the lever, C, about its pivot, A, it produces an intermediate resultant at the pivoted joint, E, in the direction shown in Figure 212, forcing the pivoted joint, E, downward, caus-

ing the primary lever, C, and the floating lever, D, to either line up straight, or tend to line up straight. This pushes part G away from pivot A.

Customarily, a small spring tension, or other tension, is continuously applied to lever C in such a way as to cause it to maintain whichever position is desired, while making the arrangement unstable in intermediate positions. Some form of stop is usually provided at the limit of travel positions for either lever C or D or both.

If levers C and D are moved to the point where they form a straight line, then slightly beyond, and up against a firm stop, while maintaining a holding tension of lesser intensity but of the same direction as the applied force, F, the action will effectively prevent any force, opposite in direction to the resultant, R, from moving part G from its fully extended position.

Knob Cams

The so-called knob cam, or hub, shown in Figure 213, may be thought of as a cam or as a lever, since it has characteristics of both. A characteristic is present for a camming type of operation in that the distance from the pivot to a given point T outside of the cam remains constant as the hub is rotated, but the distance from the extreme radius of the knob cam to point T will vary as the hub is rotated in one direction or the other.

Since the knob cam makes no pretense of sliding along an object at point P as it is rotated, but rather puts a direct pressure on that point without slippage along its periphery, it lacks this qualifying characteristic of a true cam. Actually, it much more closely approximates the function of a lever of the type shown in Figure 213 (a). The eccentric cam shown in Figure 213 (b) is representative of a true camming action; that is, the periphery of the cam slides across the surface of the object which it actuates. In this illustration, it is easy to see that as the cam, C, is rotated in the direction shown, B will be pushed away from the cam pivot.

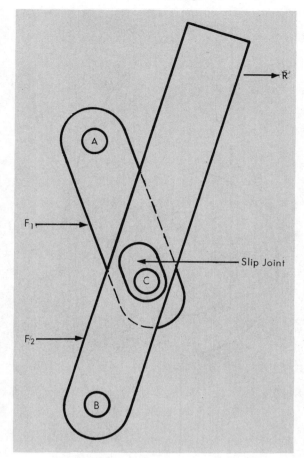

Figure 211. A compensating lever system.

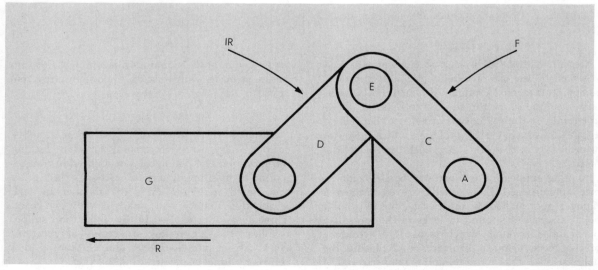

Figure 212. The floating piston type of lever system.

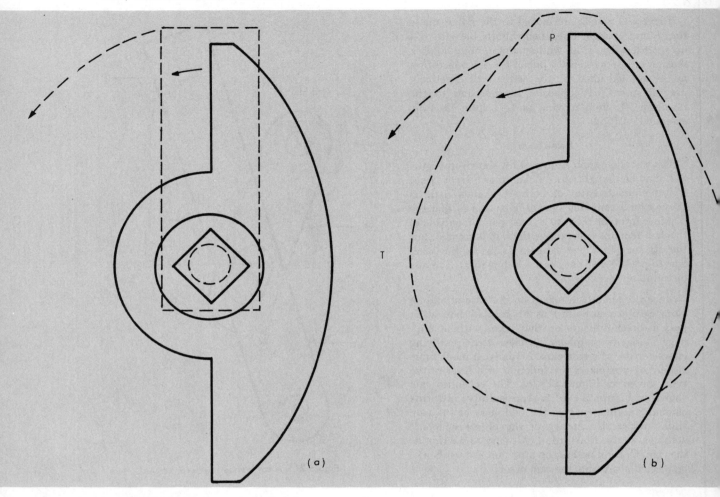

Figure 213. The knob cam as a lever (a), and as a cam (b).

Helix

Certain motions in a lock mechanism are best achieved by the use of a helix. Probably the most commonly accepted example of a helix is an ordinary machine screw thread or a coil spring. These are examples of but one type of helix. Other helical applications include the spiral helix which increases or decreases in diameter along with the lead of the helix.

Another application is illustrated by the common mechanical pencil where a helix is used to move a plunger back and forth inside of a tube. A similar helix is sometimes found in locks in the form of a push and turn knob which is used to actuate a locking mechanism. In this instance, the lead of the helix is usually not uniform, and toward one end of the helix, the lead may actually become slightly regressive in order to maintain the plunger (or turn knob) in position against an opposing spring tension.

Gears

Although there are many types of gear applications, only two are commonly found in locking mechanisms. The plain gear system used to transfer motion is the simplest of these. Although three gears are shown in Figure 215, the number in lock applications may be, and usually is, only two gears. Three gears are included in the illustration simply to show the alternating directions of rotation, clockwise, counterclockwise, and clockwise, in a series or *train* of gears operating together.

In lock applications, instead of a gear having

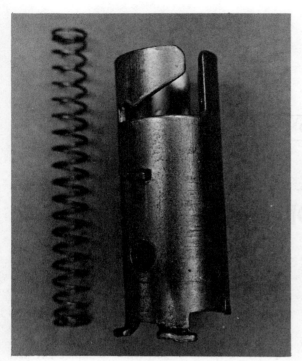

Figure 214. The helix. A true helix is represented here by a coil spring. When used in other parts of a lock, the helix is often less recognizable in form, as seen in the cylindrical knob spindle at right.

The principal difference between the gear train shown in Figure 215 and the rack and pinion is that the gear train transfers a rotary motion into another rotary motion which is opposite in direction to the original motion; whereas the rack and pinion changes a rotary motion into a straight line motion in the same direction as the original rotating motion.

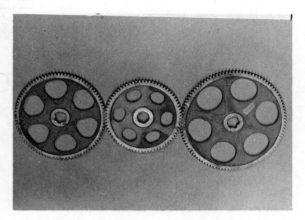

Figure 215. A train of gears. Lock applications often consist of a mere fragment of gears such as a single tooth or notch because the rotation of the parts is severely restricted — but this makes no change in principle except to provide such rotational limitations.

teeth all the way around, as in this illustration, it is quite likely that a gear will have only one, or a very few teeth as may be required to fulfill its function. Thinking of a rotating deadbolt cam as being a gear having one notch into which a gear tooth may enter, and thinking of the mortise lock cylinder cam as the gear tooth which enters that notch, will assist in understanding the adaptability of this mechanism to lock functions.

The Rack and Pinion

The rack and pinion is the second type of gear mechanism commonly found in lock mechanisms. The mortise bolt mechanism shown in Figure 216 is a true rack and pinion, the straight part being the rack and the true gear being the pinion. Once again, consider the mortise lock cylinder cam as one tooth of a pinion, and the notch in a sliding deadbolt as one notch in the rack. The usefulness of this type of gear mechanism in lock design thus becomes immediately apparent.

Figure 216. A true rack and pinion, exemplified here in a mortise bolt. Consider the mortise deadbolt and the cylinder cam which operates it as a one notch rack actuated by a one tooth gear.

LESSON 33

Design for Details

It has been said many times: "Take care of the small things and the big ones will take care of themselves." While this is not entirely accurate insofar as lock designs are concerned, it is true that many a good basic design has been spoiled by inattention to the design details required to make it an outstanding mechanism. Some of the details which commonly create problems in locks are enumerated here. They do not include all the problems of lock design, but they do point out the difference between a good design and an outstanding one.

Breaking of Square Corners

It is considerably more simple, when creating a design, to draw it showing square, right angle bends on parts which are to be stamped from sheet metal; and the dies are often easier to make, too. It is easier to fit parts into square corners than into curved ones. The fact remains, however, that the breaking of square corners into curved ones has two desirable results.

Square corners in a part which has been bent to form the square are not nearly so strong as corners which have been curved slightly; corners bent into some parts will often show the separation of metal on the outside of the square corner. Obviously, such a part is weakened. Not so obvious, is the situation where a square corner is bent into a part and internal stresses have been set up in the area of the corner which will lead to crystallization and stress corrosion. Both of these factors are much reduced by the use of a slightly rounded corner.

Square corners on the cut edges of working parts should be either rounded off or beveled slightly to allow the parts to work together more smoothly. A square edge on a flat part tends to abrade any part which it may slide across if either of the parts is tipped to any degree whatsoever, unless the square corner is on a much softer metal than that across which it is sliding. In addition to the abrasion factor, static electricity has a tendency to accumulate on sharp points. For this reason, square corners become a focal point for electrolytic corrosion, even though the juncture of two dissimilar metals may not be present.

Curved Surfaces

The design and construction of a lock requires the use of a great many curved surfaces. The thing to remember in the designing of lock parts is to keep the design simple and the curves of such a nature that they may be readily machined through the use of standard cutters, should machining be either necessary or desirable. Subsequent hand filing or hand fitting in production or assembly should be avoided since this process is both expensive and seldom satisfactory.

Curved surfaces often provide greatly improved smoothness in the operating characteristics of certain parts. When possible, a curved surface should be provided. Such curves, if kept simple, are not particularly expensive to produce; however, compound curves, parabolic, hyperbolic, or three dimensional curves are often quite expensive to produce and should be avoided wherever possible.

318

Where a curved surface is provided on the corner of a working part to allow the corner to thrust against the flat surface of another working part, it should be in a *stabilized form;* that is, the positioning and configuration of the curved surface should be in the form and dimension that the same part would tend to assume through a process of wear and fretting corrosion. The initial use of such a stabilized form tends to minimize subsequent wearing away of the part.

Space-Strength Relationship

The space to strength relationship of various lock parts is one of the most troublesome details of lock designing, and since each part has its own requirements for both the space it occupies and for the strength characteristics which it must exhibit to withstand the stresses which are placed upon it, it is probably *the* one most complex aspect of designing a serviceable lock.

A great many factors are involved in the allocation of the proper space and strength to a given part. One factor which is often neglected in lock designs is that when two parts are locked together by the use of a third part, the two parts to be locked together should be as close, in terms of physical proximity, as possible. Unless this is done, an entirely unacceptable strain may be placed on the part which locks them together. This particular problem is especially troublesome in the locking of the outside knob.

Other places where the space to strength relationship often becomes critical include the attachment of the latch bolt or the latch tail to the latch shaft. Since the latch shaft diameter is pretty well standardized at 1/4 in., and the latch shaft diameter must be reduced somewhat for making most types of attachment to it, the effective diameter is all too often little more than 1/8 in. at this point. This is utterly unrealistic, considering the strains to which the latch bolt is subjected, and is especially so when the latch in question is to be used with a panic exit device.

The sudden and severe shocks and strains to which a panic exit device is subject make requirements for latch bolt and latch tail strength particularly demanding. Even though such an attachment may be very firm, it does little good for it to be so secure if the shaft itself is going to break at that point.

The smaller and more compact the lock, the more acute becomes the strength to space relationship, especially with the cylindrical lock. In situations where there is not enough space to achieve the required strenght of parts, this deficiency may be made up by the use of special metals or alloys which have greater strength characteristics for a given amount of cross section.

It is sometimes possible to change the direction of the forces involved in the production of a given strain on a particular part; and where this is possible, it should be remembered that an end pressure on a given part will produce far less distortion of that part than will cross pressure. If the part is positioned accordingly, the cross sectional requirements will be reduced for that part.

Since much of the operation of locks is based upon lever systems, a lever which may be subject to some strain in the operation of the lock should be reviewed in design, not only at the point where the actuating force is applied and where the resultant force is expended, but throughout the entire length of the lever, and at the pivot pin as well. Too often, applied pivot pins will loosen just because the physical size of the pin or its mount is not adequate to handle the forces which are applied to it.

Sliding vs. Rotary Friction

Quite often in the design of a lock, the designer has an option for a particular part. It may be possible for the part to slide or to rotate. In this situation, it should be remembered that rotary friction is less than sliding friction. Consequently it produces a smoothly operating and usually a more durable assembly of parts. This simple fact is easily demonstrated in the sliding of a large, heavy divan, across the floor. If one attempts to slide both ends of the divan at the same time, it may be an almost back-breaking process, particularly if the floor is not waxed (and has a higher coefficient of friction). If one pushes first one end of the divan and then the other, it becomes a relatively easy task.

Why is rotating friction less than sliding friction? Because if one end is moving to accomplish a function while the other end is stationary, the friction is covering only about half the area which it would if the lever were moving the same distance on both ends. This demonstrated in Figure 217 showing an end pivoted lever and a sliding part, with the area which each covers in applying a given amount of motion at their ends. The same illustration shows the increase in the amount of area covered if the same part were to be used in a sliding motion.

Specify Parts Finishes

All too often the difference between a fine, durable lock and one which is plagued with operating difficulties is not a matter of the size, shape,

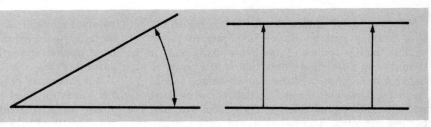

Figure 217. Rotating and sliding friction compared. The amount of area covered and, inferably, the amount of friction generated by a sliding part, is double that for a rotating part operating under similar conditions.

or design but of the finish of those parts. If an operating surface must move across another surface which is as rough as a coarse file, an effect characteristic of a filing operation will be exhibited on those parts. Not only will the operating part be rapidly worn out, but its operation will be erratic as the rough surface alternately grabs and releases it. The use of sand castings is particularly troublesome in this respect. For this reason, manufacturing processes by which these parts are produced should be specified and a minimum acceptable finish for them should also be specified.

The minimum finish specification should be in terms which recognize the manufacturing process to be used, since a No. 100 finish in one process may display utterly different frictional and abrasive characteristics than will the same number of finish when applied to a different process. This is due to the difference in the physical shape of the irregularities produced by the different manufacturing processes.

ANGLES OF ROTATION

Angles of rotation in such matters as the amount of rotation of a door knob are usually well taken care of in lock design, but are frequently troublesome in situations where the segment of arc described by one part may, in its progression through that segment of arc, be called upon to operate two or more additional parts. In this situation, the physical size of the parts assembly must be considered, and the angle of rotation adjusted accordingly, in order to produce smoothness of operation and synchronization of parts, even after some wear takes place.

SPRING TENSIONS AND FORCES

Probably every locksmith has observed at one time or another that the springs in certain locks seem to last forever, while others are almost literally eaten up, even though they do not rub another part to make them weak at the point where the break seems to consistently occur.

The reason for this is comparatively simple, but one which is often overlooked in the design of locks.

Figure 218 shows the cause. The life of a spring depends upon the material of which it is made, and the temper, of course; but just as importantly, it depends upon the amount of strain which is placed upon it. This may be thought of in terms of the amount of curvature required between its anchor point and its point of contact with the moving part which requires it to flex; in other words, its angular distortion from at rest to full flex position in relation to its length.

It can be readily seen that when part A rises from its at-rest position, spring B is going through a great deal more angular distortion than is spring C. Thus the tensions and internal stresses generated in spring B are considerably greater than they are in spring C; therefore, the life of spring B will be shorter. This is one reason for the widespread acceptance of the helical coil spring and the spiraled flat spring. These types of springs permit the distortion from the at-rest position to be absorbed gradually over a longer length of spring wire (in contrast to concentrating this strain in an extremely short spring).

Occasionally a coil or spiral spring will give consistent trouble, and when this happens, it is usually because the total length of spring stock used in making it is too short. The addition of a few coils or spirals to the spring will usually correct a situation of this type.

TOLERANCES

Wherever parts are brought together to form an operating unit, certain tolerances, or permissible limits of measurements, must be observed if the components are to fit together and operate properly. While a quarter-inch shaft may be forced into a quarter-inch hole, it will not move freely. Either the hole must be made slightly larger, or the shaft must be made slightly smaller if motion of the shaft in the hole is to be the objective.

The same thing is true of all operating parts. They must clear all of the stationary parts by a given amount before smooth operation can be maintained. If the operating parts clear the stationary parts by an excessive amount, it may cause them to lose

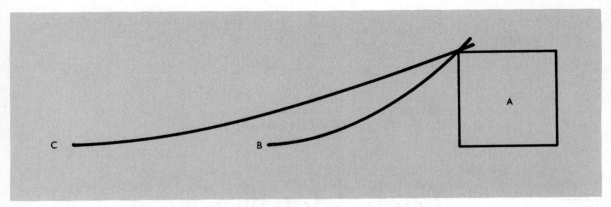

Figure 218. Angular distortion of springs. The life expectancy of a spring is largely governed by angular distortion during its flexing. The larger angle of distortion as in spring B may be expected to contribute to its deterioration through stress corrosion, crystallization, and metal fatigue to a much greater extent than will the smaller angle of distortion in spring C.

their positioning and bind because of an angular displacement. Various types of tolerances are required to maintain the proper timing, or sequencing, of various operating parts of a lock. Some tolerances may be in terms of linear measurement, or of diameter, or of angle (stated in terms of degrees or minutes of angle). Some of these tolerances are extremely critical; others are less so.

It should be remembered that tolerances which are correct and produce an excellent, smoothly operating device may prove inadequate in use due to extremes of temperature and blowing dirt or sand, or because of other severe operating conditions. A design should be created with these operating conditions in mind; and after design, the pilot model should be tested under operating conditions considerably in excess of the degree of severity expected to be encountered under field conditions. Revisions in design should be made to meet those conditions.

If timing in a lock is off, the tolerance of a part may need to be changed; furthermore, it may be necessary to adjust interrelated operations to compensate for this adjustment.

LESSON 34

Lock Metallurgy

THE metallurgy involved in the process of lock designing and manufacturing is an extremely important part of good lock design. It is not our intention to go into the metallurgy of lockmaking and to discuss things like coefficients of elasticity, yield point, and shear point for specific metals and alloys. This information is readily available from other sources, but a brief discussion of some of the metals and alloys used in the production of architectural hardware, in a rather general form, is in order.

Certain types and grades of cast metals are brittle when subjected to considerable strain, and many of them are difficult to machine. For this reason, castings of certain metals, whether poured or pressure cast are, as a general class, to be avoided in situations where strain or shock incurred in severe operating conditions could build up more force than a part can absorb without yield or fracture.

Production controls over the quality of sand castings are often difficult to maintain. This, too, should be taken into consideration as a design factor, since surface quality control of this type of casting is often more theoretical than practical, and warping and internal stresses are perpetual problems.

Sheet and strip metals and extruded shapes are often preferable to castings because they avoid or lessen many of the problems associated with the latter. They also lend themselves very well to mass production and automation techniques, as shown in Figure 219.

Stainless steel is one of the fine materials from which locks may be made, but just as it presents some unique advantages, it also presents some unique problems. It is difficult to machine and is more expensive than other materials for which it might be substituted. Certain problems are also presented by the tendency of many grades of stainless steel to gall where working parts are brought together. Its corrosion resistance does make it highly effective, and improved production techniques are making its use more feasible. Nylon bearings can also be used to avert many of the situations where galling might occur.

Copper alloys in the form of brass and bronze are readily workable and reasonably resistant to ordinary atmospheric corrosion. In the presence of chemical corrosives, however, most of the copper alloys deteriorate badly. Wear characteristics of the copper alloys and their friction characteristics are very good in many applications. These generally desirable characteristics account for the considerable use of such alloys in lockmaking, not only at the present time, but for centuries past.

Steel and the ferrous alloys are, of course, widely used in lock making to achieve the strength and rigidity required in specific applications, and do it with minimum bulk. They lend themselves readily to all of the common production techniques as well. Also, modern steels are available in degrees of hardness which can be used for saw and even torch resistant applications.

Aluminum and its alloys do not currently enjoy widespread popularity as an engineering material for locks of the better grades, largely because of its traditionally low strength characteristics. Although

Figure 219. Automatic production from metal strips. Illustration courtesy of Eaton Yale & Towne, Inc.

modern aluminum alloys are available which have very good strength and reasonably good wearing characteristics, fretting corrosion is still a serious problem in many applications. Anodized aluminum, which has had the surface treated to improve its corrosion resistance, currently enjoys considerable popularity as an attractive finish for exposed parts.

When die castings are mentioned, many locksmiths develop a sudden case of collar strangulation because they have encountered severe problems with parts made of the inferior grades. The facts are, however, somewhat different today. Die castings are no longer limited to inferior grades which practically crumble of their own accord, nor are they limited to zinc or aluminum.

A relatively new branch of metallurgy is currently devoted to *powdered metal technology*, and many of the products are just as sophisticated as the term implies. Some of these types of *pressure castings* have already made their way, unobtrusively and reliably, into lock construction. Even the types which the locksmith is prone to identify with disdain as "die castings" have been upgraded in quality and are acceptable engineering materials in lock construction if they are used properly and within their limitations.

Properly used, such materials will last for many years without a hint of trouble; but improperly used, they may break almost immediately. Nor are pressure castings necessarily brittle; many of the later developments in this field have found their way into lock construction without creating breakage problems. Indeed, some of them have hardly been identified as pressure castings; the fact is that good pressure castings, like anything else, are the result of good materials and good workmanship.

The Need
for Standard
Testing Procedures

CERTAIN design standards do exist for locks. Many of these are dimensional in nature; some are architectural standards; and others are governmental. On the whole, they cover the size, shape, and position of strike plates, the backset dimension of the lock, and the various sizes of lock front; in some instances, the length of throw of the bolt and latchbolt are also covered. Others specify the presence or absence of a guarded stopworks.

The governmental specifications are covered in "Federal Hardware Specifications FFH-106-(b)" and a few related specifications; however, the federal specification makes no reference to the workmanship or finish of internal parts. The Federal Hardware Specifications does give certain dimensional standards and provides a quite useful chart with it, illustrating the various functions of locks. Mechanical tests are specified for certain types of locks, but are restricted in scope. The author considers them somewhat less than realistic, although they call for anywhere from 200,000 to 800,000 test operations.

The British Standards Institution has a somewhat less rigid requirement, as set forth in two of their publications: BS: 455, entitled "Schedule of Sizes for Locks and Latches for Doors in Buildings"; and BS: 2088, entitled "Performance Tests for Locks." A separate marking is granted, under license, by the British Standards Institution under certain conditions.

Certain of the lock functions under test by the British Standards Institution are required to perform 200,000 to 300,000 operations successfully;

these operations are to be performed by various mechanical contrivances under laboratory conditions. Even these tests, as desirable as they are, are not truly indicative of the performance of the lock in the field because no provision has been made for the introduction of a certain amount of dirt or grit into the lock while it is operating; nor is there a stipulation as to whether the lock under test shall be dry or lubricated; or if lubricated, with what type of lubricant.

Some 200,000 or even 300,000 operations of a lock are not necessarily realistic; for example, locks on certain public or commercial buildings frequently take as many as a thousand operations a day. In the course of a year's time, this amounts to 365,000 operations. Very few owners of such buildings would regard a single year of service as acceptable performance from a lock. For these reasons, then, a test procedure of merely 300,000 operations is not realistic, particularly when it is not coupled with duplication in the testing laboratory of actual inservice conditions to the greatest possible extent.

Locks for entry doors of large office buildings and the like would require considerably more operations than would locks made for residential buildings. Techniques for picking and bypassing locks are developing rapidly, due to the superior technological development to which the average man, and certainly the average burglar, is exposed.

In addition to performance standards, other standards such as the proper spacing and location of screws for the attachment of such accessories as escutcheon plates are highly desirable. Performance tests for the smoothness of operation of various types of locks and/or lock components would seem to the author to be highly essential to a realistic testing technique. Such smoothness of performance could be stated and tested in terms such as: "The amount of force required to move (any lock part) through its area of performance shall not vary within any ten percent of that area by an amount in excess of _____ percent of the required motive force for other portions of that same ten percent."

The Japanese have evolved a quasi-governmental organization for the purpose of testing cameras destined for export. The cameras are put through an extremely rigorous testing procedure before they are released for export with the JCII (Japan Camera Inspection Institute) seal of approval.

Japan today is achieving worldwide recognition as a leading producer of fine cameras. Surely something on this order would benefit American export sales of locks, as well as result in superior locks for domestic consumption. Such an organization operated by the government or by the industry, would necessarily be free of obligations to any firm for which it provides this service. The tests and inspection methods for working parts made by that organization should be uniform and realistic.

Repair Techniques and Procedures

X

Parts Replacement

WHEN the locksmith is called upon to repair or replace a lock which has malfunctioned or broken, his first step is to get the door open so that the lock can be removed for evaluation. Ample techniques have been given for accomplishing this in most cases. The next step is the evaluation, based on what is learned when the lock is opened for inspection and the cause of the problem determined.

When the defective parts have been located, the big decision is whether to repair the parts, replace the parts, or replace the entire lock. A number of factors are involved in this evaluation. First, is the old lock worth repairing, that is, is it worth the price of the new parts and cost of the labor required to repair it? The basis for making this determination involves the protection furnished by the old lock and that to be provided by a prospective replacement. Did the old lock give adequate protection when it was in good operating condition? The amount of protection is relative to not only the requirements of a particular locking situation, but also to the requirements of a particular community, or even to a district in that community.

Second, what are the problems which will be encountered if the lock is to be replaced with a new one? The factors to consider are whether or not the new lock will fit into the same space on the door; and if not, whether the door is wood or metal and can be modified to permit the installation of a complete replacement (assuming that an identical replacement is not available). This brings up the question of whether an adequate replacement is available and what effect it would have on a special key system of which the old lock might have been a part.

Repairing of parts or making replacements for unavailable parts is likely to be an expensive process and should only be considered when proper repair parts are not available. Based upon whatever the locksmith charges for his time, it is usually more practical to replace defective parts than to repair them. Let us assume that the lock is of current production and parts are available. In order to insure that a workable supply of repair parts is not only available, but *on hand*, the locksmith should know how to shop for them.

PARTS SHOPPING

A great many parts, such as knob screws, snap rings, machine and wood screws, coil springs, and flat spring stock are available from the locksmith supply houses as assortments. Certain of these are standard to a locksmith's stock of supplies, and not only are they available in assortment form, but refills are usually obtainable on a single item basis.

Certain other assortments available from locksmith supply houses are slightly more specialized. Lock tumbler assortments for pin and disc tumbler locks are an example. Their use is restricted to a certain class or type of lock, but serves for specific replacement parts for several manufacturers. This type of an assortment is somewhat less universal than such items as coil springs and flat spring stock; and its selection depends to a greater extent upon the lock usage pattern of the community. Some assortments are almost universal, but others are used very little in certain areas, or are in demand in a narrower range, which makes stocking a complete assortment of parts, such as thumblatches, valuable in one community, but not in another.

Yet other assortments of repair parts are available for specific makes of locks, mostly of the cylindrical type. The decision of whether or not to stock a particular assortment will depend entirely upon local usage of the specific lock.

Certain lock parts are often carried in stock by locksmith supply houses, even though they may or may not be cataloged. Springs and small parts having a high incidence of breakdown may be more desirable to stock in parts of the original manufacture if their usage in the community warrants their purchase in larger quantities than assortments will provide. Parts required for a particular make of lock should be ordered through a locksmith supplier who stocks locks of that manufacture, as it is often difficult for him to obtain parts from manufacturers with whom he has not done business.

Parts stocked should be limited to those which cost less to install than a reasonable fraction of the value of the entire unit. Of parts of this type (having value greater than screws or springs), only those which are commonly subject to failure should be stocked. If limitations are not made, the shop inventory will get completely out of hand. Avoid a situation where the customer might say that repair costs are too near the replacement cost and that he prefers a new lock instead.

The next question that naturally arises is how does the locksmith know which parts will cause trouble? For the man who is moving into a new community, the answer depends almost entirely upon his knowledge of locksmithing. He can examine a lock of a given make, and can tell with a reasonable degree of accuracy which parts are going to stand up in service and which are not. Experience will help you decide which parts to stock. Experience in a particular community is even better, because then the locksmith knows what locks are in use. He understands local conditions or idiosyncrasies of the populace which have a bearing on lock performance. These considerations can cause a substantial revision in stocking requirements.

The best advice for a locksmith beginning business in a particular community is that he carry a reasonably good selection of the most common parts, primarily those sold in assortments by locksmith supply houses, and very few, if any, specialized parts for particular locks until the need has asserted itself (as it soon will). This will mean more repairing of broken or damaged parts while the need for that type of part is making itself felt. But it is sound economics, even if it may be necessary temporarily to undercharge for the labor used in repairing parts which could otherwise be replaced.

One system which will work until a need for certain parts is established is to carry a few of the predominant types of locks used in a community in stock to cannibalize for parts. There are some precautions to be observed in such parts stocking: Be sure that the replacement parts *are available*, either from locksmith supply houses or from the manufacturer, because if the locksmith sells a $3 part from a lock and it costs him $20 to replace it because the missing parts are not available, the entire transaction is likely to prove unprofitable.

Another way in which the locksmith can be caught short with this system of parts stocking is if the manufacturer changes models so that repair parts formerly available become unobtainable. This method presupposes that the locksmith has a reasonably accurate idea of what the replacement cost of parts used from locks in stock will be, but such information is not always easy to obtain. Many manufacturers do not maintain a parts price list and the locksmith's pricing to his customer in these situations must depend entirely upon past experience; it is, however, safe to assume that prices for repair parts, even in moderate quantities, will be highly disproportionate to the value of an entire lock. For example, a part having an estimated valuation of contributing 5% toward the cost of a lock will probably cost on the order of 15% to 25% of the price of a lock when buying that part separately.

Another source of parts is from salvage. Salvaged parts are obtained when a lock is replaced and the customer does not wish to keep the old one. This lock is brought into the shop and put into a salvage bin. If the lock is of an old, obsolete type and not particularly popular in the area, it is usually better to preserve it intact. If, however, the lock enjoys widespread popularity in the area, it may be well to disassemble it and store the parts in a parts cabinet where they are readily available.

At times it may be necessary to make unavailable parts, even though the cost of doing so is far in excess of the actual worth of those parts. This may occur when an exact replacement is not available and the substitute cannot be made to fit a rather expensive door, such as some metal doors; or when the lock is valuable as an antique. Even then, the job of making parts should not be attempted unless the locksmith is properly equipped with tools and the ability to fabricate such a part.

One of the most common parts which may be unavailable is a spring, and it is the outstanding exception to the rule that making parts is not economically feasible. Most lock springs are readily made in the shop, with the exception of the flat spiral spring; however, these are usually available in sufficiently wide assortments to fill most needs. Coil springs may be made by hand, by winding a piece of piano wire around a steel rod. The rod must be considerably smaller in diameter than the inside diameter of the desired completed spring. In general, the smaller the diameter of the spring desired, the more closely can the rod approach the finished diameter. One precaution to observe, other than achieving the right diameter, is that on some coil springs the direction in which the spring is wound (clockwise or counterclockwise) may be important.

Various spring makers are available from locksmith supply houses. These, with a small supply of assorted piano wire, can be an excellent investment, particularly if the locksmith makes up many springs at one time to use on such things as latchbolts. In this case, he can avoid buying assortments, or he can make his own assortments, thus amortizing the cost of the spring winder.

The making of flat springs calls for an assortment of flat spring stock and may be very simple or quite complex. It may be that the making of a flat spring for a lock requires no more than cutting off the right length of the proper size material. It may be that it will require a couple of angle bends which may be formed with pliers. It is when one of these flat springs requires various curves that problems can arise.

The bending of the curves is done around a metal bar or pin in much the way that a coil spring is wound, but it may be necessary to heat the spring to allow it to assume the shape of the required curves. In this case, careful retempering is necessary. Some experimentation in just how hot to heat the spring before quenching it in water or oil is required. If the spring is not heated sufficiently for tempering, the spring stock will loose its elasticity and take a permanent set when it is placed into service, instead of providing the return function as intended. If the spring is heated too much, excessive hardening of the metal will result and the spring will break rather than flex. Spring makers for making flat springs of complex form were available many years ago, but they were complicated and unwieldy to use. It has been many years since the author has seen one.

LESSON 37

Repair Methods and Skills

Whan parts in a lock are broken or worn, there are a number of repair methods to use and each requires a certain amount of skill. It is not the purpose of this book to teach these skills, but rather to inform the reader as to the means by which they may be adapted to the requirements of locksmithing.

WELDING

One of the common methods for correcting the condition of worn, broken, or cracked parts is by welding. The first step in a welding operation is the preparation of the surface to receive the weld. The surface must be entirely free from dirt, paint, grease, oil, and water. Lock parts should first be run through the normal cleaning procedure, then buffed on a very fine wire wheel all around the area which is to be welded. Next, the part should be subjected to a bath in a volatile solvent to remove any residual particles of grease or paint which the wire brush may have left.

The next step is to align carefully the two parts of the broken surface, joining them as firmly and closely together as possible. If a break has torn metal in a way that prevents the parts from fitting together smoothly, it will be necessary to file the metal to allow a close fit. Only that amount of metal required to make the break join smoothly should be removed. The break should then be "tacked" together with a small spot of weld at each end in order to maintain the alignment.

The break, which will now show up as a fine crack between the two tack welds, should be ground

out on the corner of an emery stone until it assumes a V notched shape; but, it should not be ground completely through. The V notch should then be filled with weld, and if its location is such that a small protrusion will not interfere with the function of the part, it should be built up slightly above the surface, and lapped over on the original surface of the two halves of the broken part.

When this has been done, the two tack welds may be ground away into a further extension of the original V notch. If the weld metal has not penetrated completely through, the reverse side of the crack should be V notched and filled with weld. If this is carefully done, very little refinishing or resurfacing of the welded area will be required; however, the locksmith can cause himself many extra hours of work if he drops excess metal where it should not be. It is of the utmost importance when making the initial tack weld that the parts be held firmly in alignment until the process is complete and the tack welds have cooled.

The welding of cracks follows a procedure which is nearly identical to that for breaks. The part is cleaned, buffed, washed in volatile solvent, and allowed to dry; the tack welds are made while keeping the tacks slightly away from the point where the crack leads into solid metal. The latter is done in order to keep the heat from causing the crack to creep and possibly become a break, with consequent alignment problems. If one end of the crack penetrates entirely to the edge of the part, this is the end which should be tack welded first to prevent its creeping. The part may then be V notched and welding may proceed just as for a break.

The building up of worn parts to restore them to their original dimensions consists of thoroughly cleaning the part by the same methods used for cracks or breaks, followed by the welding on of successive layers of weld where needed. A precaution to be observed in this building up is that after one layer is applied, it must be cleaned of the residue of the welding flux and metallic oxides which have formed before the next layer can be applied. Failure to remove this material between layers may cause successive layers to separate.

When the worn part has been built up to just a little more than the dimension required, it should be trimmed to the specified dimension and shaped with a grinder or file, at which time it may be found that the corners are not sharp and square. Some of the raw weld may still be showing; if so, the raw weld should be covered with another layer of weld and retrimmed.

At times, and with certain metals involving a peculiar shaping, it may be very difficult or nearly impossible to bring the round shoulders up entirely square without creating a huge glob that must be cut away. Under those conditions, consider the use of a lower melting point of metal to complete the build-up; for example, if the initial build-up has been made with iron, it may be better to use brass or bronze for the final build-up, and crack and pit filling layer. If the initial metal is brass or bronze, the final build-up of corners may be made with

silver solder in order to keep the parent metal from melting and losing its shape before the final build-up can be completed.

Although building up of parts with weld is the traditional method, there is another process which may be used, and it usually produces better results. This involves cutting away the worn portion of the part, usually an end or corner, as shown in Figure 220, and making a new inlay to replace the worn portion. The inlay is made as nearly as possible from the same type of metal as the original part. For example, if the original part were of cast iron, malleable iron, or steel, the inlay would be of mild steel. If the original part were of brass or bronze, the inlay would be of brass. The inlay is then silver soldered into the notch which has been prepared in the original part. Very little filing or trimming is necessary, and if the silver soldering has been well done, the part will be as strong and as symmetrical as it was when new.

Metals Used

The metal used in welding lock parts is often at variance with those used by commercial welding shops in similar situations. One reason for this is that the metal used is dependent upon a number of factors, only one of which is the type of the parent metal. Another reason for this variance is the small size or the thinness of the part to be welded. If the metal being welded is so small, or so thin that it will melt either before or at the same time that the welding rod or metal which is being applied to it

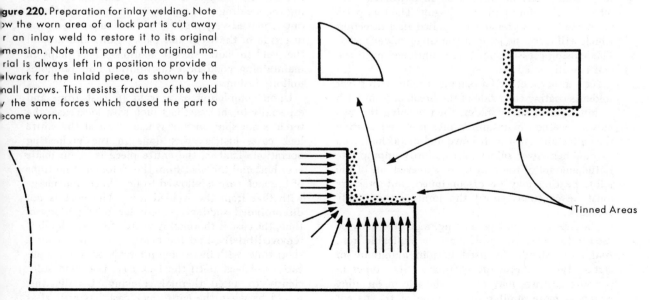

Figure 220. Preparation for inlay welding. Note how the worn area of a lock part is cut away for an inlay weld to restore it to its original dimension. Note that part of the original material is always left in a position to provide a bulwark for the inlaid piece, as shown by the small arrows. This resists fracture of the weld by the same forces which caused the part to become worn.

Tinned Areas

melts, then the entire part will melt into a useless glob before the weld can be made.

Yet another factor in the selection of a weld metal is the amount and type of strain which will be placed upon the finished weld when the repaired part is placed back in service. This differs considerably in type and degree from the average welding job. For example, the commercial welding shops usually weld cast iron with a ferrous rod intended for welding cast iron, but in larger castings. For thin material of cast iron, such as a lock case, it is far better to use bronze; bronze melts at considerably lower temperatures than cast iron rod. Cast iron is a porous metal; the lower melting point of the bronze will permit the weld metal to penetrate the porous structure of the cast iron. The lower heat required also lessens, to a degree, the danger of thermal cracking induced by the heat of the welding process. Fewer internal stresses are generated in the part during the welding process, and consequently the relief of those stresses by the post-weld heating for slow cooling is usually more complete.

Where dimensions are the critical factor in a brass or bronze part to be welded, silver soldering of the break should be carefully considered as the method of joining, as little or no melt is produced in the parent metal. Silver soldering of a break or crack does not require the grinding out of the crack in order to be effective, because the silver will penetrate and flow completely through a crack or break in metal which is not too thick. In addition to this, silver solder may be introduced in powdered form into a break before joining is begun. The two parts are heated while pressure is applied in a direction which will close the joint as the silver solder melts. This closing pressure must be maintained until the part is well cooled.

Yet another method of making a joint with silver solder is coating each side of the break or joint with a thin layer of silver solder, then pushing the two halves together, maintaining pressure, and heating the entire unit, just as if powdered metal had been used. When silver solder in a quantity appropriate to the amount of the coating is squeezed out of the joint, heating may be discontinued, but pressure must be continued until the joint is thoroughly cooled.

The size of the part permitting, mild steel may be used to join cast iron, mild steel, or malleable iron. Aluminum should be used to join aluminum or certain types of pressure castings. Other types of pressure castings may require the use of an alloy which is more similar to the type of the parent metal. One should be cautioned that magnesium is a metal which closely resembles aluminum in appearance and weight; it has the characteristic, however, particularly in small pieces, of burning with an unbelievable degree of intensity when it is subjected to excessive heat, such as in welding.

If there is any doubt whatsoever as to the possibility that the part to be welded is of magnesium, a thin sliver of it should be cut from the parent metal and tested in a safe place with a flame to determine if it will burn. *Do not hand hold this sliver while making such a test!* If burning occurs, the only method of welding which can be successful and safe with small magnesium parts is the *heliarc* method; even then, the welding of magnesium must not be attempted by the inexperienced welder.

Welding Methods

There are several different methods of welding metals together, but the one most commonly used in lock work and generally most satisfactory involves the use of the acetylene torch. In torch welding of nearly all metals, particularly cast iron, it is either absolutely essential or highly desirable to preheat the parts. Preheating is done very slowly, with the flame of the torch moving closer to the material as it begins to get hot.

The parts to be welded should be quite hot before even tack welds are made. This prevents the cast iron from cracking and stops thermal conduction, which prevents the parts to be welded from getting hot enough to arrive at welding temperature. During the welding process, the torch is not held still, but is moved very slightly all the time, along the direction of the weld and back again, and around the weld in slow, smooth motions. This helps in maintaining good heat distribution and promotes uniform fusion.

Upon completion of a weld in cast iron, and especially in thin castings such as a lock case, the torch is not shut off, but is used to heat the entire lock case, just as was done in the preheating operation. That is, the entire piece is again made very hot, and the heat from the untouched portions of the lock case is allowed to flow back and merge with that from the welded area. The heat is not discontinued suddenly except for brief periods of time; the case is thoroughly heated first, the torch is removed briefly, and the heating is resumed for a short time, with increasing intervals between applications of heat until the lock case is cooled sufficiently to avoid thermal cracking. Usually this would be when the entire lock case has arrived at about the temperature of a hot skillet. The entire cooling process to this point should take at least 10 minutes. Slow cooling in this fashion helps to relieve any internal stresses present in the lock case.

The author's own preference in this matter of welding is the Eutectic process, which gives excellent metal flow characteristics and permits the welding to take place at somewhat lower temperatures than do many other types of welding. For further information on the Eutectic welding process, either torch or electric, write to: Eutectic Welding Alloys Corporation, 40-40 172nd St., Flushing, N.Y. 11358, or contact your local distributor.

Electric Welding

Electric welding may be used with some success on larger parts made of steel or iron, or cast iron, and with a certain degree of success for brass or bronze; however, in the case of brass or bronze, the weld is usually not as smooth as the acetylene weld would be. The author's experience with the electric welding of cast iron lock cases has been highly unsatisfactory unless the case is stress relieved in a refractory furnace after welding, due to the thermal stresses set up by the intense and localized nature of the welding heat of an electric weld. Often a cracked lock case will break further during the process of making an electric weld, unless it has been well preheated before welding begins.

Heliarc Welding

The heliarc welding process is comparatively new; essentially it consists of electric welding inside a blanket of inert gas. The gas blanket contains no oxygen and will not support combustion; usually it consists of helium and argon. Welds produced by this method are usually somewhat smoother than the more common electric weld, but not quite as smooth as a good torch weld.

Because of the inert gas blanket surrounding the weld, the material being welded is not subject to oxidation as it is with other methods, and the resulting weld is comparatively bright and clean. This is of particular importance when metals such as aluminum and its alloys or magnesium and its alloys are welded.

Furnace Welding

Furnace welding can be done in a restricted number of instances in a refractory furnace for a very limited class of applications. It requires that welding metal in a granular or powdered form be mixed with flux and piled on top of a crack. The furnace comes up to temperature to melt the welding metal (usually silver solder). The weld metal continues getting hotter, and the metal which when first melted was a thick viscous liquid will become more runny and will penetrate into the crack, welding it together. The furnace is then left to cool rather slowly so as to avoid setting up thermal stresses in the repaired part.

Furnace welding may also be used for inlay welds and overlay sweats, in which instance the granular metal and the flux are placed between the areas to be joined. It is necessary to provide a means for keeping the parts in firm contact and in their proper position as the metal granules melt; otherwise the parts to be inlaid or overlaid are very likely to slide off to one side or fail to make a firm bond.

Furnace overlay and inlay welds may also be made in a manner very much like the method used for making the same type of welds with a torch. Both surfaces are first "tinned" with the weld metal, a very small amount of flux is added to the tinned surfaces, and they are placed in the furnace for joining the tinned surfaces. It is necessary to provide positional support to avoid sidewise sliding during any such operation in the furnace, and some pressure is desirable in a direction which assists in seating the applied part. Here again it is especially important that the weld metal be of a substantially lower melting point than that of the parent metal.

Forge Welding

The forge weld is rapidly approaching the status of a lost art in this country, but for those locksmiths who are occasionally called upon to repair antique locks, the process is described here. The metal to be joined, usually a malleable iron or mild steel, is heated to white hot for some distance back from the point where the junction is to be made, on both of the parts to be joined, by using a welding torch, a refractory furnace, or a blacksmith's forge. When the metal is white hot, and nearly molten, the parts are quickly withdrawn from the heat source and placed on an anvil with the two hot ends slightly overlapping one another. This lap must be as great as the size of the parts will permit and yet allow the metal to be worked sufficiently before cooling sets in.

Figure 221 shows two square bars laid out in position for forge welding. The metal is just right for joining when the heat is white and the surface appears to be flowing just a trifle. At this time, small "freckles" will be evident on the surface of the metal and appear to move about. The quicker

Figure 221. Forge welding. The bars shown on the anvil are in position for forge welding. The amount of overlap will depend primarily upon the size and shape of the bars to be joined.

the parts can be positioned on the anvil, the better. The overlapping parts are struck repeatedly with a heavy hammer, which forces the two metals together. This produces the initial weld, and the part is usually flattened in this position until it is about one-third greater in thickness than the metal on either side of the weld (if the two ends are the same size).

By this time, the weld is partially joined; the now one part is flipped over 180° and the hammering continued for three or four quick blows, then turned 90° for a few more.

The shape of the bar is now roughly square, but larger in the weld portion than on either side. Preferably all of this should have been done on the initial "heat" of the parts; this means real speed for this stage, because it is the crucial phase of forge welding.

If the first joining is firm and quickly done, the weld should be a good one; if not, it may fail to ever join properly. For this reason, it is a good idea to preheat the anvil so as to avoid drawing the heat from the weld in this crucial stage. This does not mean that the anvil should glow, or even nearly so, but if it is quite hot to the touch, it will be a big help. After the first joining has been made, the joint is reheated, and hammering is resumed. Reheating and hammering are alternated until the weld is hammered to the shape and size of the original bars joined.

The reheats should not be quite so hot as the initial heat. When the metal has lost all of its red color in the area of the weld and just turned white, it is about right. During all of the hammering

processes, the metal should be hot enough for the hammering to produce a significant result (usually not much colder than a dull red glow); and the heat at the beginning of each hammering cycle should be at a *welding temperature*.

During the later stages of making the weld, the weld is shaped to conform with the shape and size of the original material. This stretches the joined parts out, allowing them to resume substantially their original butted-together length. If properly done, a weld of this type can be very strong. Goggles and appropriate protective clothing are required, as in other welding operations.

Riveting

Several types of riveting are used in lock construction, and all have their particular usefulness. First, and perhaps most common, is the *swage riveting process*. In this the rivet comes prepared with a head of the appropriate style on one end. The opposite end is allowed to protrude through the two pieces to be joined by a distance which is governed by the type of head to be formed on this protruding end of the rivet, and by the size of the rivet.

The term swage riveting comes from the process by which the head is formed. When the rivet is in place between the two parts as shown in Figure 222, a head is formed on the length of straight shank protruding from the side opposite to the original head by hammering in such a way as to give it the desired shape. This is often done with a *header;* that is, a tool with a cup the size and shape which the new head is to assume. Small rivets, such as most of those found in locks, may commonly be worked cold (without heating), regardless of the metal of which they are made; but large, steel rivets must be heated to effectively form a good head and join the parts tightly.

Pressure Riveting

Pressure riveting is very much the same as swage riveting except that the head is formed by applying

Figure 222. Rivet after swaging. Such riveting can hold parts firmly together when properly done.

pressure to distort the tip of the rivet into the proper shape, rather than by hammering. This is usually done with a press or a squeeze riveter. A small assembly type press is often sufficient for heading the smaller sizes rivets commonly used in locks. For special purposes, such as attaching a part to a shaft, either the heading tool or the shaft may be rotated as pressure is applied. This method insures a high quality head on the rivet and offers both strength and appearance.

"Pop" Riveting

Pop riveting may be used where joints do not require exceptional strength. These rivets consist of a hollow tube with a rivet head on one end. A pin much like a finishing nail passes through the rivet. The pointed end of the pin is left protruding from the head end of the rivet as shown in Figure 223.

The rivet is inserted through the parts being joined, and a special riveting tool is used to press on the head at the same time that it pulls the pin

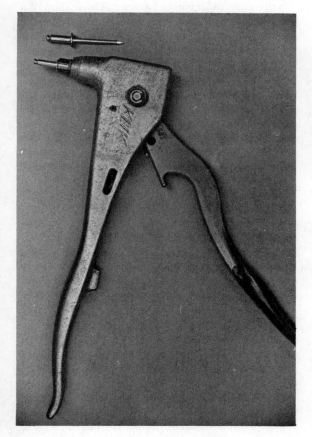

Figure 223. Pop rivet in tool, ready for insertion into the parts to be joined.

through. When the ball (represented by the head of the finish nail) reaches and enters the tube, the tube is forced to expand. When the ball has penetrated approximately to the position of the surface of the near part, and as pressure is applied to pull the pin through the rivet, the ball will break off. This permits the pin to be removed from the rivet, completing the task. This particular type of riveting is especially useful in those situations where no tool may be used to provide backing for more conventional rivets. It has the additional advantage of being fast and convenient.

Expansion Riveting

Expansion riveting, although similar in principle to pop riveting, more or less reverses the technique. The tube of the rivet shank becomes solid, having little or no hole at the point (inserted) end of the rivet. The tube is split into three or four longitudinal sections for part of its length, beginning at the tip. Inside the tube of the rivet there is a solid pin, which, before the rivet is used, protrudes slightly above the surface of the head.

Figure 224. Expansion rivet.

The opposite end of the pin is somewhat pointed, much like a key pin for a lock. In use, the rivet is inserted through the parts to be joined. The pin is struck with a hammer to force it through the rivet. The split tip is forced apart, and in effect it forms a head on the tip of the rivet.

PINNING

Pinning is used in locks in situations where it is necessary to prevent motion at right angles to the length of the pin. Such usage is found in the attachment of the two parts of a two-piece deadbolt to one another. The attachment of the latch shaft to the latch bolt is another instance where pinning is commonly used. Properly used, pinning is one of the fastest and most effective means of joining two parts or in maintaining the alignment of two parts which are held together by screws.

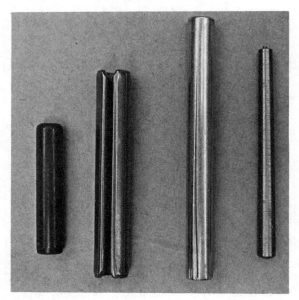

Figure 225. Pin types. From left: straight, roll, sliced, and taper pins.

The types of pin most commonly found in lock construction are the straight pin and the roll pin; also used, primarily in filing cabinet locks, is the sliced pin. Taper pins are rarely specified in locks, but are occasionally found in safe and vault work. These various types of pins are shown in Figure 225, enlarged in order to make the details of their construction more apparent.

All of these types utilize a straight hole into which the pin is driven, except the taper pin. The latter requires the use of a special *taper pin reamer* in the hole before the pin may be utilized. When drilling holes, both parts must be drilled at the same time. In order to do this, the alignment must be very carefully made before drilling begins and must be maintained throughout the entire drilling process.

Drilling is done initially with a drill which is slightly undersize compared with the size of the hole which is finally to accept the pin. The reason for this is that the nominal size of a drill will often produce a hole which is slightly in excess of its rated size, because of irregularities in the sharpening of the point, even in the case of a new drill. If the hole is first drilled undersize, and a drill or reamer of the finish size is used to bring it to the proper diameter, the fit of the pin will usually be quite close. Conversely, if the full size drill is used to begin with, and there is some wobble, the hole will be oversize and the pin will fall out.

Certain special situations, such as those where alignment of the two parts is difficult to maintain during the drilling process, will require that one part be drilled separately. When this situation prevails, and one part is drilled first, the second part is brought into proper alignment with it, and the drill is run through the hole. Then the drilling is completed. This drill should be undersize compared to the final size desired, just as in the previous example.

Drilling and alignment of parts for taper pinning is done in the same way as for straight pins, except that instead of using a second drill for bringing the hole to the finished size, a taper reamer is used. The size of drill to use for a particular size of taper pin is given on charts. It is of vital importance that the depth to which the reamer penetrates the hole be carefully controlled if proper fit of the pin is to be maintained.

A properly fitted taper pin does not come flush with the surface of the material being pinned until it is driven in flush with a hammer. The amount of a pin which is left to be driven in depends upon the size of the pin and its taper. In lock sizes, this would usually be about 1/8 in. In larger sizes, such as for safes and vaults, 3/16 in. is usually better.

Sliced pins are straight pins which have a slice lengthwise along the side very much like the scoring of a chisel mark; they are not made of as hard a metal as the special hardened metal used in commercial straight pins. The ridge raised along the edge of the slice helps to take up excess space in the drilled hole; consequently, the full sized drill may usually be used in drilling for the sliced pins. This eliminates the need for undersize drilling and subsequent reaming, as for the straight and taper pins. Sliced pins are primarily used in pinning aluminum or brass parts, as these metals are sufficiently soft for the slice to bite into the metal being pinned. This helps the pin to be firmly retained, even though the hole might not be tight enough for a straight pin to be effective.

Roll pins have very much the same use as straight or sliced pins. They, like the sliced pin, do not require reaming; but, in most applications, they do not have the strength of either the straight or sliced pin. In applications within their strength limitations, they do not work loose in the hole any more readily than might other pins under conditions of vibration, unless the force of the vibration is great enough to induce compression of the roll of metal of which they are formed. If compression of the roll pin occurs, the pin will rapidly work loose.

Draw Swaging

Draw swaging can sometimes be used when a part is too short to accomplish its proper function. It involves a thinning out and lengthening of a part (usually a flat part). It is done by hammering the part in such a way as to cause the lengthening process to extend in the desired direction. A given volume of metal is very nearly incompressible; hence, if part of that volume of metal is forced to move out of the space it occupies, it will occupy the immediately adjoining space. Stated otherwise, if the thickness is decreased, the width and length must increase proportionately. This is the basis of draw swaging.

Referring to Figure 226, we find that the direction of the blow struck by the hammer can, to an extent, control the direction of distortion or the direction of

Figure 226. Controlling the direction of a draw swage. Blows which apply their force from the direction shown will tend to increase the length more rapidly than the width; conversely, blows applied from a position 90° different from that shown will tend to increase the width more rapidly than the length.

draw. The direction of draw tends to be as nearly as possible away from the direction from which the blow is struck. For this reason, a blow on a part such as a cylinder cam, delivered from the direction shown in Figure 226, will tend to increase the length of the cam at the same time that the cam is made thinner and wider. Delivering the blows from the direction shown will make the cam longer with less increase in width than will blows from any other direction.

When the cam has been made sufficiently long, filing may be done to restore the correct shape to the distorted end or other part. Draw swaging with the subsequent "trimming the draw" operation lessens the cross section of a part and consequently weakens it; therefore, this process should not be used if the full strength of the component is essential. It is, however, very useful in situations where two dimensions of a part are vital and the required strength is not too great.

Refinishing

Since there are more means of refinishing than there are metals involved in the exposed surfaces of locks, it would be a difficult task to describe, or perhaps even to enumerate them all here. Certain techniques, however, are basic, and these will be described to a limited extent. The first important step in refinishing any metal part is the removal of the old finish; even though the part may be of solid metal or plated, the factory usually applies a final coating of clear lacquer to protect the finish.

Removal of this coating may be done with paint remover or prolonged soaking in carburetor cleaning solution. Sometimes a very strong solution of detergent is effective. Nitrocellulose base lacquers are usually the most difficult to remove. On removal from the cleaning solution, any solid coating remaining on the surface, or clinging to the irregularities of a surface such as a lock case, may be removed by buffing with an extremely fine wire brush. Sometimes, a stiff bristle scrub brush, soap, and water will do this job effectively. For some solid metals, this may be all the refinishing required.

As an alternative or supplement to the method described, a sandblast using very fine grit may be employed. Some solid metals, notably brass and bronze, may be refinished beautifully by dipping them in a commercial brass brightener solution which many cities use to renovate their water meters. This solution is sometimes called "meter brightener."

Extreme care must be taken in the application of these solutions to avoid contact with skin, and particularly the eyes. A full face shield and rubber gloves are advocated by the author as a minimum safety requirement; however, safety measures recommended by the manufacturer for his specific product may require more. Most of these solutions are damaging to clothing also, and clothing protection may be advisable. *Always check the manufacturer's label before using a chemical product.* A solution of this type does a beautiful job, is convenient, and is not at all dangerous if proper precautions are taken. After taking the brass or bronze parts from the solution, thoroughly wash in hot water to remove all remaining traces of the solution.

Plated metals may be refinished, as may solid metals, by first removing the lacquer coating and scrubbing the parts with soap and water or scouring powder and a brush. If plated metals *must* be buffed, it should be very carefully done; otherwise the plating may be worn through, irrespective of whether buffing is done with a wire or cloth wheel.

If care is taken, plated or solid metals may be further enhanced by buffing with a cloth wheel to which jewelers' rouge has been applied. Such buffing produces a finish which is mirrorlike, and most attractive if the part is in good condition. If the part is bent or scarred, the finish tends to magnify defects; a less extreme or a duller finish will conceal them better.

A surface that has been washed with water after the final finishing should be dried immediately with a cloth or paper towel. Water marks will show if a surface is allowed to air dry.

When refinishing is completed and considered satisfactory, a coat of clear lacquer should be applied with either a spray gun or a pressurized can. It should be pointed out that two types of lacquer, acrylic and nitrocellulose, are available for spray gun use; however, pressurized spray can manufacturers offer only acrylic lacquer. The nitrocellulose lacquer is somewhat more protective to the finish under most corrosive atmosphere conditions than is the acrylic type.

Painted surfaces may conveniently be repainted with spray can paints after cleaning and removal of the old paint by the method previously outlined. They are satisfactory for this purpose.

The refinishing of plastics is another matter entirely; and since many plastics are damaged by immersion in organic chemical solutions, such as solvents, paint removers, etc., the only practicable way to refinish them is by soaking for a period of from a half hour to two hours in a very strong solution of detergent. Liquid dishwashing detergent is very good for this process.

At the end of the soaking period, if the surface of the plastic has not cleaned up sufficiently, brushing vigorously with the same detergent solution will often accomplish the desired results. Sanding plastics and then coating them with a clear acrylic lacquer can sometimes be effective, providing the sanding is done by hand; but some of these plastics will not tolerate the lacquer or the propellant, or both, and their finish will be permanently damaged. This course of action, therefore, is reserved as a last resort and should be tested carefully on an inconspicuous part of the plastic before coating surfaces which are exposed to view.

SMOOTHING PARTS

Quite frequently new lock parts from the factory, or old ones subjected to wear and tear, are rough enough to require some smoothening. This may be done by grinding, or by hand or power filing.

Hand filing of parts to make their operation smooth is a process requiring considerable dexterity if one is to keep the parts true and meeting one another properly. This requires that the file move precisely and smoothly across the part being cut. If you need practice handling a file, such practice can be obtained by taking a good, wide ruler and setting a plastic (be sure it is unbreakable) water tumbler in the middle of the ruler. The tumbler is filled with water and the ruler (simulating a file) is passed across an imaginary part to be filed, without spilling any of the water from the tumbler which is kept balanced on the rule while the simulated filing is being done. This is the best practice the author knows to teach one's hands and arms the correct motions for properly filing a part and keeping it square.

Power filing may be done with a rotary file, either plain or end cut, mounted in a quarter-inch electric drill or in the chuck of a drill press. The rotary file is moved over the area to be smoothened, or vice versa. It is important to keep the file moving at a moderately rapid and uniform rate, or gouges will result. If the part to be filed is recessed, a rotary file having an end cut should be used.

Two sizes of rotary files in two or three different grades of coarseness are ordinarily quite adequate. The sizes preferred by the author are 3/8 in. and 1 in. cylindrical, all having the end cut feature. Nicholson File Company and others make them but they are available from most industrial supply firms. In addition to being handy for refinishing lock parts, they are also handy for adjusting the size or position of holes for lock cylinders or cylindrical locks in a door.

Grinding of parts to smoothen them is done in much the same manner as when the rotary file is used; that is, the grinder is moved continuously back and forth across the entire area at a uniform rate of speed. The part should not be held against the grinder too long or overheating of both the part and the stone may result. The author prefers to use a fine stone for the grinder.

One should be sure that the stone is rated for at least as many revolutions per minute as the grinder is capable of producing with no load; otherwise, the wheel can explode with disastrous results. Resin binders in the stone are more susceptible to heat than are ceramic binders, but will withstand more shock. A grindstone should be kept well dressed and true. A cracked stone or one which is defective should be immediately discarded. An exploding grindstone can expend the force and violence of an artillery shell.

Another method of smoothening such parts as lock cases which are excessively rough, and which cannot be smoothened adequately by filing or grinding without serious compromise to tolerances or strength, calls for an acetylene welding torch.

Use it to spread a thin film of some low melting point metal, such as silver solder, over the entire affected area where operating parts must come into contact with the rough surface. Very little metal should be used and the rough surface must be thoroughly cleaned before attempting to do this.

The flow of metal will be uniform if the material upon which it is being applied is extremely hot first. Both the parent metal and the weld metal must be kept hot enough to run freely throughout the entire operation. What smoothening remains to be done should be minor (if the heat has been maintained properly) and can be completed with a file or rotary file.

At times, paint can be used to help cover surface irregularities of castings, provided that they are not too severe. Here, again, spray paint in a pressurized can is usually satisfactory. A precaution to be observed is that parts or levers which function on a pivot which is permanently attached to a part to be painted should have the pivot masked off with masking tape if the tolerances are close enough for a coat of paint to interfere with their operation. This method does not, of course, have the durable qualities of other methods of resurfacing, but can often produce acceptable results in short term or light duty situations.

Another method of smoothening parts in those instances where a part has been scored and a ridge raised from the parent metal is by peining. This involves light, quick blows with a ballpein hammer, applied directly along the raised portion, to drive it back into place in the parent metal. A small amount of filing is usually beneficial after peining.

If the blemish is small, and the parent metal is soft, it may be possible to burnish the blemish back into the parent metal. This is done by rubbing it briskly with considerable pressure, using a smooth, steel bar of appropriate shape with a highly polished finish (if further scratching or scarring is to be avoided). For less critical applications, it may be possible to use either the ball or face of a ballpein hammer (if the surface used is smooth enough). Burnishing with a smooth tool produces a pleasing finish on the affected part.

Synchronization of Parts

Very seldom are any two problems in the synchronization of parts the same, therefore no effort will be made to cover all of the possibilities involved in retiming the sequence of various functions in a lock. Instead, the locksmith will have to use his own mechanical ability and ingenuity to reason what the timing is supposed to be for a particular lock.

In arriving at a decision on cutting down or building up various parts to change a synchronization, no single part should be selected for modification without first considering the effect on other parts. In other words, each part which forms a portion of a synchronized operation is dependent upon some other part. While a change may be beneficial at one point, it may be deleterious at another; therefore, all of the possibilities should be considered *before* any changes are made.

Bushings

Frequently, difficult problems in bearing areas or pivoted levers caused by excessive tolerances may be compensated for by the use of bushings. Often bushings may be cut from standard sizes of tubing; or, if the locksmith has a small lathe, they may be made on a lathe. The tubing or material for the bushings may be steel, stainless steel, aluminum, brass, copper, oilite (oil impregnated porous bronze), or plastic.

An assortment of tubing for use as bushing material can often be obtained by arrangement with a local machine shop, since most machine shops have a certain amount of tubing of suitable sizes in short lengths and left-over pieces. If the bushings are to be of nylon or some other plastic, the most common way of making them is by turning them from solid bar stock in a lathe.

OVERSIZE PARTS

Another means of correcting improper pivot tolerances is by the use of an oversize part. A shaft or pivot pin is made with a somewhat greater diameter than the original. It is possible that the lever or part which pivots may not be worn entirely round. In this case, it may be necessary to ream the hole in addition to supplying the larger pivot.

GENERAL CHANGES IN TOLERANCES

General changes in tolerances sometimes take place in a lock, just in the ordinary process of wear and tear. When this occurs, the best practice usually is to replace the lock with a new unit; however, in the case of some particularly expensive locks or

collectors' items, it may be desirable to either bring these tolerances back to the original or to change one or two essential parts by building them up. Rather rarely, the tolerance adjustment may take the form of cutting something down.

When either of these operations is undertaken, it should have the objective of changing the tolerance requirements to the point where the unit will function properly without adjusting an entire series.

This work may be physically accomplished by any of the processes which have been covered for the building up of parts; but, as is the case in a synchronization problem, care must be taken to ensure that the change being made is beneficial.

This means a process of pure mechanical reasoning on the part of the locksmith, and he must be able to see through and follow a chain of interrelated actions.

REPLACEMENT OF ENTIRE LOCK

When, based on the economics of the situation with regard to the lock's value and the cost of repair, it appears to be desirable to replace the entire lock with a new one, certain factors must be considered before a decision is irrevocably made. One of these factors is the dimensional requirements of the door or other unit in which the lock is installed. For example, certain older doors have a

door stile which is too narrow to accept most modern locks.

The thickness or thinness of the door is often a deterrent to the replacement of a lock, as opposed to its repair. It may be impossible to obtain a lock thin enough to operate in the door without cutting the stile in two or seriously compromising its strength. Conversely, certain locks, especially of the cylindrical type, are not produced in ordinary production for use in doors which are substantially thicker than 1-3/4 in. to 2 in. In situations of this type, a lock would either be a special order or completely unavailable.

Another factor to consider in the determination to replace a lock is the adequacy of any lock which may be used to replace the one which has broken down. For example, the broken lock may have been specially built, or be of an exceptional strength which is no longer produced, or for some other reason, the adequacy of an available replacement is subject to question. Conversely, the broken lock may be extremely deficient in these qualities; drastic changes on the door may be necessary in order to install a lock of superior quality.

The adaptability of the door to accept a replacement lock is a very important factor in making the repair-replace evaluation, based on the cost of doing the work and the appearance of the door after such a replacement has been made. This is especially applicable to metal doors where the broken-down lock has nonstandard dimensional requirements, or where the configuration of the lock front requires the door to be cut out to accept a form of that particular shape. If a door has been cut in this fashion, replacement of a lock can be extremely difficult. If a panic device is involved, the problem is further compounded by the addition of requirements for screw holes, pull handles, and so on.

Keying requirements can often be a problem area in the search for an adequate replacement for a particular lock. If a lock is part of an extensive key system and replacements for it are no longer available, it may be more economical to repair it at a cost which is not economical for no other reason than to not upset an extensive master key system.

All of these factors must be weighed into the decision of whether to repair or replace a lock. Most customers will appreciate a full and frank evaluation of their own particular situation, and the majority will readily accept the advice of the locksmith, provided he takes the time and trouble to explain the basis of his recommendations.

XI

Electric and Electronic Locks

Operation and Control

Electric locks and locks based on electronic systems are rapidly growing in both usage and importance, thanks to their remarkable versatility and adaptability. The author heartily recommends the study of electronics, particularly solid state electronics, to all locksmiths. These lock types, although just out of the developmental stages today, will almost surely have a profound influence upon the locksmithing profession within a period of a few years. The locksmith who is prepared and equipped will be in great demand, and will receive commensurate rewards.

Electric and electronic locks are being used to replace both key and combination locks on such things as doors, gates, and equipment. For example, unmanned space vehicles may be said to be equipped with electronic locks, since they are triggered to perform certain operations upon receiving certain combinations of radio signals. Indeed, some of the technology which has made the electronic lock possible is a by-product of the space program. This technology is now being adapted to the operation of such prosaic things as a lock bolt.

Electrically operated locks are currently of two types: one involves the use of an electric solenoid (an electromagnet used to operate a plunger); the other uses an electric motor to accomplish a similar purpose, with automatic cutoffs at either limit of its function. These electrically operated locks may be controlled by a device as simple as a push button or a switch. Others, more complex, may be operated by push buttons and relays, with the relays operating in such a manner as to require a certain sequence of operation of the push buttons.

The relays may be replaced by a transistorized, or, more accurately, their semiconductor equivalent. Such a lock is currently marketed by Sargent & Greenleaf under the name of Cypher Lock. It is a remarkably versatile device, involving such things as penalties for pushing the wrong button, manifested in terms of a time delay or an alarm. It also provides the locking arrangement. Such devices can also be made to respond to electrical impulses from sources other than push buttons. These other sources act to replace push buttons and often consist of electronic sensing devices of one sort or another. They may be sensitive to magnetism and be used in conjunction with a magnetic tape or thread incorporated in such a device as a card key or an industrial or governmental badge, or they may be set to respond to some entirely different magnetic situation.

Radio waves constitute such a magnetic situation and have long been used in a rudimentary form on locks for such purposes as opening garage doors and the like. Other sensors, either possible or presently in use, include photo cells, which make the lock respond to various light patterns or wave lengths or combinations; sonic sensors, which respond to sounds of certain intensity or frequency; and the Geiger-Mueller tube, which may be used as a sensor to permit the lock to respond to certain types of radioactivity, either in general or by a set pattern.

With the microminiaturization of semiconductor devices, whole circuits involving several semi-conductor functions may be embodied in a single tiny device. This means that as far as locks are concerned, it is rapidly becoming possible to have electronic sensors, which were previously much too bulky to be practical, but which are now definitely within the realm of practicability. It will soon be possible and practical to have a lock whose sensing unit is a television camera that can match up the fingerprint of an individual with one on a pre-recorded magnetic tape. It would either accept (open) or reject (remain locked) on the basis of whether or not the fingerprints matched.

The greatest obstacle to this at the present time is that even a slight deviation in something as complex as a fingerprint from the identical positioning pre-recorded on the tape would require a computer of respectable size to erect and align the opening image with the pattern image; but computers are getting smaller and more efficient.

The following illustrations show methods by which semi-conductors can be used in directing and controlling the flow of an electrical impulse to lock or unlock an electrically operated lock.

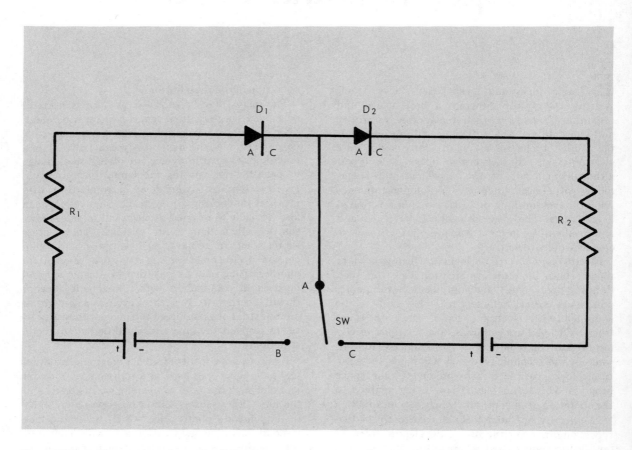

Figure 227. Diodes used to direct the flow of current. In addition to rectification of alternating current into direct current, diodes may be used as shown to channel direct current signals into predetermined paths depending upon the polarity of the applied signal current. When the switch (SW) is closed to terminal B, negative voltage is applied to the anode of D2 with no result; but this same negative voltage is simultaneously applied to the cathode of D1, and current will flow through D1 and may be used to operate other devices, represented by R1. Although the negative voltage applied to the anode of D2 produced no effective conductance of current through D2 because of the blocking action of diodes to inverse currents, by moving the switch position to terminal C, the applied voltage will become positive and D2 will conduct; D1 will not.

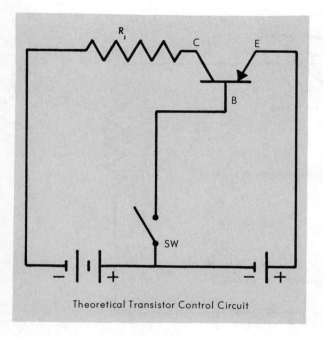

Theoretical Transistor Control Circuit

Figure 228. Transistor control circuits. Transistors, because of their versatility, are the workhorses of the semiconductors. Their usefulness in control circuitry is based upon the fact that when the proper voltage is applied to the base, current will flow for as long as the voltage is applied. Control of the base voltage, either by electrical or mechanical means (or by both in combination) establishes control over the flow of current through the transistor. The voltage for the control function after the initial application may be supplied by a feed-back arrangement or solely by outside current. Either of these arrangements may be modified in a number of ways. The base, emitter, and collector of the transistors are marked B, E, and C respectively.

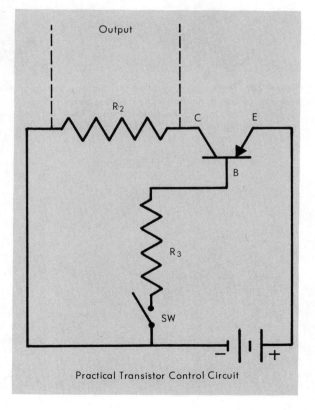

Practical Transistor Control Circuit

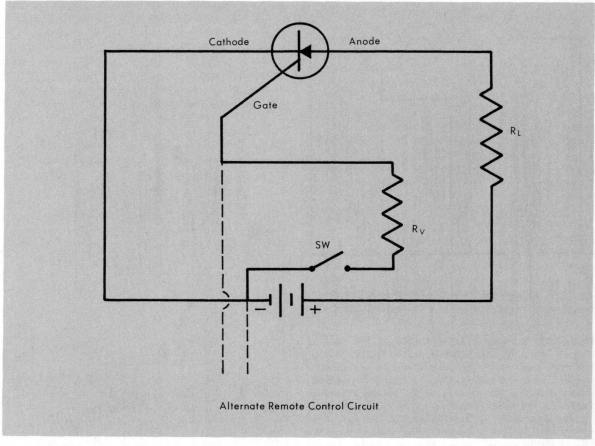

Alternate Remote Control Circuit

Figure 229. The silicon controlled rectifier has the unique property of being non-conductive until a small voltage is applied to the gate. It then conducts until the flow of current is interrupted, ever after the exciter voltage is discontinued. As with the transistor circuit, the exciter voltage may be applied remotely, either by electrical or mechanical devices.

XII

The Locksmith and the Locksmith's Shop

Other Sources of Income

THERE are a number of almost untouched areas in locksmithing which can be taken advantage of by an alert and aggressive locksmith.

RESIDENTIAL REKEYING

The average locksmith can do residential rekeying, but does he advertise it? Does he let his own customers know that he can do this work and is equipped for it? A great many people, particularly in some of the older homes and in a large number of the newer ones, have a hodge-podge collection of locks. One key fits the front door, another the back door, another the side door, another one or two keys fit the garage, another a tool shed, etc. The end result of all this is that the average householder is fortunate if he can locate keys for all of the doors of his own home.

The market for the rekeying of locks is usually for locks keyed all to one key, rather than master-keying, and is just about as simple a job of keying as a locksmith can undertake. Many times, a house is keyed all around with locks of the same make, model, and keyway. In this type of situation, the job is simplicity itself. Before advertising such work, the locksmith should either visit his supply house or check his catalogs very carefully for all of the types of locks that he can use in setting up a complete residential keying system. Usually this will include mortise, rim, and cylindrical locks, the cylinders for them, and matched keyway padlocks. All of these items must be available in one common keyway, and preferably with the same number of tumblers.

Several manufacturers can supply all of these locks complete, or just the cylinders for them in a one key system, but the locksmith must either have these things on hand or know where to obtain them quickly when the customer wants the job done. Customers tend to become impatient with the three to six month deliveries which many manufacturers seem to regard as their reasonable prerogative. For this reason, it is advisable for the locksmith who plans to do this work to carry all of the necessary components for a key system as stock items in his shop, preferably in the form of unkeyed cylinders. These unkeyed components should be carried in the dominant manufacture of residential locks in his area (if that type is of reasonable quality and suitable for adaptation to the key system).

If cylinders are available for all types with matching padlocks, they should be stocked (unkeyed, if possible); if a better deal can be had on keyed locks, they can often be used, but at a sacrifice of the time it takes to remove the existing keying. This time is usually not too significant except for padlocks in which disassembly procedures may tend to become tedious.

Assuming that the locks are of a five pin tumbler construction, they may be keyed in advance in three of the tumbler positions, leaving the last two tumbler combinations available for key changes from one customer to another. When the key changes are used up on the two tumblers, the next tumbler may be changed and the same combinations used on the remaining two again, just as in the development of any other code series. It is important in this type of work to be systematic in choosing key changes so that one customer does not have the same key as another. To do otherwise would be cheating your customer and destroying your reputation in the process. This means keeping records of combinations used, and storing them in a safe place to assure maximum protection for the customer.

Once the locksmith has surveyed the potential for residential rekeying in his area, he should work up a loose-leaf brochure which he takes with him when interviewing prospective customers and checking the rekeying potential of their homes. If the customer's present key system is adequate in quality and capable of being rekeyed into a one key system, this should be done. It may make a bit less money for the locksmith at the time, but it pays big dividends in customer satisfaction and word-of-mouth advertising.

If the customer's locks are not suited to rekeying, by reason of different keyways or the inadequacy of the locks, the situation should be briefly outlined to the customer. By showing him in your brochure the locks available, and the exact price of each lock, a firm quotation may be offered on the spot. It is surprising how often a sale can be made if one is prepared with necessary illustrations and prices.

Apartment and Rooming House Masterkeying

Apartment and rooming house masterkeying, particularly for small apartments and neighborhood rooming houses, offers the opportunity for a comparatively small, simple job of masterkeying at a good profit. Most locksmiths are content to leave apartment masterkeying to the construction contractors who build the apartment house. But what of the small apartment house which may have been remodeled from a large residence? Here is a field which is virtually untouched and is wide open to the locksmith.

The same conditions prevail for rooming house masterkeying; indeed, such masterkeying may be more lucrative in the long run because the houses are divided into individual rooms and there are usually more of them to a building. The locks required for rooming houses are virtually identical to those for residential rekeying, and the same locks stocked for one purpose will fill the other.

For either of these systems, the locksmith should be prepared to replace an old bit key mortise lock with a modern cylindrical unit, using the mortise lock replacement kits which most manufacturers of cylindrical locks offer in an attractive form.

Farm Key Systems

Farm key systems are another potential source of revenue for the locksmith. In most rural areas today, the advent of improved automobiles and roads has not only made the city more accessible to the farm, but the farm more accessible to the city and its vandalism, burglary, and other types of crime. More and more farmers who had never locked their houses before are locking them now.

The modern, progressive farmer is likely to have machinery worth tens of thousands of dollars in addition to his home. Livestock may be highly valuable, and fence gates should be locked when not used. But how is the farmer to lock all the gates on his farm, lock doors on his equipment shed, and lock two or three on his home, without spending half his time searching for keys or sorting them? Farmers are wide-open prospects for a one or two key system. The padlocks for such a system should be very carefully selected by the locksmith to withstand the conditions of blowing dirt, snow, ice, water, etc., to which they will be subjected in farm use in that particular area.

Once the locksmith has selected and organized a system of locks which are directly and specifically oriented for farm use, an advertising compaign can be launched to bring it to the farmers' attention. They should be made to realize that they can save time and prevent the loss of valuable farm equipment by replacing inadequate padlocks with a modern locking system. This may be pointed out in a mailing brochure, or by radio advertising on a farm oriented program.

If the locksmith is interested in selling safes and fire resistant file cabinets, the prosperous farmer is a good prospect. The locksmith should by no means try to high pressure the farmer or "sell him a bill of goods." It should be understood that the locksmith is merely offering the farmer a service for minimizing his exposure to theft, and at the same time saving his time in a way which is convenient and practical.

The amount and type of advertising that the locksmith does will naturally depend upon the number and type of farms within his working area. Naturally, a brochure should be prepared solely for farm needs, even though the locks shown may include some of those used for residential or other work. This does not mean that a brochure presented by the locksmith has to be printed to his order; on the contrary, it can be composed from catalog pages of the various manufacturers, or excerpts from those pages.

Prices for each unit should be stated clearly in the brochure, and the customer should have an understanding that should a complete lock change be necessary, installation charges would be extra. The farmer is just about the forgotten man when it comes to lock service, and anyone offering it usually receives an enthusiastic welcome.

BURGLARY ALARM SALES AND SERVICE

The sale and servicing of burglar alarm systems can be a highly lucrative adjunct to locksmithing work. The locksmith is in a position to supply a complete line of burglar protection for business and industrial premises, and even to an occasional private home.

Burglar alarms vary widely as to the type of sensing equipment used to detect burglarious entry. The response to the alarm initiated by the sensing system differs widely too. For example, sensing systems may consist of simple door and window contacts; they may be of the capacitance type, operating from radio waves; they may be sonic detectors; the detectors may respond to infrared waves; they may operate through the breaking of light beams, visible or otherwise; or they may be actuated by some mechanical action which the burglar either performs or does not perform.

The response to these various alarms may include almost anything which can be electrically actuated or controlled. It may be a simple bell, horn, or siren on the premises, with or without a flashing light on the roof. It may give only a remote alarm, sounding in the police station or the owner's home, or both. It may take the form of a prerecorded telephone call to the police department. It could even involve such actions as sliding gates across the doors and windows to catch the burglar. All of these and many more are included in the burglar alarm potential.

In order to sell and service alarm systems, the locksmith should be intimately familiar with everything about the types which he offers, and have a good working knowledge of other kinds as well. He must be prepared to make any needed repairs and to make routine service calls and tests of the systems. Not only must he be prepared to do this for the systems which he sells, but he must also be ready to cope with other systems upon which he may be called to work. Here, again, the study of electronics is a highly desirable thing for locksmiths.

If a locksmith wishes to sell burglar alarms, then safe and vault sales and service are also lucrative additions to his activities. It is not, however, the author's intention to go into safe and vault work in this text as that is a highly specialized and diverse field of study, just as is burglar alarm electronics. It is sufficient to say that at the time the customer is negotiating for locks and alarm systems, a sale of a safe may be included if the locksmith is prepared to take advantage of the opportunity.

Safes and burglar alarm systems should be sold with discretion; where the customer is under-protected, he should be so informed and given the opportunity to bring his equipment up to date. Recommendations by the locksmith can come to carry a great deal of weight in the community if he goes about it in a thoroughly honest and ethical manner. He should see that the customer has as much protection as he needs, but is not over-protected merely to make sales.

SECURITY AND BURGLARY PREVENTION COUNSELING

Burglary prevention counseling is a field which takes cognizance of the many facets of a good locksmith's abilities. Merchants who recently have been burglarized are ideal prospects. The locksmith goes over the entire building from basement to attic, around the building walls, inside and out, including the roof, in order to locate every possible means by which a burglar might enter; he then recommends protective measures to the merchant to minimize the possibility of burglarious entry. This may include the installation of bars on windows (other than show windows), which should be bolted or welded on so that they cannot be removed; special glass, such as the Amerada burglary resistant plate glass for show windows; special locking arrangements for doors; protective measures for skylights, coal bins, and utility entrances of all kinds; even the possibility of a burglar cutting a hole in the roof or walls should not be overlooked.

Door locks, if otherwise adequate, may be upgraded by the installation of spool or mushroom tumblers. Other situations may require the replacing of the lock cylinder or the entire locking unit with one of a radically different type.

A sound way to organize the locksmith's burglary prevention counseling operation is to charge a flat fee of so much per 100 sq. ft. of floor space for all floors occupied; or, the fee can be individually negotiated if the situation does not seem applicable to this type of charge. It is suggested that part of the fee for counseling be applied against the purchase price of materials or the fee charged by the locksmith for implementing his recommendations should they be accepted by the customer. Burglary prevention or security counseling is not, however, ordinary locksmithing work; it is based upon highly technical and professional knowledge. Therefore, fees charged should be proportionately greater.

Security counseling is by no means limited to the physical measures which may be taken to protect an establishment against burglary. It may include such things as shoplifting prevention, either by gadgetry or procedures revision, or both. It may deal with theft by employees, in which case recommendations would include a variety of measures such as gadgetry, inventory control, spot checks, operating methods, bookkeeping and accounting procedures, clearing of new employees through the police department, and a review of policies and procedures for both supervisors and employees.

The revision of the interior of the building may be helpful in solving security problems. These and many other problems must be coped with by one who is to engage in the security counseling field. As can be seen by the bare enumeration of some of its aspects, this subject is too diverse and complex to cover in this text.

Architectural Counseling

Architectural counseling is another aspect of locksmithing which is strictly professional and is based upon sound and encyclopedic knowledge of locks, and their strengths and weaknesses, and includes many of the aspects of burglary prevention counseling as well. To do this counseling, one should have knowledge of all types of door hardware, including hinges, door closers, mechanical and electrical hold-open devices for doors, and wide knowledge of the reliability of the various makes and models of all types of door and architectural hardware. One also needs to know such things as window construction and locking methods for windows, alarm systems, safe and vault requirements, and have at least a basic knowledge of general construction practices.

When the locksmith has this knowledge and can demonstrate it, he can go to a local architect and explain what he can do for him in the way of recommending manufacturers and models of hardware for any building that the architect may plan. He can offer assurance that the hardware will perform to the satisfaction of the architect, the construction contractor, and especially the owner of the new building. This can be done at a fee which will interest the architect, and may be coupled with a mutually advantageous advertising program to the effect that hardware for the building is to your specifications. An unobtrusive label could be worked out which would be advantageous to both the locksmith and the architect, and affixed to the principal entry doors.

Another part of a package deal could include the environmental servicing of every lock and latch in the building before installation, and either making the installation or checking it out after it has been made. The sale of the entire hardware package for the building could be made to the construction contractor, if the locksmith is able to offer it at a sufficiently attractive price. The author has seen many buildings, well designed and built, in which door hardware caused trouble even before the construction period was completed because the architect drawing the specifications had no way of knowing whether or not the hardware was entirely reliable. The building contractor could make the same mistake without the expert advice of a reliable locksmith.

The locksmith who is looking forward to entering this field should, whenever possible, attempt to secure dealerships for both good hardware and locks and for inexpensive but highly serviceable models suitable for residential construction. With such dealerships, the locksmith should be able to sell large volumes of door hardware to construction contractors in his area at a substantial reduction from retail price and yet make a good profit on the sale as well as on the servicing of these locks prior to installation. One of the fringe benefits of such an operation is that the locks which the locksmith sells at retail out of his shop will cost him the jobber's price, rather than the higher wholesale price, thus increasing his profits.

Custom Lockmaking

Building or modifying locks to meet special performance criteria can be both lucrative and satisfying to a first-rate locksmith. The range of lock functions in normal commercial manufacture is great, it is true; however, certain situations arise from time to time wherein normal commercial practices prevent the lock from performing in the desired way. Some of these situations will require only a very simple modification to an existing commercial design. The addition of a screw to immobilize one of the parts, or the removal or addition of a part, may be all that is required in some cases.

Other situations may require the building of a lock from scratch to fit a given space requirement or to perform certain functions not ordinarily expected of the type of lock in question. For example, it may be required to modifty a lock so that it registers with an audible or visual signal at a remote location whenever the door is unlocked or opened. This type of operation is usually taken care of with micro-switches, either in the strike or on the door frame. Occasionally it is necessary for the switch to be in the lock itself, but this is less desirable, due to the flexing of the wires as the door is opened and closed.

It may be desirable to make the lock operate part of the time from a remote electric push button, and, at other times, to work by manual operation completely independent of the electrical function. The list of special modifications is almost endless, and the job of building custom-made locks to suit a particular application involving electricity is equal-ly diversified. A good background of general lock-smithing knowledge is adequate for most such purposes; however, in order to meet all of the requirements of this type of work, a certain knowl-edge of electricity is also required.

When one has made a few of these locks to suit special purposes, one may find that word-of-mouth advertising results in more jobs of this type. Have a few demonstration models where customers can see them. Commercial and industrial users are the primary source of this type of work. And, frequently, the locksmith who is prepared to do this work will find that other locksmiths will refer customers to him or will negotiate to have the work for one of their customers.

LESSON 40

Shop Location

THE first step in establishing a locksmithing business is choosing a location. All of the factors involved should be considered as fully and carefully as possible. Although there are too many to enumerate, we will attempt to list some factors which *must* be considered.

The population of the area is important when considering the establishment of a locksmithing business; it must be sufficient to make such a business a realistic venture. The size of a city is less important than the population of the trade area. The city itself may be rather small, but it may have a number of satellite towns which increase the population of the trade area; or the farming population in the vicinity may be divided into small farms with a high population density. A smaller city, therefore, can support a locksmithing business even though the population of the city itself may be comparatively small, say 12,000 to 15,000 people. Conversely, a city of this size may be unable to support such an establishment if the population of the trade area is small.

The competitive factor is another aspect of selecting the location. If a small city already has a locksmith's shop, in all probability, a new shop would have rough going for some time. Dime stores and hardware stores which cut keys, while competition, are not difficult to overcome. Most of these places carry a very limited selection of key blanks. Their key machines are usually lacking in accuracy, perhaps not because the machines are inaccurate, but because they are not properly adjusted. Also, key duplication is not one of the more lucrative

operations in locksmithing; but it serves a valuable purpose in bringing customers into the shop and makes a small profit besides.

A locksmith who has to depend upon key duplicating alone will quickly be out of business. Key duplicating is an asset. It is both valuable and important. But it is not a decisive factor. It can build or destroy a reputation and it helps pay the rent, and as such it should be sought assiduously, but it will not make a locksmith rich.

Another factor to be considered in the acceptance or rejection of a location is the economic orientation of the area. This may not be an impediment to the establishment of a business, but it certainly will affect the type of business which is done. For example, an agriculturally-oriented trade area usually means that the majority of the locksmith's customers will be the merchants, distributors, and others who serve the farmers of the area, and many of his customers will be farmers as well.

Conversely, an industrial area means that the bulk of a locksmith's business will be residential work performed for those who work in the factories. Business people, their homes, and the industries themselves will be somewhat secondary. Each type of economic orientation poses its own factors, and these have to be analyzed with regard to the orientation of the locksmith, his work, and his merchandise; indeed, even his merchandising philosophy and merchandising methods will be affected by this area orientation.

Yet another factor is the personal suitability of the area to the locksmith. This includes both tangible

356

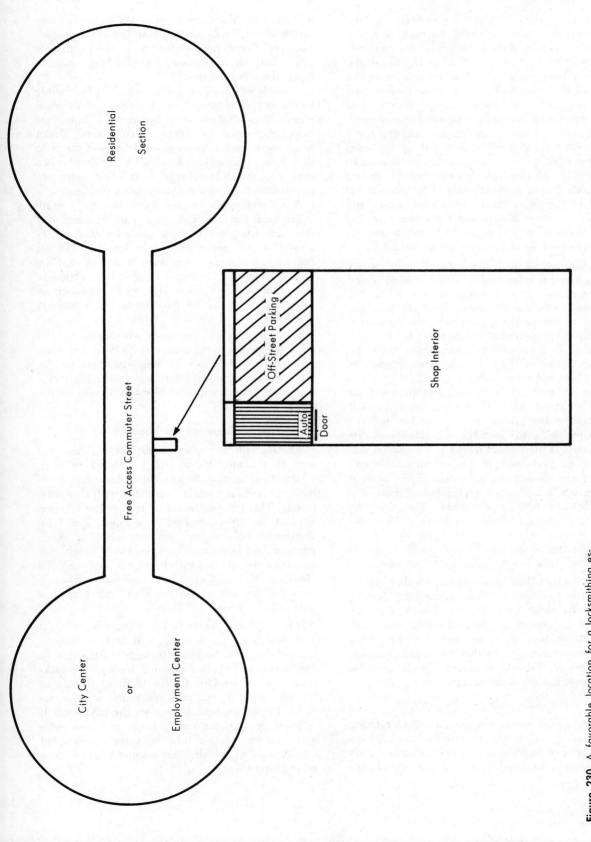

Figure 230. A favorable location for a locksmithing establishment. If on a two way street, the establishment should be located on the street side which favors access to residential bound traffic, as shown. If the street is one way, the side chosen is a less important factor.

and intangible factors such as whether or not he likes the community in which he must live, the physical condition of the community, the way it is governed, the tax structure, the climate, the people, and an indescribable thing best referred to as the "feel" of the community. All of these factors, and more, are to be considered in the choice of a location in which to begin a locksmithing business.

Several things can be helpful in weighing these factors. One is information provided by the local Chamber of Commerce. An interview with some of the officials of the local government is always beneficial. One of those officials to be interviewed should be the police chief. He can be particularly helpful in matters concerning the crime rate, the types of crime in the community, and what preventive and enforcement measures are available.

In addition to this, the establishment of an amiable working relationship with the local police department is nearly essential to the successful operation of a locksmithing establishment. Working together in the proper atmosphere, the locksmith and the police department can do much to lower the crime rate in a community, enhancing both the efficiency rating of the police department and the locksmith's financial condition.

If one is buying out an existing establishment, he should investigate all of these elements and weigh them against the present owner's profits and the way in which he fits into the organization of the community. If the present owner is acclimated and is showing a good profit, then a new owner who can adjust easily should do well also. If the present owner is not in harmony with local conditions, and his profits are small or nonexistant, then one who can attune himself should do better. Often the purchase price of such an establishment is much lower than that of one which is doing well; it should be, as it will take time to build up the business.

It may be that there is nothing much that can be done with a business when community feeling toward it has deteriorated. Such a shop is better left alone, or purchased for the equipment alone and moved to another, more receptive location. This would mean moving to another street and changing the name of the establishment. Or it may be necessary to move to another city.

CHOOSING AN AREA OF THE CITY

Assuming that one is not purchasing an existing establishment, obviously it will be necessary to choose an area in which to start. The same factors which were considered in the choice of the city

again prevail. For example, who will be the customers? Will they be factory workers? Business people? Farmers? Commuters who work or live in another city? Will the customers be the wage earners themselves, or their wives?

Once these questions are answered, it is possible to determine by geographic location and personal observation what areas of the city (or town) are most frequented by potential customers. Under some conditions, a location on the line of travel to and from a particular place can be desirable. But make sure that if travel is by automobile, entry into or exit from the traffic pattern is not a problem.

A difficult problem for most business establishments is the lack of adequate parking space. If it does not exist, can parking space be made available? If so, how much will it cost? Leasing of lots or buildings to convert to parking is expensive. (The locksmith may at least take comfort in the thought that thousands of successful retail and service businesses exist where parking space is severely restricted.)

Another factor to consider is whether or not the other businesses in the area are of a type that draw people who will be locksmithing customers. Or, for example, a locksmithing business in the heart of a wholesale district will attract few customers off the street.

CHOOSING THE BUILDING

Whether you rent space or buy the building, the cost is a major factor. As a general rule, in establishing a new locksmithing business, it is better to rent for a while, preferably with an option to buy. This permits an actual trial of the location without the risks involved in a large capital investment, such as the purchase price or down payment on a building. These costs vary widely, not only from one city to another, but from one area of a city to another, and must be considered in relationship to the advantages or disadvantages of a particular location.

It is possible to locate in too cheap a building; if it is shoddy or run down, it will repel customers. Conversely, if a building is highly attractive to customers, and costs too much money, the locksmith will be working for the landlord instead of himself, since it will take too much of what he makes to pay the rent, or to meet the payments. If he buys the building outright, the purchase price may represent a loss of interest on the money spent for the building, which might amount to as much as an exorbitant rent.

Aside from cost, the building selected must provide adequate floor space for the type of operation in which the locksmith is going to be engaged. If he plans to handle just locks, the floor space requirement would be relatively small, but by no means cubicle size. If he wishes to handle safes, he will require more floor space.

If he is to do automotive work, still more floor space will be necessary, and a drive-in door is highly desirable; in some parts of the country, outdoor space is acceptable, providing rainfall and other climatic conditions do not make this impractical. If he does not intend to run purely a locksmithing business, and plans to have other activities as well, the nature of the latter will create an additional set of requirements for floor space.

Window display space should be adequate. Here again, the amount and type of window space required will depend almost entirely upon the type of operation which is contemplated. Should the locksmith have a shop which handles locks, alarm systems, safes, builders' hardware, or other merchandise, then the display requirements for the window should be commensurate with the size and variety of the products in which he deals.

The building selected should be adequate to take care of a reasonable degree of expansion. Since people expect stability in any business, and perhaps even more than usual in locksmithing, it is not good business practice to change buildings to find more room every time the locksmith gets in a shipment of merchandise.

If the locksmith plans to do a considerable amount of automotive locksmithing, it is preferable, particularly in a cold or wet climate, to be able to drive the customer's car into the shop to make keys, to thaw out the locks, or to repair them. This is desirable not only for the comfort of the locksmith, and because a heated building is of material assistance in, say, thawing out a frozen lock, but is also desirable in those rather numerous situations where the locksmith does not want the general public to observe just what he is doing or how he is going about it. A building with a door large enough to admit an automobile, either from the front or from the rear, is highly desirable.

With the present profusion of service stations, a

Figure 231. A combined locksmithing and bicycle shop. Note the off-street parking and drive-in door for automotive and safe work.

great many are finding competition too severe, and they can't make the grade. So, there may be a service station available in a desirable area which would make an admirable facility for handling auto locksmithing work. The provision for parking space is also attractive, and if two or more stalls are available, it would provide additional space for safe or alarm system work, if these operations are to be part of the establishment.

Lighting, or the potential for lighting is, of course, an important condition because of the close nature of much of the locksmith's work. Inadequate lighting may be rather costly to modify. The general condition of the building itself is also a consideration, aside from its attractiveness (or lack of it) to the public. Factors to consider are: The condition of the heating plant, heating costs, insulation, air conditioning, fire resistance (since welding must be done), adequate wiring for power tools, sufficient floor strength (for safes and cars), and insurance rates (certain types of construction assure favorable fire insurance rates). Liability insurance is not likely to vary much, provided that the building is structurally sound.

These factors, then, all go into the selection of the particular building which the locksmith will occupy, and must be considered in light of the requirements of the type and extent of business in which he proposes to engage.

Shop Layout

HAVING selected the building, the layout of the shop is the next item for consideration. One of the most important parts of the layout of the shop is the window, because it is the window which says to the customer: "Please come in. We will try our best to please you." Without an attractive window, the locksmith is handicapping himself by losing some of his best advertising; or he may be throwing it away if he has a good window and isn't using it properly.

A window display should be simple to be effective. It should not be cluttered with a hodgepodge of material which is so indiscriminate and disorderly as to confuse the eye and mind of the potential customer. Above all, the window, and anything visible through it, should be kept scrupulously clean and neat at all times. This is a big "plus" factor in customer attraction.

The glass should be cleaned regularly and the material displayed should be chosen with two thoughts in mind: (1) keep the window simple, with only a few items which will display well; (2) the merchandise selected for the display window should be compatible in size, shape, and color. The articles should be either in wide acceptance, or unique so as to arouse curiosity.

Try to be original when "dressing" a window. You might show a number of related articles in a harmonious, integrated arrangement. Then, as a change of pace, set up a display of unrelated but interesting mechandise. The unexpected makes people curious.

Choosing the window display from merchandise which the businessman wishes to "push" is a good idea, assuming that the requirements of a good window display are met. If an item does not make an attractive window display, it is better left out of the window. Selling a particular product can be done in another manner once the window has been made sufficiently attractive to bring the customer into the shop.

There is a technique in the selection and arrangement of display merchandise which approaches the artistic. It involves the use of symmetry and asymmetry relating to the physical size and angles of the objects to be shown. Normally a window should be made to appear symmetrical; for example, if one end of the window has something which is tall, then the other end should likewise have something tall, with a lower object or a yet higher object in the middle.

A saw-toothed pattern in a window is to be avoided if the height of the objects is random or alternate, because it creates chaos to the eye and detracts from the effectiveness of the display. This is not to say that a window should display material of the same height; or low, high, low; or high, low, high. It may be desirable for it to be high, medium, low, but the arrangement should be both systematic and simple in its physical configuration, and it should not disrupt the visual pattern of the window as a whole.

The principle of not disrupting the visual pattern also applies to colors. A conglomeration of garish colors is offensive to the eye and should be avoided in a window display. A single brightly colored object may be used in the center of the window to attract the eye. One on each end may be used, or

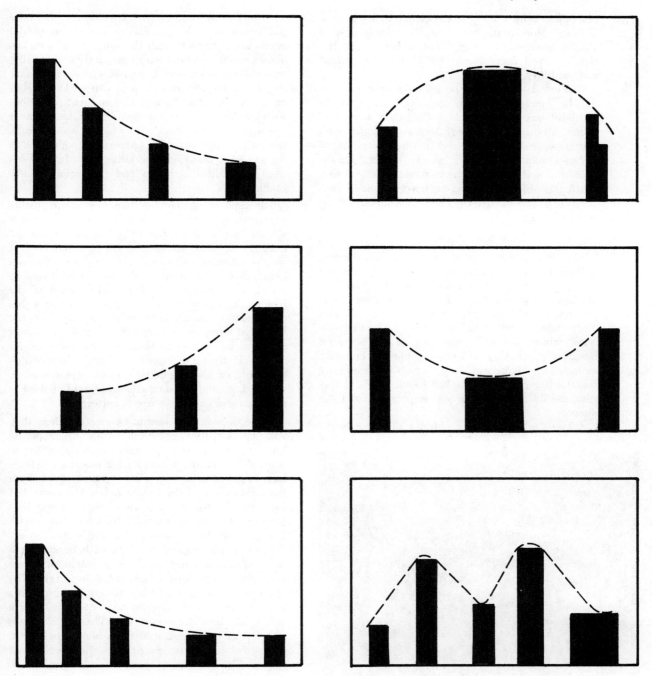

Figure 232. Configurations of desirable symmetry in window displays. Dotted lines represent the imaginary curves presented to the eye by the height of various displayed items. Contrast these with the illustration at bottom right which fails to attract the viewer's attention.

one on each end and one in the center if the window is large. More than three brightly colored objects should never be used, no matter how large the window; and they should be chosen for mutual compatibility.

Window lighting may be effective, not only to display merchandise in general, but also to spotlight one particular object to gain attention. Other attention directing devices include such things as color wheels and multiple strands of ribbon radiating from a single object. Raised platforms and special colors to direct attention to a particular object may also be tried, provided that they are in accord with the planned symmetry of the window.

THE KEY-MAKING AREA

The key-making area is usually located directly behind the window, where it will be visible from the street. Choosing this location is simply good psychology, provided that the area is kept neat. Passers-by may look in, see the duplicating machines and a wide selection of key blanks kept neatly on hooks along the wall, and stop in. It is sound business psychology to keep key blanks on hooks mounted on a large wall rack. Even if the beginning locksmith must space his key blanks widely, this rack should appear large and impressive from the street.

A large selection of key blanks distinguishes a

Figure 233. The key making area from the customer's viewpoint. Lee Hickenlooper explains the merits of a lock as his father, Glen, and brother, Jay, look on. Glen Hickenlooper is the first double winner of ALOA's lock-picking championship (1956-1960).

locksmith from the dime store key duplicators in the public's mind, and presents a more professional appearance. Even though the same number of key blanks could be stored more compactly and feasibly on revolving key racks, they make a more significant impression on the customer if he can see them all at once, spread out and apparently in great profusion. Another device which can be used by the beginner to make his initial stock of key blanks appear to be somewhat larger is to devote more than one hook to the same key blank, consolidating them later as his need for additional hooks and his supply of key blanks grows.

The duplicating machine should be readily visible to the customer while the key is being made, but it should be placed so that the customer cannot get cuttings from the machine in his face and eyes. A safe arrangement is to mount the key machine on a bench directly in front of the key blanks, and to install a counter or low partition in front of the bench. The partition will keep the customer a safe distance from the machine and protect him from the hazard of flying metal chips. A partition or counter may also be shelved on the back side and used for overstock storage of boxed key blanks from which the hooks on the key board may be replenished as required. A small vacuum cleaner should be used to keep machines and floor free of cuttings.

The code machines should be located where they will be less visible and less accessible to the public than the duplicating machines. It is important to conceal codes from which keys are being cut for a large keyed-alike or master key system, not only from the observation of the public inside the shop, but from people passing by, and from unscrupulous persons who may be watching from a distance with high powered binoculars or telescopes.

For example, should the locksmith be cutting a series of keys in a new master key system for one of the local banks, anyone interested in burglarizing the bank could conceivably benefit by observing even the individual code for one of the locks for a series, let alone the master key. These precautions are part of the specialized work of the locksmith; ethics and good business practice dictate that not too much of this type of information be available to the general public.

The cylinder disassembly area may be located in either the front or the back portion of the shop. Some locksmiths prefer it one way and some the other; there are good arguments for either point of view. Those who favor the location of the cylinder disassembly area in the front of the shop feel that it

Figure 234. Inside view of the same area shown in Figure 233. The hexagonal key display revolves to allow easy access while working at any of the key machines. Extra key blanks are stored in boxes on shelving at the right. Shelving above, below, and to the left of the window is for locks, parts, and accessories. The customer area on the other side of the "sales and service" window is equipped with a comfortable seat and annunciator chimes on the doors. A safe display is in a separate room.

puts the locksmith where he can be seen working from the street a greater percentage of the time.

It is, of course, true that the activity going on is a good advertising feature. It is also true that the area where things are cleaned and prepared for servicing, rekeying or masterkeying will be in the back of the shop where the equipment for cleaning is available. It resolves itself more or less into a dilemma of keeping equipment in the back where it is more efficient, or up front for display purposes. No matter how this is resolved, by preference or the physical layout of the building, it should be emphasized that the visibility of the work area be restricted so that the key codes to which a lock is being keyed are visible to absolutely no one but the locksmith. This is necessary not only to protect his customer, but also to protect himself should a key be made from codes by someone who has observed them and who might use the key for burglary.

A separate workbench should be provided for all keying work. This bench or table area need not be particularly large, but it should be equipped with its own tool rack where the tools needed for rekeying are at hand and neatly stored. The locksmith should be able to reach for a tool without having to look for it or search for it among a clutter. The table top should be kept scrupulously clean all the time; the smallest particle of dirt in a pin tumbler lock cylinder can cause it to malfunction.

This area should be equipped with bins for the storage of incoming locks which have not yet been worked on. Locks should be separated according to which customer they are for. Completed work should be placed in out bins, then transferred to an area for storing until the customer picks it up, or until it is reinstalled in the customer's door. A junk bin for unsalable, unusable, and unclaimed lock cylinders should be immediately at hand and used *only* for lock cylinders. Often these cylinders, even though damaged in some respect and uneconomical to repair, may be used for parts to repair other similar cylinders.

The lighting of the working areas where close work such as key making or cylinder keying is done, should be adequate but not so bright as to make it difficult to work in over protracted periods of time. The cylinder disassembly area should be equipped with a special light for occasional use in making critical inspections; this light should *not* be of the nonglare type. One of the modern, small, high intensity lamps is excellent for this purpose. It may not be used very often, but when it is needed, it is badly needed.

The key-making and cylinder disassembly area (if

it is located in the front of the shop) should maintain a high level of sales appeal. This means that it should be kept clean and neat, the colors should be pleasing to the eye, and more than a token respect to the principle of symmetry should apply, just as in the window. Lighting of the remainder of the front of the shop should be adequate, without excessive glare, and arranged to minimize shadows.

THE LARGER WORK AREA

The larger work area is almost always at the rear of the shop. Sometimes it is visible from the front of the building and sometimes not, depending to a large degree upon how many people are employed in the shop. For example, if the locksmith is working alone in the back area, and there is an intervening partition, he may not know when a customer comes in. The customer also may not realize that the establishment is open for business if he can see no one. A door chime, arranged to give an audible signal when a customer enters, is only a partial and none too reliable solution.

Unless the shop is sufficiently large to keep someone up front all the time, it is usually better if the work area at the back is arranged in such a way as to permit the locksmith to see the front of the shop and the passing public to at least be able to see that there is someone inside and ready for business. The back work area, just as the front, should be neat and free of grease and dirt. The wall paint should be clean as it is up front, particularly if the work area is visible from the front of the shop.

Workbenches should be large. Even in a one-man shop there should be at least two workbenches. Each bench may have its own rack of tools; or, a common tool rack may be provided between benches which are located back to back. The reason for two benches is that if the locksmith has a large job spread on his workbench, and a rush job comes in, he can do the special work on a second bench.

A third bench or sturdy table should be provided for power tools, such as a bench grinder which has a grinder on one side and a very fine wire brush on the other. A drill press is desirable, and if the locksmith has a lathe, it should be mounted on the bench, assuming it is small enough. All power tools (especially the lathe) should be mounted well away from the grinder because of the emery dust. The electrical circuit leading to these power tools should be adequate to operate them all at the same time, otherwise a provision should be made which will permit an electrical load no greater than the capa-

bility of the electrical circuit at any one time. This may be done by having a limited number of sockets, or by a somewhat more complicated switching arrangement. It is, of course, highly preferable to have adequate electrical circuits to carry the entire load.

The area where locks are brought to be cleaned prior to actual repair work should be separate from the workbenches and, in fact, separate from any place where other work is done. The solvents, and the dirt and grease which they take out of a lock, will quickly deface and spoil the appearance of both workbenches and tools, and will make the area where they are used very distasteful to work in. This is an excellent reason why the cleaning area should be separated from all other operations; the extreme fire hazard presented by some of the solvents and by the grease that is removed from the locks by those solvents is another.

The cleaning area should also be well ventilated, since a number of solvents are volatile. It may be desirable or even necessary to have an exhaust fan in this section. Just because it is an area for cleaning locks and parts does not mean that it should be untidy. To help assure cleanliness, provide metal drain trays or sinks, and table space with a removable paper covering. To do this, attach a roll of brown paper to one edge of the table and affix a cutting edge to the other. Unroll the paper across the top. As the paper soils, tear it off and replace it with a clean section. This method will keep the working surface fresh in appearance and pleasant to work on.

A number of the cleaning solutions used may be both volatile and flammable. It is essential that the place where they are used and stored be located in the part of the building which is most fire resistant. Volatile or flammable liquids should be stored in approved safety containers. Since solvents and cleaning solutions are the principal flammable materials in the shop, the same area should be used for the storage of surplus supplies of other flammables such as oils, greases, alcohol used for thawing locks, paints, lacquers, and thinners.

If the shop is to include a section for automotive or safe work, it also should be at the rear of the shop. While the workbenches will serve for all mechanical work, a certain amount of floor space in the back part of the shop should be reserved for the exclusive use of these activities. If the locksmith does his own welding, the area should be provided with a steel topped table on which to work.

Such work should be done in a very highly fire

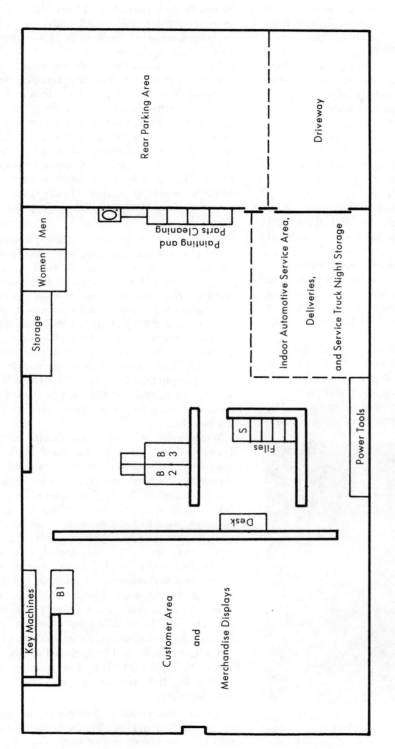

Figure 235. A shop layout. Note the use of zoning for all operations. The key blank display is behind the machines and boxed blanks are stored under the counter top surrounding that area. B1 is the cylinder disassembly bench. Shoulder-high partitions for the customer area and the office are actually door-covered shelving for storing merchandise, parts, and supplies; the customer area partition also contains some glass-covered individually-lighted shelving which serves as a showcase. More storage shelving, broom closet, etc., are provided in the work area. Bench B2 is for small jobs and B3 is for larger jobs; cabinets for small parts are provided at the end of each of these benches.

resistant area, well removed from the cleaning and flammable storage area, and as far as possible from the place where automotive work is done. Since sparks are consistently produced in a welding process, and flammable or explosive vapors from gasoline or volatile solvents will travel, the fire hazard from having these two types of operations is extreme unless the most rigid safeguards are maintained at all times.

TOOLS

Just as the furnishings and equipment of the shop may vary from that of the initial set-up, so also may a locksmith's tools. Certain tools are basic necessities for the beginning locksmith. There are other desirable tools for the beginner, and these will become necessary or even mandatory later on. A full complement of tools will make it possible to cope with any problem of any nature with which one might be confronted in locksmithing. The desirable tools and equipment make their need felt as time goes by. When sufficient need is manifested to warrant their acquisition, the alert locksmith will acquire them.

Utterly basic to the locksmith's needs are the tools which he will carry out on the job. These tools should be neatly arranged in a tool box, and left packed up at all times. These tools should not be used in the shop (with some rare exceptions). The first group of tools to be considered comprise the

Figure 236. Box top key vise. A special shop-made vise for key making is attached to the top of the key making kit. It eliminates the need for carrying a separate vise, with its bulk and weight.

locksmith's all-purpose tool box. This is carried for general purpose lock work only, and does not include specialized tools such as are required for lock coding and key making.

Of all the tools the locksmith uses, his screwdrivers are the most important and most used. The carry-out tool box should include about four assorted screwdrivers with the regular point, and with tips of all different sizes. Except for the smallest size, the 6 in. blade length is the most satisfactory, as a general rule. There are a few situations where the shorter 4 in. length is the more desirable; so, if one wishes, an additional medium-sized tip, 4 in. screwdriver may be carried. Three screwdrivers with assorted sizes of the Phillips-type point are also desirable.

Four pairs of pliers are suggested, assorted thus: water pump, 9 in.; slip joint, 8 in.; needle nose with wirecutter near the pivot joint; Super Champ electrician's terminal pliers (used for the screw cutting feature). The terminal pliers do a neat, clean job of cutting machine screws, and the threads require no point trimming or cleaning to make them work. Two ballpein hammers, 12 ounce and 4 ounce, and one plastic-faced hammer, lightweight with replaceable tips; one prick punch; one center punch, medium size; three pin punches in sizes 1/16 in., 3/32 in., and 1/8 in.; three hacksaw blades, one coarse and two fine with only the edge hard, also go into this carryout kit.

When using the hacksaw blade without the hacksaw frame, the fully hardened blade tends to snap easily; therefore these blades should be of the type which is hardened only on the toothed edge. Also included in this kit are three plastic or metal boxes with at least six compartments each. The boxes are used to carry assorted screws for escutcheons, knobs, cylinder set screws, and other common lock screws. Some of these common screws may be carried in longer lengths than ordinarily are required and the electrician's terminal pliers used to cut them to the proper length. This does not mean that screws should not be carried of the same diameter and thread in lengths which are just right for the majority of locks, but in addition to these, a few longer ones should be included so that they can be cut to any length. It is wise to avoid hardened screws for this purpose as they may ruin the terminal pliers.

A small assortment of flat and round spring steel in three or four sizes is a highly desirable item in the tool box. An assortment of taps is necessary. Only the plug taps are carried in this box in sizes: 6-32,

8-32, 10-24, 10-32, and 12-24, and are minimum basic needs for this item, together with a small T-type tap handle and five number drills in the tap sizes mentioned. Experience may also dictate carrying a few of these taps in the taper and bottoming style as well.

Two rotary files are highly desirable in size 1 in., coarse tooth, end cut, cylindrical shape; and 3/8 in., medium tooth, end cut. These sizes are adequate for most purposes. Mill files should be carried in 6, 8, and 10 in. flat; 8 in. square; 8 in. triangular, extra slim taper. Bastard files are carried in 12 in. half round, and in 4 and 10 in. round (or rat tail). In Swiss pattern files, the following are carried: 1 warding, 6 in., No. 2 cut, .040 or .050 thick; 1 assortment of "needle files"; and 1 hand (flat), 4 in., No. 2 cut. In addition to the box tools, a 1/4 in. electric drill should be carried in the car or truck for use with the drills and rotary files when necessary.

Another carry-out unit is the keying kit. These are available in kit form, or the locksmith can assemble his own. If he chooses to assemble his own, the basic tool box is, of course, essential and should be large enough and suitably arranged to be adaptable to other needs which will, no doubt, develop as time goes by. A plug holder affixed to the top of the keying tool box is a necessity. It should be the type which will mount firmly on the box without requiring a vise, and also should be of the rotary type so as to accept different plug sizes. The various sizes of followers are needed for use in removing the plugs from the lock cylinders and for reassembling them. The plug followers may either be purchased or shop made.

Assortments of pins, upper and lower, in .115 and .095 diameters, together with springs for both of these sizes, should be carried. A few mushroom drivers in assorted sizes, as made by Yale, or the spool tumblers made by other manufacturers, should be carried for security keying purposes. If one is assembling his own kit, it is desirable to fit the tumbler assortments according to size, into very small plastic tubes of the type that small items of fishing tackle are sold in. Small medicine tubes will also do for this purpose. A holder should be made to keep these tubes in place and in order, and they should be labeled with their proper size.

A pin tray is an essential item in the box to keep the tumblers in order during keying operations. A pin vise is frequently needed when the tumblers carried do not exactly fit the situation, and a slightly longer tumbler must be cut down or rounded a bit in order to work properly. A 4 in. and

a 6 in. Swiss pattern hand file should be included in the kit, in about a No. 4 and No. 2 cut, for use in making adjustments to the tumbler length. Also, Swiss pattern round files of 8 in. length, in No. 2 and No. 4 cuts, should also be carried.

A pair of tweezers for handling tumblers and springs is also required; the author's preference in tweezers is a curved dissecting forceps in 5 in. or 5-1/2 in. length, of the type used by biology students. A few disc tumblers in the Briggs & Stratton and a few in Yale types should also be carried in this basic keying kit. Later, others may be desirable, including some of the less common Yale disc tumblers, the Briggs & Stratton side bar tumblers, and the Rockford disc tumblers. It is wise to provide in the initial kit for this expansion, in order to avoid having to assemble a completely new tool box and accessories later.

The carry-out key making kit consists of a large, hip roofed fishing tackle box with 50 or more compartments, which are to be used for storage of an assortment of key blanks, the Curtis Key Clipper, such code books as the beginning locksmith can muster (primarily for automotive and disc tumbler locks), a very small, inexpensive clamp-on vise (or a box top vise of a type similar to that shown in Figure 236) for use in making keys by one of the impression methods. Of the two types of vise, the box top model is handier, since there is not always an available place to attach a vise of the clamp-on type, and a flat bench vise (or machinist's vise) is a bit heavy to carry in a kit of this type.

Also desirable in the kit are: 2 broken-key extractors; 3 assorted picks; 2 tension wrenches; and a straight pick for sight reading and occasional emergency picking jobs; a penlight; a pair of round 8 in. Swiss pattern files in No. 2 and No. 3 cut; a wiggle or impression system tool; a set of car opening tools; and a small piece of piano wire bent into an L shape, with one end filed to a flat-sided point, to be used for removal of disc tumbler lock plugs. The key blanks carried should include a few of each of the more popular types in the area, and an especially wide selection of automotive key blanks.

A very few of the most commonly used blanks should be carried in larger quantities to avoid having to continually replenish the box. Usually some 8 or 10 different key blanks will account for 75% to 80% of the total keys cut for other than automotive locks.

These three carry-out kits will comprise the basic necessities for out-of-shop work. It will be found that surprisingly few additions to the basic kits will be necessary as time goes by, with the possible exception of some specialized tools which may be required for the disassembly of particular types of cylindrical locks. Certain tumbler lengths for a particular manufacture of lock may be needed too, in any given community, and these may be carried in the keying kit, either in addition to, or in preference to, the original assortment of tumblers.

The code books, as dictated by the economics of the situation, must be used both in the carry-out keying kit and in the shop — as will the Key Clipper and the electric drill. After each use in the shop, the code books and the Key Clipper should be returned to the kit, so that it is ready to go at all times. Other kit tools should not be removed at all so that no time is lost in gathering tools, and none of the essential ones are forgotten.

The tools which are for use in the shop, and which remain there are very similar in nature, although of a somewhat broader selection, to those in the carry-out tool kits. For example, in the line of hammers, the shop tools may include: 1 to 3 pound blacksmith's hammer; ballpein hammers in 16, 12, 8, and 4 ounce sizes; 1 to 12 ounce riveting hammer; and one medium and one lightweight plastic faced hammer with removable tips.

A good set of lock picks should be kept in the shop and taken out on the job whenever there is a possibility of their being needed. These should be carried in the locksmith's pocket and returned immediately to a securely locked safe or depository upon his return.

Basic needs in shop key machines are: a machine of the capability of the Ilco "Minute" machine for general duplicating; and a code machine of the capability of the Ilco "Universal" for coding. Other machines may be necessary at a later date. You might want a machine to cut Chicago "Ace" keys, and another for the duplication of keys using precut blanks. A reliable automatic duplicator can be an attractive addition to the locksmith's complement of key machines.

If an automatic duplicator is purchased, it is important from the standpoint of customer psychology that it does not resemble those used by dime stores and hardware stores. If too close a resemblance is apparent, the customer will assume that, because of the similar appearance of the machine, the keys he gets from the locksmith are no better than those from the dime store, no matter how closely the locksmith's machine may be set. An automatic duplicating machine should be purchased on a specified trial period basis, or with a

strict, written guarantee of accuracy within the limits which your shop allows for duplication. Although a number of automatic machines will maintain good accuracy, many will not; hence the precaution.

The quarter-inch electric drill may be used both in and out of the shop until such time as the locksmith decides to purchase another. The shop equipment should, however, include an electric drill having a chuck capacity for a half-inch bit and turning at no more than 550 revolutions per minute (to avoid burning the larger size drills). A bench grinder and buffer are basic requirements.

A key micrometer or dial type caliper are necessary to set and check key machines. Of the two, the author prefers the dial type caliper; a high quality model is available currently at under $20. Parts cabinets and tool racks may be small at first, but should be capable of expansion as the locksmith's further needs are established. Approved cleaning tanks for solvents, and a sink and table for parts cleaning operations are necessary. Workbenches, with the exception of those used for power tools, should be low enough for the locksmith to draw up a chair and sit down to work. Standing in front of a workbench for long periods can be much more tiring than sitting.

If the workbench is made at a height convenient for sitting, the chair should be of about the regular height, thus eliminating the expense for special seating. Power tools are more efficiently used while you stand than sit; therefore, the bench should be higher. If an air compressor, even one of the type commonly used with paint sprayers, is not included in the initial shop equipment, it should be one of the first additional items purchased. Its usefulness in cleaning parts, particularly intricate parts such as lock cylinders, serves to make it nearly indispensable.

The service station type of air compressor will provide more compressed air, and has wider uses, but the paint sprayer type (small and portable) compressor is good for most purposes. Adequate eye protection is essential when air is used. Horseplay involving compressed air can be extremely dangerous, or even fatal.

The next addition to shop equipment should be a drill press. This need not be new, but should be of a sturdy type having a very rigid, firm spindle, even when fully extended. A machinist's vise for use on the table of the drill press is also highly desirable. Later, particularly if custom lockmaking is to be done, a milling table like that manufactured by

Palmgren, is a valuable addition for the precision locating of holes to be drilled.

The mill table may also be used with the drill press for light milling operations, if the table is kept rigid and the drill press spindle is sufficiently free of wobble. Ordinarily, in a drill press, it is preferable to use a rotary file in the milling operation and to take very small cuts because of the wobble of the drill press spindle. Clamping of the drill press spindle at the preselected height will help in maintaining its rigidity when used in the milling operation. A large bench vise and a small clamp-on vise or two are basic shop equipment and should be obtained at the time the shop is opened.

As the shop becomes better established, and the need develops through making locks to order and custom rebuilding, it may be desirable to add a small lathe. It should be of the screw cutting variety, and above all, it should not be too large. A large lathe is time consuming and unwieldy to operate, regardless of the purchase price. A quality used lathe is just as acceptable as a new one, but it should be kept clean and neat. The lathe, like any power tool, must be used with care because it can be dangerous.

Gloves should never be worn when operating power equipment. Loose clothing is to be avoided as it can become entangled in the equipment. A special word of caution for the lady locksmith: Long hair has been known to become entangled in power equipment, resulting in an instant customer for a wig, with considerable pain to boot! Eye protection is mandatory before operating power equipment, *including key machines.*

Full equipment for the shop, naturally, will include a certain number of specialized tools, and additional key machines and key blanks. Of course, the ultimate nicety of a fully equipped shop is a service truck equipped with two-way radio communication! It is just about a shop on wheels, and includes such conveniences as a key duplicating machine, cleaning equipment, specialized tools, a wide assortment of key blanks — in short, almost anything a locksmith needs to do any job anywhere.

A word about the quality of the locksmith's tools: The basic tools should be of the finest quality that the locksmith can afford, since they will be the means to his livelihood for a number of years. Unusual tools, bought for a specialized job which might or might not recur for a number of years, need not be of such high quality, if cheaper tools are available and will do the job. These tools are not part of the locksmith's permanent kit to be routinely carried out on jobs, or even displayed in the shop.

The tools which are carried on the job, or which are displayed in the shop because they are continually used, should be of adequate quality to do the work well and to do it for a long time. Their appearance and serviceability should be maintained, and even enhanced a bit. This is particularly true of tools which the customer sees. For example, the painted parts of hammer heads may be repainted to change the usual drab black to bright blue, or another color.

Other tools may be painted with a neat band of the same color, not only to preserve the appearance of an integrated set of tools and to create a favorable impression on the customer, but also to distinguish the locksmith's own tools from those of another craftsman who may be on the job at the same time. The color scheme may be carried out on all the shop tools and those in the carry-out kits.

Another idea which usually makes a fine impression on the customer is to carry a sheet of plastic cloth and to spread it on the floor under the lock so that no dirt, oil, grease, wood cuttings, or the like will get on the customer's floor. An alternative is to carry a roll of plain wrapping paper, which may be torn off in suitable lengths and disposed of after each job. This is small item, surely, but it is one which customers appreciate.

THE MERCHANDISING AREA

The merchandising area of the shop has a number of uses. One of its functions is to keep the customer from wandering casually back into the work area, thereby getting himself injured, or just generally making a nuisance of himself. The merchandising area in all too many shops is practically indistinguishable from the junk bin. This is not good merchandising practice. The area should be kept orderly and clean.

Individual samples of the locksmith's products should be displayed in a showcase; but when he has more than one of an item the surplus should be stored in shelving which has hinged doors. The doors may be either opaque or glass fronted, at the owner's option. If they are opaque, they are somewhat easier to maintain, whereas if they are glass, the rules of color and symmetry when arranging items on the shelves must be observed. Merchandise stored in closed shelving does not require tedious and frequent dusting as does open shelf storage, and is much more impressive in the long run than a large quantity of material that is piled in a "where it'll go" fashion. The doors should swing-out rather

Figure 237. A modern service truck. This truck, like many of its kind, is equipped for two-way radio communication with the shop. Communication of this type saves mileage and driving time, and provides speedier, more efficient service to the customer.

than slide open because if anyone opens them, he is more inclined to close them, too. This keeps the contents clean and avoids a shopworn appearance. Showcase items should be plainly marked as to price and code marked as to cost. An additional marking, either a number or letter, should indicate where the stock supply of that particular item is located.

Certain companies occasionally supply cut-away samples or see-through models of their locks for use as "sales-pushers." These items are fine and should

be put to use providing they do not show so much as to compromise the protection which would otherwise be afforded by the lock. Any sales-pusher which reveals information negating the protection offered by that or any other lock should never be displayed because, after all, the locksmith's profession is one of trust. Displaying anything which will compromise the protection given to any customer by his lock is a violation of that trust.

The type of sale known as an "impulse sale" can be profitable in any business, and locksmithing is no exception. The item purchased on impulse is usually in an eye-catching display. It should be placed near the cash register where every customer will see it as he steps up to pay his bill. The item may be something such as an exceptionally fine key case, a matching key chain and tie clasp, an unusual lock such as one with key operation for window sash, or a locking electric plug.

Whatever the item is, it should offer a large profit, and be readily salable in that particular community. What will sell readily will vary from one community to another; trial and error based on experience provide the only answer to what to display. As has been mentioned, "impulse items" are often in an eye-catching display; sometimes garishly so. There is a common tendency to over-do a good thing. The author feels that only one item should be displayed

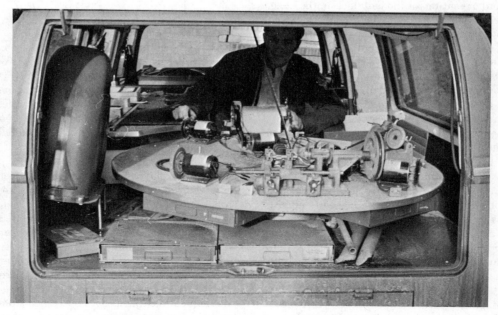

Figure 238. Key machines in service truck. Key machines are battery-powered and mounted on a revolving table in this truck.

Figure 239. Interior of service truck. The top of the work bench in this service truck is carpeted to keep parts and tools from falling off. The uncarpeted areas are the tops of tumbler storage boxes which are set flush with the bench top. Notice the easy-to-reach tool rack. Except for necessary knee space under the vise and the key machine in use, all available space is utilized for storing key blanks, parts, locks, tool boxes, and service kits.

at the cash register at a time; if it is felt that there are two items worthy of display at the same time, the second item should be in another part of the shop, possibly on the counter by the key machines. Too many of these displays create an unfavorable impression, however.

THE CUSTOMER AREA

The customer or waiting area is separated from the work space by the merchandising section with its showcases and storage shelving. The customer should be made as comfortable as possible while waiting for work to be completed. This is particularly true if the locksmith wants to attract women customers. Since women take care of the household business while the men are working, this is an extremely important factor. Keeping customers comfortable means many things. It means choosing pleasing colors. It means comfortable chairs, and perhaps a sofa.

The author has seen comfortable and attractive sofas which have been improvised from seat and back cushions of wrecked automobiles. These are easily reupholstered in any color, and a little ingenuity can provide a framework to convert them into attractive seating for waiting customers. A little music may be provided by a tape recorder, or radio, and is often the difference between a patient customer and an impatient one. The placement of the seating should be such that the merchandise for sale is clearly visible. This factor alone will often result in sales which might otherwise be missed. A soft drink machine will often add not only to the cutomer's sense of well-being and patience, but also to the merchant's profit.

The author can envision a good many of the old-time locksmiths snorting at the idea of an appealing customer-waiting area, but he asks them to consider what percentage of his customers are women, and what percentage of the net annual income they spend; and to consider if the waiting area in their shop is conducive to women shoppers coming in to buy a padlock or to wait for work to be done.

Figure 240. An attractive and effective merchandising area in a combination type shop.

LESSON 42

Merchandising

Advertising and the "gimmicks" associated with successful promotion are sales aids which the locksmith often fails to utilize properly. Several media are available for advertising. Chief among these are television, radio, newspapers, handbills, and the telephone directory.

The problem of selecting the proper medium is important. The decision may depend upon the habits of the local population; upon the availability of the medium at the right time (day and hour); the cost of the medium; and the material to be advertised. For example, the locksmith does not buy costly television time to advertise his key duplicating facilities. Television time may be desirable for advertising systematic protection against burglary, which includes such things as burglar resistant locks for doors, protection for windows and building openings, and alarm systems.

This would make a good subject for television advertising in the proper community at the right time. It would probably be successful immediately after a wave of burglaries in or near the community. Radio or newspapers may be used to advertise less expensive items or services.

A key to successful advertising is to offer something "free." It is surprising how many people will use an extra fifty cents' worth of gasoline to get a twenty-cent item free. For example, it does no harm to advertise that a free key ring (costing several cents) will be given with each duplicate key next Saturday, or a free key case will be given with each service call. If the key case is inexpensive but durable, and has the name of the locksmithing establishment discreetly on it, it is excellent advertising.

Free coffee while waiting for work to be completed is an excellent and relatively inexpensive way to promote good customer relations. Be sure that the coffee is good and is served in disposable cups. Perhaps a free costume pin for all women customers on certain days of the week will bring in many who would not otherwise come in. Quite often these items may be bought at a reasonable price in quantities. They should be relatively inexpensive compared to their return, but should never appear cheap or shoddy. Whenever practical, they should carry advertising.

Experience with the various media and careful records of sales before and after advertising in a particular medium will enable the businessman to evaluate the actual value of the promotion. Once evaluation has been made, advertising a product or service, if successful, should be continued. However, the same item should not be continually advertised. Some variety in advertising is not only desirable, but necessary. If some form of humor can be injected into the message, it may attract more attention.

CUSTOMER RELATIONS

Customer satisfaction should be striven for with diligence, not only in the work which the locksmith does, but also in the attitude with which he meets his customer's problems, and in the truthfulness of his advertising. No advertisement should promise more than can be delivered.

Customer satisfaction with advertising is only one aspect of customer relations. The customer should be met with a casual and natural courtesy. He should never be treated with either arrogance or subservience. The courtesy extended should not be artificial or strained. What constitutes courtesy varies from one community to another and from one part of the country to another — even from one individual to another. Courtesy to one customer may be ''Hi, Bill, come on in!'' and to the next, it may be ''Good afternoon, Mr. Jones. Please come in.'' Either form of greeting may be considered courteous by one man and discourteous by the other.

Another matter of customer relations is the warranty on the locksmith's work. Warranty work is always an expense for the locksmith because he cannot charge for it, whatever the cost to him. It is usually better to lose money on one transaction than to have a dissatisfied customer who may be vociferous, and cost the locksmith other customers as well. For this reason, it is strongly urged that the locksmith never haggle with his customer over warranty matters if it can be avoided.

To avoid a dispute as to the warranty on duplicate keys, the following suggestions are offered: The original key which the customer brings in may be either one of the original keys which came with the lock, or a duplicate. If his original key is worn, or if it is obviously a poorly-made duplicate, the locksmith should make it clear to the customer at that time that no warranty can be made on the new duplicate key except that it will measure within, say, .001 in. of the sample. The key should then be measured before delivery to the customer and marked unobtrusively to indicate to the locksmith that he has not given it a warranty. In such cases, it is also a good idea to mark the pattern key inconspicuously to avoid a claim later that another, better key was the one from which the locksmith made his key.

The keys manufactured by the locksmith should always be stamped with the name, initials, or symbol of his establishment in order to prevent warranty problems; and keys which are made from worn or poorly-made keys, after being measured for accuracy with the customer's sample, should be inconspicuously stamped, perhaps with a small N (signifying ''not warranted''). The customer should not be left with any illusions that the duplicate of a worn or defective key is guaranteed to work any better than the pattern key he submits for duplication.

Questioning Legitimacy

At times, there is a question in every locksmith's mind as to whether or not the person who is asking him to open a locked door, or make keys for a house, automobile, or business firm has a legitimate reason for his request. As a minimum precaution to avoid innocently becoming an accessory to a crime, the locksmith should ask the customer to show him some proof that the home, or whatever, is his; and this proof should be recorded. For example, the locksmith could keep a prepared work order form requiring the signature of a customer, the number of his driver's license, and his address. If any of the information appears suspicious, police assistance should be requested to establish the right of the customer to enter the locked item or building.

For example, if the locked object is an automobile, the customer should be able to establish his identity with his driver's license, and the registration receipt for the automobile. The number of the registration should be recorded along with the driver's license number, and the license tag number should also be recorded if it is different from the number of the registration receipt. All of this can be done in a casual and business-like way. The procedure should be followed for every job of this type, no matter what the circumstances.

In instances where a serious question exists, or where some of these factors are incompatible, a telephone call, explaining the situation to the local police department, should resolve the matter. Some communities may require, by law, a register or other record of all duplicate keys made, and the signature (and address) of the customer.

Ethics for Locksmiths

The locksmith, being in a profession which involves the trust of the public, is particularly sensitive in his relationship with the community to any touch of scandal or illegal activity. Any precaution, and every precaution possible should be taken to insure against being involved in any type of illegal activity.

In the course of his work, or in the course of making a sale, the locksmith often has the opportunity to advance constructive suggestions which will result in added work or sales for himself and additional protections or convenience for his customer. No opportunity should be missed to offer these sales and service suggestions; however, these suggestions should always be of true benefit to the customer from the standpoint of efficiency, convenience, reliability, or increasing his protection.

The Nosy Customer

Locksmithing is one of those professions which seems built to order for the nosy customer. Quite often, the author hears the question: "How do you open a lock like that?" It can usually be passed off with a remark such as: "Well, I do everything but taste it." Everyone grins, and that is usually the end of the matter. If the customer persists, he should be told that it would be a violation of professional ethics if this information were divulged.

The Arrogant Customer

Occasionally a customer is encountered who is arrogant and dictatorial. He tells the locksmith how to do the job, or that he is going about it the wrong way. The response to this should not depend upon the boiling point of the locksmith; every effort should be made to be courteous. If, however, the situation passes the point where the locksmith feels that he can no longer tolerate it, the customer should be told as gently as possible to take his business elsewhere.

Holdup, Armed Robbery, and Burglary

The locksmith makes his living by preventing three specific crimes: holdup, armed robbery and burglary. Therefore *extreme* precautions should be taken to avoid these things happening *to the locksmith himself.* Not only will the locksmith suffer the cash loss involved, but it will reflect upon his professional competence to protect his customers from the same thing. Inferably, the locksmith's business will suffer from the loss of the public's confidence. No criminal expects to make a million dollars out of the holdup or burglary of a locksmith, but, the locksmith's tools and picks are a tempting bait for burglars who may want to use them in a much bigger robbery.

In addition to this, the locksmith's knowledge may pose some hazard to him or to some member of his family. He might be kidnapped to force him to gain entry to, say, the local bank for a burglar (or a ring of burglars). Therefore, the locksmith's shop and home should be protected to the greatest possible degree.

The possibility of being held up while on a phony service call is a matter for careful consideration. Someone should always know where the locksmith is going and when he expects to be back. In addition, if he gets a service call, and the situation is suspicious, he should ask for police protection. In some communities, a locksmith may arrange with the police department for a permit to carry a gun on such calls; but this should never be considered unless he is skilled in the use of a gun.

If he has suspicions when he reaches the site of the call, he should simply drive past without stopping and either return to the shop and wait for the call to be repeated, or ask the police to investigate, or both. It is true that this does not always promote good customer relations if the call is legitimate, but it may save his relations with the rest of the people in the community, or even save his life! The author has had personal experience with this type of situation and is well aware that it is not melodramatic or fictional, but can happen to any locksmith. With the crimes against property having nearly tripled over a period of thirty years, this type of situation can be expected to increase also. Locksmiths should be wary of being duped.

Service Calls

When making service calls, the locksmith should extend every courtesy and consideration to his customer. If the customer's property becomes dirty or greasy as a result of the locksmith's work, he should clean up the mess.

Service calls are usually based on a minimum fee plus time and parts costs. Giving the customer a small memento, such as a key chain or small key case (with the locksmith's advertising), builds good customer relations and helps to soften the impact of the minimum fee if the job takes only fifteen minutes or so.

When a customer brings a lock into the shop for which keys are to be made, it is likely to have a code number from which the locksmith can cut keys. When this situation exists, it offers an unparalled potential for building good customer relations. The locksmith can charge the customer the fee for a code key and point out to him that since the key was made from this number, it is not as expensive as it would have been had there been no number.

It is true that the locksmith will make somewhat less money at the time of this particular transaction; but, if he wants that person for a steady customer, there is no better way of insuring that this will happen. The cost, in comparison to the value of having repeat business, is quite small. In addition to this, it is a situation which lends itself effectively to word-of-mouth advertising which is such a vital factor in the continuing success of a locksmithing business.

Competition

The locksmith is subjected to a number of sources

of competition; for example, dime stores, hardware stores, service stations, and garages — any or all of which may have key duplicating machines and key blanks. When a customer inquires about this competition, a good response is: "Our key machines are set to duplicate keys with a variation of less than 1/1,000 in. I do not know what the accuracy of our competitors' key machines is, but an inaccurately cut key can destroy a lock cylinder."

It is a good idea to have a lock cylinder available which has been used with a misfitting key to the point where the holes for the pin tumblers are egg shaped. The tumblers may be left in the holes and the key in the plug so that the customer can see for himself precisely what a misfitted key does to an otherwise good lock. There is seldom any quibbling over competitive pricing of duplicate keys after the customer has seen a ruined lock. He becomes aware that the locksmith is guaranteeing the accuracy of his keys to a point which prevents this type of damage from occurring.

Another phase of competition is in the sale of locks. Hardware and chain stores may have locks in competition with the locksmith's own. If this is the case, identical items should be priced as economically as the locksmith can afford, irrespective of the competition's pricing. The trouble with a locksmith's dealing in a cheap line of locks is that people expect his products to be good, better than his competitors'. It does nothing for the public image the locksmith so badly needs if his products fail to live up to expectations. However, if the locksmith acquires an inexpensive line of locks at a price which will allow him to compete successfully, and the product represents a fair value to the public, then he should make the most of it, by all means.

LEGAL AND MORAL PROBLEMS AND RESPONSIBILITIES

The locksmith is in a unique position in the community insofar as his ethics are concerned. The trust which the community places in him prohibits his being completely frank and open in certain transactions. That is, he cannot share specific knowledge of his trade with the general public and be considered ethical, or even self-respecting; divulging information can result in the loss of property by burglary. If necessary, this should be pointed out to customers. They usually respect a locksmith for this type of reticence, once they understand the reason behind it.

Another peculiar aspect of locksmithing is that it is easy to become involved in an illegal operation. It behooves the locksmith to know and abide by all of the laws which govern his trade and to be aware of those situations which might lead to difficulties. A frank talk with the local district attorney would be helpful and should include such matters as what the locksmith must do to operate safely within the law and what precautions should be taken to avoid becoming an unwitting accessory to the criminal acts of others. These requirements vary considerably among states, counties and cities.

Answering public inquiries is a sometimes difficult and at other times satisfying experience. If the questions can be answered in a general way, without divulging specific information which may pose hazards to the public, then the locksmith not only has the privilege but the obligation to answer in such a way that the public will be aware of just how valuable a good locksmith is to his community. This can often be done by citing some phase of the history of locks and their use, or by relating an anecdote which is couched in terms sufficiently general that no one can recognize who the people involved may be. Such anecdotes should never expose a member of the community to shame or ridicule. Remember, the teller of tales is subject to slander suits, and many people will realize that they, too, could become the butt of a joke if they patronize that particular locksmith in a like situation.

In situations which tend to become awkward because a customer asks for specific information, often the best tactic for the locksmith to employ is to divert the customer's interest to other channels. For instance, one could start by saying: "You know, that reminds me of a story I heard the other day . . . ."

A locksmith may enhance his position through his cooperation with civic authorities. This means observing zoning requirements, helping in community projects, and acquiring membership in the chamber of commerce; or, it may take more concrete forms such as working with the police department on burglary investigations. In fact, it would be helpful to the locksmith if an arrangement could be made with the police whereby they would notify him when there is a burglary in his area, and allow him to help determine what means of entry was used. This would, of course, enable him to acquire knowledge which would be useful in preventing the same type of entry from happening again, or happening to those who depend upon the locksmith for protection against burglary.

A locksmith's relationship with churches can contribute to or detract from his position in the community. It is suggested that, when called upon for his services to a church, the locksmith should not charge for his work, whether the church is of his own faith or not. If it is possible to do this without financial strain, the donation should include both labor and merchandise, within reason. In those instances where the locksmith feels that he cannot afford to donate both materials and labor, substantial reduction from his usual charges should be made.

If a new church is being built in the community, it is an excellent idea for the locksmith to donate the door locks and latches. A discreet way usually can be found to let the public know that the donation was made by the locksmith. It is usually not necessary to have this announced with fanfare; word gets around pretty well in a community. Such an act enhances the locksmith's standing in the community, unless he is too eager to point out his generosity.

The locksmith's private life, while technically his own, is a matter of public concern. How he conducts his private affairs, his relationships with other people, and his business practices, must be beyond reproach. Bad personal habits may quickly destroy the standing of a locksmith.

The locksmith should keep records, not only those required for purposes of income taxes, business taxes, etc., but also, records of his operations about which there might be a question of legality, in particular, those situations which call for opening houses or automobiles. These records should be kept systematically on file for a given period of time before they are destroyed. So, the locksmith should have a desk and filling cabinet in the shop for maintaining these financial and other records. Record-keeping involves a great many things; among them is the computation of the fees which the locksmith will have to charge for his work in order to show a reasonable profit.

The factors which go into these calculations include such things as rent; utilities; the ex-

penditure on shop supplies such as cleaning fluids, lubricants, etc.; the cost of customer services and conveniences and their maintenance, the costs of give-away or premium items, window displays and other advertising, and other items incidental to the maintenance of the establishment.

All of these items should be figured on an annual basis, and apportioned per hour during the time that the establishment is open; do not include days closed, such as Sundays and holidays. This apportionment is the *break-even figure,* an hourly average of expenses. Then the locksmith decides how much per hour he feels he has to make to live on, and this is added to the break-even figure to find what he must average per hour to make his business worthwhile. He cannot hope to keep busy at that price every hour he is open, so an adjustment upward must be made to compensate, and a further upward adjustment must be made for that time which he spends in keeping up the appearance of his shop.

Now, if the locksmith merely invested money in a savings account, instead of in his business, it would draw interest; so, in order to be realistic, the loss of the interest which his capital investment would otherwise draw must be figured into the overhead costs. This is the only realistic approach to computing what he must charge his customers for the time he spends on their work.

Again the author urges the locksmith not to allow the customer watch what he does. This is another form of sharing confidential knowledge with the public and is actually against the public interest as well as contrary to the locksmith's own interests. It is simply another rule of ethics for the locksmith; and ethics also dictate that excluding the public from watching the locksmith at work must be done as politely and as unobtrusively as possible.

It is the sincere hope of the author that, by writing this book, he has contributed to the profession of locksmithing, and in so doing, has increased and improved protection of the property and lives of the people of the United States.

Recommended Reading for Locksmiths

The National Locksmith (NLA)
The Locksmith Ledger (ALOA)
Reed General Code Set (E. D. Reed & Son)
Security World (magazine)
Catalogs of lock manufacturers and locksmith supply houses
The Corrosion Handbook (Ulig)
A good high school physics text

Recommended Courses for Locksmiths

Elementary bookkeeping
Elementary accounting
Electronics (including solid state)
Basic machine shop
Inorganic chemistry
Organic chemistry
Engineering drawing
Psychology
English
Business law
Criminology
Speech
Philosophy
Government

Glossary

ACTION. Pertaining to the construction and resulting operating characteristics of a lock or latch. *See:* Plain action; Easy action.

ACTIVE END. Pertaining to that end of a panic bar which operates the lock.

ALLOY. A mixture of metals; or iron to which a small amount of carbon has been added (carbon steel).

ANNEAL. A form of heat treatment to remove hardness from metal alloys (so that cutting or cold forming operations may be performed). Often the cooling process in annealing takes hours, days, or occasionally even weeks.

ANODIZED. A corrosion-resistant finish produced on metal (usually aluminum or one of its alloys) by electrolytic means.

ANTI-THRUST SLIDE. A British term for a deadlocking plunger.

ARCHITECTURAL COUNSELING. Advising an architect on the proper selection of locks and latches, or specifications for them, in the design stage of a building.

ARMORED FRONT. A two-piece front for a mortise lock which provides protection for the cylinder set screw or cylinder clamp by covering it with the armor front or scalp.

AT REST POSITION. 1. The normal position for a unit, module, or interacting group of parts. 2. The position in which all spring-operated parts rest with the spring tension at a minimum. 3. The position of least stress.

AUTOMOTIVE LOCKS. The types of locks used on automobiles.

AUXILIARY LATCH BOLT. A secondary latchbolt used in deadlatch mechanisms to control the operations by which the main latchbolt is deadlocked.

AUXILIARY LATCH BOLT ASSEMBLY. The latchbolt and those related parts which are more or less permanently assembled into a single unit of which the auxiliary latch bolt is the only portion visible during normal operation of the lock.

AUXILIARY LATCH BOLT MODULE. The equivalent of an auxiliary latch bolt assembly, but with parts which remain individual rather than part of an assembly.

AUXILIARY LATCH LEVER. Part of an auxiliary latch bolt module which, though pivoted, fulfills the function of an auxiliary latch bolt shaft.

AUXILIARY LATCH LEVER SPRING. The spring which causes the auxiliary latch lever to accomplish its automatic return function.

AUXILIARY LATCH MECHANISM. A group of parts that operate together to perform a deadlocking function on a latchbolt.

BACKSET. The distance from the high edge of the door to the center of the knob and/or lock cylinder.

BACKSET EXTENSION LINK. A device available for some cylindrical locks to allow the knob to be set further from the edge of the door than is standard for that particular lock.

BARREL. The tube of a lever lock. Sometimes the term "barrel" is erroneously used for "cylinder barrel" in referring to the rotating plug of a disc or pin tumbler lock cylinder.

BARREL AND CURTAIN. A warding arrangement in the barrel or tube of a lever lock which closes off part of the keyway as the key is turned. This type of warding is very rarely used on pin or disc tumbler locks.

BARREL BOLT. A surface-mounted, small, round bar, equipped with a hand knob, used in a sliding action to bar or bolt a moving part to an immovable object.

BARREL KEY. A pipe key (Br.). A bit key having a drilled (hollow) stem.

BARRON, ROBERT. The Britisher generally credited with being the inventor of the lever lock.

BASE FRONT. The undersection of an armored front; an inside front.

BASTARD BLANK. A key blank not of the same manufacture as the lock it is intended to fit. The term does not necessarily denote inferior quality.

BEVEL. The amount by which a part's surfaces deviate from the norm of being at right angles. (For example, the bevel of a lock front with respect to its case.) Colloquially, the term is often used merely to express the existence of such a deviation from the right angle configuration.

BI-CENTRIC CYLINDER. A pin tumbler lock cylinder having two separate plugs, one operated by the change key, the other by the master key.

BIT KEY. A key having a bit extending at right angles from a solid, round shank. Keys for mortise locks of the warded or lever types are usually bit keys.

BOLT. That part of a lock or latch which does the fastening. *See also:* Barrel bolt; Claw bolt; Deadbolt; Hook bolt; Rotary bolt; Spring bolt (or springbolt).

BOLTWORK. That portion of a lock, latch, or locking device which secures, but which is not actually part of a combination-carrying mechanism such as a lock cylinder (although commonly associated with it). Boltwork likewise does not include trim, escutcheons, accessories, or ornamentation.

BOX-OF-WARDS. A box containing the wards for a warded lock, intended to provide a realistic means of combination changing. Warding provided in a box-of-wards is often quite elaborate.

BRAMAH, JOSEPH. The British inventor (in 1784) of a lock which bore his name and utilized notched tumblers.

BRIDGE WARD. An internal ward which is not attached directly to the lock case or cap, but indirectly through a bridge. A bridge ward meets the key in the center of the bit rather than at the ends (as would a case ward), or at the sides as would keyway wards.

BRITISH STANDARDS INSTITUTION, 2 Park Street, London, W. 1. Establishes standards and makes performance tests for locks, other durable goods, and chemicals for Britain.

BROACH. A tool which is pulled or pushed (usually through a hole or notch) to form a particular size or shape.

BUILDER'S HARDWARE. A general term referring to locks, hinges, door closers, other window and door hardware; and other non-plumbing and non-electrical hardware for houses and other buildings or structures.

BURNISH. To polish by drawing a smooth, hard object across the surface, while applying sufficient pressure to obtain the required finish.

BURGLAR ALARM. A device producing an audible or visible signal, locally or remotely, as a warning that an intruder has entered a protected area. Burglar alarms are of many types and are based upon many different principles. The selection of the proper type in accordance with its objectives is especially important.

BURGLARY PREVENTION. A system used to protect premises against burglary. It includes the use of adequate locks, protection for windows, burglar alarms and stratagems intended to increase both the time a burglar is exposed and his risk during that time.

BURR. A sharp piece or ridge of metal adjoining an edge which has been cut or machined. A fray (Br.).

BUSHING. An inserted shim or bearing, usually, but not always, for a round shaft. A method of compensating for worn parts.

BUTT. A common hinge, both sides of which are actually or potentially set into mortises.

CABINET KEY. A key, intended for warded or lever locks, and having a stem which is drilled to fit over a pin in the lock. Sometimes called a pin key on this account. A barrel key. A pipe key (Br.).

CABINET LOCK.   A general term which includes a wide variety of locks intended for or used on boxes, cabinets, cupboards, desks, drawers, or utility doors. This term is often used to mean a lock requiring a cabinet key.

CAM.   1. A throw piece or tongue, as in a cylinder cam.   2. An eccentric cam, as are some dead-bolt cams. A knob cam or hub (*see* Hub).

CAP.   The cover of a lock case.

CAPPED CYLINDER.   1. A pin or disc tumbler cylinder which has an ornamental metal covering crimped over the front of the cylinder. In ordinary usage, all but the plug is covered; but in automotive usage, the plug, too, may be covered — with supplemental sliding covers provided for the keyhole.   2. A pin tumbler cylinder with the tumbler cap(s) in place, retaining the tumbler springs.

CASE.   A housing which serves to contain the parts of a lock and keep them in their correct relative positions.

CASE HARDENED.   A tempering of the surface of a metal to an extreme degree of hardness, while leaving the center much less hard to avoid embrittlement of the part. Chemicals are used to induce the surface to harden in this manner.

CASEMENT.   Pertaining to a window having one or more hinged sash.

CASE WARD.   A ward attached to a lock case or cap with the attachment point within the rotational path of the key. Wards having the attachment point *outside* the rotational path of the key are called bridge wards.

CASH BOX LOCK.   A cabinet lock having a strike which partially enters the lock case and which is secured thereto or released by the action of the key.

CATCH.   Fastening devices of various kinds, usually intended to hold a door shut when closed, and commonly found on cupboard doors or gates. Common types of such devices are: ball catch, cupboard catch, elbow catch, friction catch, gate catch, magnetic catch, and transom catch.

CHAIN DOOR FASTENER.   A chain intended to be mounted on a door frame. A slide-in receptacle on its free end is mounted on the door. When the chain is in place, it is possible to open the door slightly to establish the identity of a caller before unlocking. Such a chain is not an anti-burglary device.

CHAMFER.   The bevel or radius which is put on a corner, rather than leaving it square and sharp.

CHANGE KEY.   1. An individual key for a single lock or keyed-alike group.   2. A device used for changing the combination of a key-change combination lock without having to disassemble the lock. (Not a cylinder, flat, or bit key.)   3. A special key used for changing the combination of certain types of change key (lever tumbler) locks.

CHANGE KEY LOCK.   A type of lever lock which changes its combination to correspond with that of any key of appropriate type and dimension which may be used to lock it. Only that key will then open the lock. Not a common type of lock, and very rare in the United States. The term "change key lock" also applies to certain lever locks whose combination may be changed by the use of a special "change key" to alter its keying requirements.

CLAW BOLT.   A lock bolt which is itself stationary, but which has one or more (usually two) key-operated auxiliary bolts protruding from and retracting into the edges of the stationary bolt. It is these auxiliary "claw bolts" which perform the actual locking action.

CLOSET SPINDLE.   A knob spindle for use where space is limited. A button or thumbturn on one end of the spindle substitutes for one of the knobs.

CODE.   *n.* The serial number or symbol which refers to the specific pattern to which a particular key is cut. A short term for key code or code number. *v.tr.* To key or encode a lock in conformity with code series specifications.

CODE KEY.   A key made or cut from a "code number" which stands for a specific pattern to which a given key must conform.

CODE MACHINE.   A machine expressly intended for the production of code keys.

CODING.   1. A code number.   2. A code series.   3. The process of keying a lock to correspond with a given code number; encoding.

CODING SYSTEM.   1. The order in which a code series progresses.   2. A code series having special features or adaptabilities.

COMPLEMENTARY KEYWAY.   A keyway which will accept two different key blanks, one of which is used for masterkeying a lock.

CONSTRUCTION KEY.   A key which is operable only during the construction period of a building. When the construction period is over, a simple operation is performed which invalidates the construction key. The invalidation procedure and the mechanics of construction keying vary from one manufacture to another.

CONTINENTAL ACTION. An easy action pattern employing a follower.

CORROSION. A process of chemical attrition, including but not limited to rust.

CREMONE BOLTS. A pair of bolts controlled by a central handle or knob and used to secure the top and bottom of a door.

CUP HANDLE. A handle, usually in the form of a half ring, which is free to drop into a recessed cup in a door. Such a handle is used to replace a knob when the protrusion of a knob is excessive.

CUTTER. The file or milling type of cutting tool used on key machines.

CYLINDER. That portion of a lock which contains a disc or pin tumbler mechanism and through which the key operates.

CYLINDER BARREL. A rotating lock plug.

CYLINDER CLAMP. A device by which lock cylinders are retained or secured in a lock or door.

CYLINDER CONNECTING BAR. A bar, usually of sheet metal, used in a wobbling attachment with the rear of the lock plug of a rim lock cylinder to transmit the motion of the key to the lock.

CYLINDER HULL. The body of a lock cylinder which contains the plug of a pin or disc tumbler lock.

CYLINDER KEY. A term broadly used to refer to a pin or disc tumbler key.

CYLINDER SET SCREW. A set screw used to retain a lock cylinder in its proper position in a lock (usually a mortise lock).

CYLINDER RING. An ornamental spacing washer, sometimes adjustable in thickness, which is used to adjust the effective length of a lock cylinder to match the requirements of lock and door.

CYLINDRICAL LOCKS. Locks which are customarily installed in round or cylindrical holes in a door. *Synonyms:* tubular locks, key-in-the-knob locks, knobset (Br.).

DEADBOLT. A lock bolt which is secure against end pressure on the bolt.

DEADBOLT BOLT. That portion of a lock bolt which is visible when in the extended position.

DEADBOLT DETENT LEVER. A lever which is commonly used to secure a deadbolt against end pressure.

DEADBOLT TANG. The internal extension of a deadbolt.

DEADLATCH. A lock function in which the outside knob is locked and the latchbolt is secured against end pressure.

DEADLOCKING LEVER. A lever by means of which some part of a lock is secured against end pressure. Usually used to secure latchbolts or deadbolts.

DEPTH INTERVAL. The interval between key cuts on a key intended for a particular manufacture and type of lock.

DEPTH OF CUT (of a key) 1. A misnomer, since the term refers not to the amount of material cut away, but to the amount remaining. 2. Progressive numerical symbols used to refer to the depth of the cut (or notch) in a key in reference to an original optimum full width of the key bit.

DERIVED SERIES. A code series which bears a mathematical relationship to a parent or *master* series from which it is derived.

DESIGN MODULE. A cluster of parts operating in unison to provide a single function or operational aspect of a lock.

DETECTOR LOCK. A lever tumbler lock which is so designed that the overlifting of any tumbler will cause a special detector tumbler to become engaged, thus foiling an attempt at picking. Release of the detector tumbler is accomplished by turning the proper key sharply as if to lock the lock. The rotation of the key is then reversed to operate the lock in a normal manner.

DIE CASTINGS. A loose term commonly used to describe "pot metal" castings and low grade pressure castings, usually having a certain amount of aluminum or zinc content. Although "die cast" parts may be accurate in size, they are frequently of high porosity, difficult to weld, and comparatively low in structural strength. Not to be confused with pressure castings, though often confused with that term.

DIFFERS. Two or more keys, each of which is different from and noninterchangeable with all other keys in the same set or series.

DIFFER BIT. A pin tumbler used directly above the key pin to allow two different keys or differs to operate the same lock in masterkeying. *Synonyms:* master pins, master key chips, split pins, upper pins.

DISC TUMBLER. A lock tumbler stamped from

sheet metal and designed to be entirely housed within a rotating lock plug. *Synonym:* wafer tumbler.

DISPLAY KEY. A special key used in hotel key systems to secure a lock against all other keys except the emergency shut-out key or the grand master key.

DOGGING STOP. A stop or plunger used to secure a locking knob.

DOOR CHECK. A door closer.

DOOR CLOSER. A spring operated device for closing doors. Both opening and closing speeds are regulated by means of a hydraulic or pneumatic check.

DOOR HOLDER. A door stop with a hook or other device to hold the door in the open position. Some door holders are electro-magnetic, hence are remotely as well as automatically controllable.

DOUBLE BITTED LOCKS. Those locks using a tumbler system which requires the key to be cut on both sides.

DOUBLE ACTING LEVER TUMBLER. A lever tumbler which must be neither over-lifted nor under-lifted for the lock to operate.

DOUBLE GATING. A method of masterkeying lever tumblers by providing two positions in which the fence may enter the tumbler.

DRAW SWAGING. A process of striking or squeezing a metal part in such a way as to cause it to become elongated.

DRAWER LOCK. A lock of the type used for cabinet or desk drawers; however, it may be used for other purposes.

DUMMY CYLINDER. A plug or stopper, similar in appearance to a lock cylinder, used to fill a hole where a lock cylinder has been removed.

DUMMY LEVER. A British term for a master tumbler (lever).

DUSTPROOF STRIKE. A strike having a spring-operated cover for the hole which a bolt is to enter. Commonly used with head and foot bolts and some types of vertical rod exit devices.

EASY ACTION. A lock having the latchbolt and latch lever or follower separately sprung, allowing the latchbolt to be depressed easily, while providing firm and positive return for the knob.

ELECTRIC SOLENOID. An electrical device used to produce a straight-line mechanical motion. A plunger operated by electro-magnetism.

ELECTRO-PLATED (or plated). Pertaining to an ornamental or protective coating of metal which has been deposited by electrolysis.

EMERGENCY-SHUT-OUT KEY. A key in a hotel key system which operates all guest housing, even though locked from the inside. This key is also used to lock out guests who are no longer welcome. A door locked with it will not respond to the guest's key until the emergency-shut-out key has been used again, this time to release the lock from its lock-out condition.

ENGINEERING STANDARDS. 1. Design qualities and qualifications having as their objective achievement of an acceptable quality in the durability and performance of the finished product. 2. Those modules (clusters of parts) which are standard (with minor variations) in achieving certain objectives in the design of a lock.

ENTRANCE LOCK. 1. A lock having thumbpiece operation of the latchbolt from both sides. 2. Any lock used in an entry door.

ENVIRONMENTAL SERVICING. The care and lubrication of a lock in a manner consistent with the climate and other environmental conditions in which it is to function.

ESCUTCHEON. The ornamental portion of a lock or latch which is affixed to the surface of a door; includes knob bearings, and may include a thumbturn or a hole through which a lock cylinder may pass.

EXTRUSION. Stock material which has been formed to a particular shape in a continuous strip of considerable length. Often, the manufacture of parts from such an extrusion consists of merely cutting off the proper length from it. The word is often used to refer to a finished part which has been manufactured by such a process.

FARM KEY SYSTEM. A key system oriented toward convenience of operation coupled with reliability of locks under farm conditions.

FILING CABINET LOCK. A plunger type of lock principally used for filing cabinets.

FINISH. The final processing or coating to which the exposed or visible portions of a lock are subjected, usually with the dual objectives of enhancing their appearance and of increasing the amount of protection against corrosion.

FIRE EXIT BOLT. Another name for a panic exit device.

FIXING LOCKS. The repair of locks (U.S.). The installation or fitting of locks (Br.).

FLAT STEEL KEY. A key made from flat steel; commonly, this is a key for a warded or lever lock.

FLASH. The ridge sometimes found where two dies join in the production of a part.

FLOATING STOP. A stop or plunger whose purpose is to provide a disengaging action for the digging stop.

FOLLOWER. A part combining the functions of latch lever and hub. A lock using this part is usually provided with lever handles, since the knob spindle is free to rotate in only one direction.

FORGING. v.tr. The process of forming metal to shape while it is sufficiently hot to be relatively ductile. Such shaping may be the result of dies coming together with sufficient pressure to squeeze the metal to fill them, or as the result of a controlled hammering process. n. The completed part resulting from such a process.

FROG ACTION. An action whereby the hub acts on a specially shaped latch tail or latch tail extension to draw the latchbolt. Certain types of cylindrical locks have an "easy frog action," making use of an intermediate, separately sprung, sliding part.

FRONT. That portion of a mortise, cylindrical, or tubular lock which shows at the surface of the door edge. See: Armored front, Plain front.

FUNCTION. Pertaining to the mode of operation of a lock or latch. See: Deadbolt, Deadlatch, Jimmy-proof rim lock, Panic exit device, Passage latch, Springlatch.

GALVANIZED. Zinc coated, whether by the electrolytic process, or by the "hot dip galvanize" process whereby the part to be coated is dipped into the molten zinc.

GARAGE LOCK. A rim springbolt, usually without provision for holdback.

GATING. Pertaining to the location, type, size, or configuration of the notch or gate in a lever tumbler.

GRAND MASTER KEY. A key which will operate all of the locks operable by a related group of master keys, but which prohibits inter-operation of the locks by either individual keys or more than one master key of the related group. A grand master key usually operates locks of two or more different keyways.

GREAT GRAND MASTER KEY. A key which will operate all of the locks which are also operated by two or more different grand master keys — a

level of operation nearly always achieved through the use of compatible keyways.

GRUB SCREW. An archaic term for set screw, sometimes used in reference to set screws with unusual thread configurations.

GUARD KEY. The bank's key by means of which dual custody of safe deposit boxes is maintained. Opening of such a lock requires the use of the guard key before the change key is operable.

GUARDED PUSH BUTTONS. A stopworks incorporating an auxiliary mechanism to prevent the push buttons from being moved to the unlocked position when the lock is in the locked condition and the door is closed. Synonym: guarded stopworks.

GUEST'S KEY. Operates the lock(s) for one room or unit in a hotel key system.

GUIDE KEYS. A set of keys, each of which has every cut made to the same depth of cut on any one key; but each key is different (exceptions to this arrangement do exist for double bitted locks). Guide keys are used to produce code keys through a process of duplication, one tumbler position at a time.

HALF SURFACE BUTT. A hinge intended to be mounted with one side set into a mortise and the other side surface-mounted.

HAND (of a lock). Pertaining to the direction of swing of the door on which a lock may be used. Some locks are not reversible and must be ordered according to the hand of the door. Locks (and doors) may be: left hand (LH); right hand (RH); left hand reverse bevel (LHRB); or right hand reverse bevel (RHRB).

The author's rule for determining the hand of a door is:

1. Stand on the inside, facing the door.
2. The hand with which you reach for the door knob establishes the hand of the door.
3. If the door opens away from you, add the suffix "reverse bevel."

HARDENED. Tempered.

HARD PLATE. A plate of especially hard, tool-resistant steel used to protect locks from drilling.

HASP. A hinge-like device by means of which a padlock may be used to secure two objects together. See also: Safety hasp.

HEAT TREATED. An alloy is heated to a specified temperature, then cooled under controlled conditions to obtain the desired hardness or resilience in the finished metal or part.

HINGE. 1. A simple mechanical device to permit a door, window, gate, shutter, or the like to operate in a pivoting action through a segment of arc. 2. A pivot.

HINGED. 1. An object or part which moves about a pivot in a segment of arc. 2. Having hinge(s) installed or integral.

HOOK BOLT. A lock bolt which travels in an arc to hook over its strike.

HOTEL LOCKS. Locks specifically designed for use in hostelries and having special operating features, either to ensure the privacy of the guest, to protect commercial displays, or to exclude an unwelcome guest. The mechanical methods by which this is accomplished vary widely.

HOUSEKEEPER'S KEY. Operates all locks which are also operated by several different maid's keys in a hotel key system.

HUB. That part in a lock which is used to actuate the latch lever, shoe, or yoke in the process of drawing a latchbolt by turning a knob. Also called a cam, a knob cam, or a follower (Br.).

IMPRESSION SYSTEMS. Two systems, either of which may be used to produce a marking on a key which indicates the position at which a key must be cut in order to fit. The two impression systems are known as the "pull" and the "wiggle" systems.

INACTIVE END. That end of a panic bar which is farthest from the lock.

INBOARD. Pertaining to the end of a panic bar which operates the lock; in the area of the active end mechanism of a panic bar.

INDICATOR. A special feature in some locks indicating whether the room is occupied. In some hotel locks, the indicator also acts to exclude keys which would otherwise operate the lock.

INSTALL. The affixing or attaching of a lock or lock accessories, including all preparation (cutting, fitting, aligning, etc.) of that which is to receive the lock.

JIMMY. *v.tr.* To gain entry by prying; any forcible entry.

JIMMY-PROOF RIM LOCK. A rim lock having a vertically moving bolt which locks through appropriate holes in the strike, and having

other burglar-resistant features not commonly found in most rim locks.

KEY. The instrument by which a lock of other than the combination or puzzle type is properly operated.

KEY BLANK. A key cut to the proper general configuration and size with keyway wards precut where applicable, but with either internal warding cuts or tumbler cuts remaining to be made before the key is operable.

KEY CHANGES. Pertaining to the number of differs available or the order of appearance of those differs in a code series.

KEY CODES. A systematized arrangement of the differs of which a given lock is capable without unintentional interchange of key operation between any two locks in the code series.

KEY DUPLICATING. A process whereby the ward or tumbler cuts of a key are cut on an appropriate key blank so that the duplicate key will operate the same lock in the same way as the supplied pattern from which it was made.

KEY DUPLICATOR. 1. The machine used to duplicate keys. 2. A person who makes duplicate keys, but does no other key or lock work.

KEY MAKER. A person who makes duplicate keys or keys from code, but does no other key or lock work.

KEY PIN. That pin tumbler which is in contact with the key as the lock is operated.

KICK PLATE. A metal or plastic plate applied across the bottom of a door to prevent scarring.

KEYWAY WARDS. Wards positioned in the keyhole (or keyway) to prevent a wrong key or key blank from entering the lock.

KEYING LOCKS ALIKE. Locks assembled with like keyways, tumblers, and keying dimensions are said to be keyed alike. A single key will operate all such locks. Keyed alike locks are often mistakenly called masterkeyed.

KNOB. 1. A door knob. 2. A rotating boss.

KNOB CAM. A hub. One of the intermediate parts commonly used in transferring the action of the knob to the latchbolt.

KNOB SPINDLE. A shaft, usually square, used to transfer the motion of the knob to the hub.

LANKET HOLE. A fully enclosed slot in a part which is intended to fit over a pin, post, or stump with the purpose of guiding the part and/or limiting its travel.

LATCH. A lock-type device having a latchbolt operation only. Such latchbolt operation is usually by knob, or, in the case of locking types of latches, by key.

LATCHBOLT. A beveled, spring-loaded bolt used to keep a door closed.

LATCH BOLT ASSEMBLY. The permanently assembled module (or part) of which the latchbolt is the visible portion when in service in a lock.

LATCH BOLT SHAFT. A part connecting the latchbolt with the latch tail, and used as a guide for the travel of the latch bolt assembly of which it is a part.

LATCH BOLT SPRING. The spring which actuates the latch bolt assembly.

LATCH LEVER. A lever actuated by the knob cam (hub) which, in turn, operates on the latch tail to draw the latchbolt.

LATCH LEVER, COMPOUND. A two-piece latch lever, the pieces of which are coordinated to perform the function of a latch lever.

LATCH LEVER, ROCKING. A one-piece latch lever operating in a rocking-chair-like motion.

LATCH LEVER LIFT BLOCK. A sliding part which is moved in a lifting motion by a thumblever to operate the latch lever.

LATCH LEVER LIFT LEVER. A pivoted part which serves the same purpose as a latch lever lift block.

LATCH LEVER SPRING. The spring which returns the latch lever to its "at rest" position.

LATCH SHOE. A sliding part sometimes used instead of a latch lever.

LATCH TAIL. The part opposite the latchbolt on the latch shaft, and which is operated upon by the latch lever to draw the latchbolt.

LEVER. A spring-loaded, usually pivoted, double-acting detent, operated singly or in groups by a key to provide the security function for a particular type of lock. A lever tumbler.

LEVER HANDLE. A handle whose general configuration is that of a bar bent at a right angle and used as a substitute for a door knob. The use of lever handles can throw great stress on the internal mechanism of a lock.

LEVER LOCK. A lock using lever tumblers.

LEVER TUMBLER. *See:* Lever.

LOCK. A device designed to obstruct or impede ingress or operation by unauthorized persons, and which is capable of resisting an amount of force or manipulation consistent with the requirements of its purpose.

LOCK CODING. The process of arranging or fitting the tumblers of a lock (or locks) to conform with the requirements of a given code series.

LOCKER LOCK. A particular configuration of a utility lock, usually a lever lock, which is most frequently used on clothing or storage lockers.

LOCKSMITH. One who makes keys and deals with the mechanical aspects of locks and other protective devices.

LUGGAGE LOCK. A class of locks used principally on luggage and often resembling a hasp which has a lock permanently built in. Luggage locks are often of warded or lever construction, but the use of pin tumbler locks for this purpose is widespread and growing. Disc tumbler locks are also used to some extent for luggage locks.

MAID'S KEY. Operates (usually) all guest housing on one floor or corridor in a hotel key system.

MAISON KEYED LOCK. A lock keyed to accept several different individual keys is said to be maison keyed. The process by which this is accomplished is called retrograde masterkeying.

MASTER. A short term for master key or masterkeying.

MASTER KEY (or masterkey). *n.* A key which will operate various locks, each of which is also operable by a separate and different key. *v.tr.* To prepare the internal mechanism (tumblers) of locks in such a way as to be operable by such a master key as well as by an individual key.

MASTER TUMBLER. A lever tumbler whose purpose is to operate several other tumblers simultaneously, thereby providing a master key operation. A dummy tumbler (Br.).

MILD STEEL. A steel of comparatively low hardness with excellent machining and working characteristics. Mild steel is sometimes case-hardened after working to size and shape.

MIXED PROGRESSIVE. A system of key code progression.

MODIFIED PROGRESSIVE. A system of key code progression.

MONO-LOCK. A lock designed to set into a notch cut into the edge of a door. Although the key is in the knob of such a lock, it is not a cylindrical lock set, but has features of both mortise and cylindrical locks.

MORTISE. *n.* A hole, usually rectangular, into which a lock or part may be set. *v.tr.* To make or produce such a hole. *adj.* Pertaining to such a hole or the lock or part which is designed to be fitted into such a hole.

MORTISE CYLINDER. A lock cylinder with a threaded body which is intended to screw into a suitably threaded lock case, usually of a mortise lock.

MORTISE LOCK. A lock, the case of which is enclosed by a mortise cut into the edge of a door.

MUSHROOM TUMBLER. A pin tumbler which is so called because of the resemblance of its longitudinal cross section to a mushroom; this type of tumbler is used to enhance the resistance of a pin tumbler lock to picking.

NIB WARD. An internal ward (case ward) which is formed by pressing a tab of the case metal inward.

NIGHT LATCH. 1. Colloquial term for a rim springlatch. 2. A British term for a push button controlled lock.

NIGHTWORKS. Stopworks (Br.).

NORMALIZED. A term for metal which has been stress-relieved after working or cutting operations by subjecting it to a process of annealing, tempering, or re-tempering, until the part so treated meets a standard specification over all of its area.

NOSE. 1. The keyway-containing barrel of a lever or warded lock. A nozzle (Br.). 2. The small projection at the extreme tip of most flat steel keys.

ONE WAY ACTION. A lock whose knob or handle will turn in one direction only.

ORIGINAL KEY. 1. A key supplied by the lock's manufacturer. 2. A key made from the lock according to the dimensional requirements of its tumblers or wards.

OUTBOARD. Pertaining to the end of a panic bar farthest from the lock. In the area of the inactive end mechanism of a panic bar.

PADLOCK. A lock which closes and locks upon itself in order to provide a locked enclosure for the securing of other objects, and which may be opened or removed by the operation of a key or combination.

PANIC EXIT DEVICE. A lock or latch which is always operable by pressure on a bar which extends across the inside of a door. The door always opens outward, away from the bar. A fire exit bolt.

PARACENTRIC. 1. Pertaining to a keyway containing wards which cause the body of the key passing through them to shift from one side of the keyway to the other, with the objective of preventing a flat tool from being inserted to pick the lock. 2. Pertaining to a key made to pass such a keyway.

PASSAGE LATCH. A latch with operation from both sides at all times, and having no locking function.

PEIN (or peen). The process of hammering, often with the ball end of a hammer, either to effect a hammered finish or to move a small amount of metal, for example, tapping around a hole or on a pin to make it fit tighter in the hole.

PICK. *v.tr.* To open or unlock a key-operated lock by the use of various tools to simulate roughly the action of a key. *n.* A tool used in the process of opening a lock in such a fashion.

PIN. *n.* 1. A pin tumbler. 2. A straight, roll, sliced, or taper pin. 3. A pivot, usually applied by riveting. *v.tr.* 1. The installing of tumblers in a pin tumbler lock. 2. To attach two pieces or parts together to form a whole through the use of a straight, roll, sliced, or taper pin.

PIN TUMBLER. A tumbler having roughly the form of a straight (dowel) pin. Two or more such pin tumblers are required in each tumbler position.

PIPE KEY. Cabinet key (Br.).

PLAIN ACTION. A lock which utilizes a single spring to extend the latch bolt and return the knob. Such a spring usually operates on the latch bolt, but occasionally operates on the latch lever or follower.

PLAIN FRONT. A one-piece front.

PLUG. The rotating, keyway-containing portion of a tumbler type lock or lock cylinder. A cylinder barrel.

PLUG FOLLOWER. A rod or tube of about the same diameter as a pin tumbler cylinder plug which is used to push the plug out of the cylinder hull, leaving the drivers, or topmost tumblers, in the cylinder hull. The key must be in place or the lock picked before the plug follower may be used.

POST. A pin, stump, or peg, whether round, square, or oblong, which is often constructed of the same material and at the same time as the parent body of which it is a part. A post is ordinarily used as a support member or pivot.

PRESSURE CASTINGS. Castings made in dies or molds by the application of extremely high pressure to a powdered metal matrix. Modern technology has made pressure casting of nearly all metals both practical and economical. Such castings retain most of the advantages of more common forms of the parent metal and provide a distinct advance in the formation of previously impossibly complex shapes with an amazing degree of accuracy.

PROGRESSIVE. A system of key code progression. An orderly sequencing of key codes.

PRISON LOCK. Usually, a lever lock of extremely rugged construction having heavily sprung tumblers and requiring a key which is larger than the usual lever lock key. Keyways for prison locks frequently are paracentric, a rarity in lever locks.

PULL HANDLE. A door handle having no thumbpiece or other provision for drawing a latch.

PUSH PLATE. A plate applied to the inside of outward opening doors, or to the outside of inward opening doors, to protect them from finger marks or wear.

PUZZLE LOCK. A lock which depends solely upon some secret aspect of its construction for the protection it furnishes.

RABBETED FRONT. A front with a vertical offset.

RADIATION. Subatomic particles of various charges, masses, and energy levels.

RAILROAD PADLOCK. A lever tumbler padlock taking a cabinet key and having a spring-loaded cover over the keyhole.

REFINISH. v.tr. To restore an object to its newly made appearance.

RE-KEY. v.tr. To change the combination of a tumbler-type lock.

REMOVABLE CORE. An inserted portion of a cylinder or padlock which contains the keying and which may be removed as desired by the action of a special key, usually referred to as a "control key."

RETROGRADE. The reverse of the masterkeying process, whereby one lock acts as a master receptacle for several different individual but carefully selected keys.

REVERSIBLE KEYS. Keys which are cut on both sides in such a way that either side may be inserted in contact with tumblers, which oper-

ate against only one side of the key at a time.

REVERSIBLE LOCK. A lock having a front bevel which is capable of being adjusted, even to the reverse position, and having internal parts so constructed that they may be turned over without impairing their function. Rabbeted locks are seldom reversible.

RIM CYLINDER. The detachable cylinder for a rim lock.

RIM LOCK. A lock which is mounted on the inside surface of a door, customarily along the periphery, or rim, of the inside surface.

RING HANDLE. A cup handle.

RIVETING. A fastening technique involving the use of a pinlike device which is struck or squeezed to cause it to expand in such a way that it is no longer removable. A rivet, ready for use, is supplied with one end already enlarged (or headed) in any of a number of standard shapes.

ROSE. Part of some lock trims, it contains the knob bearing, and is usually of a round or ornamental shape. Synonym: rosette.

ROTARY BOLT. A bolt having a T-shaped or knobby eccentric end; it locks by rotating the bolt after the bolt end has passed through the complementary shaped hole in its strike.

SAFE LOCK. A combination lock not of a padlock or portable type.

SAFETY HASP. A hasp which is so designed that all screws or bolts used to attach it are covered by the body of the hasp itself when it is in position to receive its padlock.

SASH. The movable part of a window.

SASH LOCK. A lock of the type used to secure a window sash, or a pair of them, in place.

SCOTCH SPRING. A spring of uniform thickness, but tapering in width.

SCRAMBLED. Pertaining to a systematic code series which has been put into a random sequence.

SCREWLESS SHANK. A knob shank which is secured to a knob spindle by means other than a set screw.

SECRET GATE LATCH. A lightweight, knobless rim latch with a concealed means of retracting the latch, usually a finger-lever (similar to a thumblever) built into the bottom of the case. This type of unit is a privacy device rather than a protective one.

SETTING-UP THIMBLE. A plug holder used in the process of keying or for other work.

SET SCREW. A machine screw which is threaded full length and is usually used in locks to prevent the turning of other threaded parts.

SHACKLE. The staple of a padlock; manacles.

SHEAR LINE. The surface of a pin or disc tumbler plug, where all tumblers must align before the plug can turn.

SHOE. A sliding part which replaces a latch lever.

SHORT TUMBLER METHOD. A method of masterkeying disc tumbler locks.

SHOULDER SCREW. A screw with a body larger than the major diameter of the threaded portion and having a head suitable to retain a pivoted part on the shoulder portion.

SHOW CASE LOCK. Any of a number of lock patterns intended to lock doors of the types used on show cases, but commonly referring to a lock for locking together two sliding panes of glass.

SIDE BAR LOCK. A lock whose locking function is controlled by the action of the tumblers upon a bar which is positioned along their side.

SKELETON KEY. A key which has been cut away or *skeletonized* to permit it to pass the wards of more than one lock.

SLIDING LEVER. A lever tumbler which is not pivoted, and which moves in a straight line, rather than with a rotary motion.

SMOOTHING PARTS. Pertaining to the removal of irregularities from the surface of a part at those places where it comes into moving contact with another part, thus reducing friction and enhancing durability. Customarily, both (or all) such parts are smoothened.

SPINDLE. That part of a lock which couples the knobs to the hubs.

SPRING. A flexible device of metal in any of a wide variety of geometric configurations intended to provide an automatic return function to a part or a group of parts.

SPRING BOLT (or springbolt). A square-ended locking bolt which is extended by spring action and which is not secured against end pressure. A springlatch (Br.).

SPRINGLATCH. A lock whose sole locking function is achieved by the locking of one or both knobs. A springbolt (Br.).

SPRING STEEL. A general term referring to any

of a group of alloy or carbon steels, specially formulated and/or tempered for use in springs.

STAPLE. That portion of a hasp through which a padlock is hooked.

STAINLESS STEEL. A general term for a large group of ferrous or nonferrous alloys especially developed for corrosion resistance in combination with other specific desirable properties.

STEEL. Any of an extremely large family of ferrous alloys of high purity, with other element(s) added in very carefully controlled proportions.

STOP. Part of a stopworks. *See:* Dogging stop, Floating stop.

STOP LEVER. The lever connecting and coordinating the dogging stop with the floating stop.

STOP SPRING. A spring used to maintain the stopworks in the position in which it is placed.

STOP YOKE. A slotted protrusion from the lock case intended to support and guide the dogging stop.

STOPWORKS. That mechanism which locks the hub and consequently the knob of a lock or latch.

STRESS RELIEVED. A heat treatment process designed to make uniform the hardness of, and relieve internal strains and stresses in, metal before and/or after working to ensure against the warping or cracking of the finished part. Stress corrosion cracking is minimized in metals so treated.

STRIKE. The plate or other part attached to a door frame with which the latchbolt and/or deadbolt engages. Strikes are sometimes used for similar reasons on drawers and windows, as well as on doors.

STUD. A screwed-in, headless pivot pin.

STUFFED SHANK. A knob shank having an ornamental covering of metal.

STUMP. A post or attached pin (Br.).

SUBMASTER KEY. *n.* A key of lower echelon than the master key and operating only a selected portion of the locks operated by the master key. *v.tr.* To key locks to operate within the framework of a masterkey arrangement of this type.

SYNCHRONIZATION. The timing or sequencing of the various operations incident to the operation of a lock or latch.

TAPPED. Having internal threads (as a hole).

TEE HANDLE. A handle shaped like the letter T. The term is most commonly used in reference to a handle to hold a tap while tapping holes, but door handles of that type are made for use with locks and latches.

TEMPERED. Another term for heat treated; often used to convey the idea of less precisely controlled conditions.

TEMPLATE. A layout pattern, usually of metal or card stock.

TENSILE STRENGTH. The strength of a material in relation to its cross section.

THREE POINT LOCK. A lock operating to secure a door at the top, bottom, and center, simultaneously. Usually found on paired doors.

THRUST WASHER. A washer used for the latch bolt spring to thrust against.

THUMBPIECE. A lever operated by thumb pressure to activate the internal mechanism which draws the latchbolt. Often called a thumblever or thumblatch.

THUMBTURN DISC. A part used in lieu of a mortise cylinder; the thumbturn disc is connected to, and operated by, a thumbturn (turn knob).

TOLERANCE. The permissible limits of dimension by which parts may vary and yet function properly.

TRANSOM. A sash, sometimes moveable, located directly over a door.

TRANSOM LOCK. A lock commonly used for securing transoms and similar to a sash lock or, more commonly, a cupboard catch.

TRAVELING LEVERS. Levers which are mounted on the bolt of a lock and travel with the bolt rather than being pivoted on the lock case as are common stationary levers.

TRIPLEX SPINDLE. A knob spindle made in three longitudinal sections.

TRUNK LOCK. A luggage lock, usually of warded or lever construction, and largest of the luggage locks.

TUBE. That portion of a lever lock which corresponds to the cylinder hull of a pin or disc tumbler lock. The housing for the barrel.

TUMBLER. A moveable part which must be in its proper position before a key will turn.

TUMBLER CHAMBER. A recess, hole, or pocket in a lock plug or cylinder hull intended to receive disc or pin tumblers.

TURN KNOB CYLINDER. A lock cylinder which is operated by a turn knob rather than a key.

TURN KNOB CAM. 1. A cam, operated by a turn knob, acting to throw a deadbolt. A type of deadbolt cam. 2. A thumbturn disc.

TYPE OF CUT. Pertaining to the configuration of the notches in a key.

UTILITY LOCKS. Locks whose use is not restricted to a single purpose or usage; locks for other than door or automotive use.

VERTICAL ROD DEVICE. A panic exit which utilizes latching operations at the top and bottom, rather than at the side of the door.

VISUALLY CONSTRUCTIVE IMAGINATION. A mental process, helpful to mechanical creativity.

VISUALLY RECONSTRUCTIVE IMAGINATION. A mental process — helpful in lock picking, the replacement of missing parts, and in the rebuilding of worn or broken parts.

WARDED LOCKS. Locks which utilize a system of baffles, rather than moveable tumblers, to achieve key differs.

WARDS, INTERNAL. Baffles intended to keep an improper key from turning in the lock.

WARDS, KEYWAY. Baffles intended to keep an improper key from entering the lock.

Index

References to charts and illustrations are printed in boldface type.